DICTIONARY OF WEEDS OF EASTERN EUROPE

DICTIONARY OF WEEDS OF EASTERN EUROPE

Their Common Names and Importance

in

Latin, Albanian, Bulgarian, Czech, German, English, Greek, Hungarian, Polish, Romanian, Russian, Serbo-Croat and Slovak

compiled by

GARETH WILLIAMS and KÁROLY HUNYADI
Working Group on Education and Training
European Weed Research Society

ELSEVIER
Amsterdam – Boston – Heidelberg – London – New York – Oxford
Paris – San Diego – San Francisco – Singapore – Sydney – Tokyo

The distribution of this book is being handled by the following publishers
for the U.S.A. and Canada
Elsevier Science Publishing Co., Inc.
52, Vanderbilt Avenue
New York, New York 10017, U.S.A.

for the East European countries, Democratic People's Republic of Korea, People's Republic of Mongolia, Republic of Cuba and Vietnam
Kultúra Hungarian Foreign Company,
P.O.B. 149, H-1389 Budapest 62, Hungary
for all remaining areas
Elsevier Science Publishers
25, Sara Burgerhartstraat
P.O. Box 211, 1000 AE Amsterdam. The Netherlands

Library of Congress Cataloging-in-Publication Data

Dictionary of weeds of eastern Europe.

Includes index.
1. Weeds--Europe, Eastern--Dictionaries--Polyglot.
2. Weeds--Europe, Eastern--Nomenclature (Popular)--
Dictionaries--Polyglot. 3. Dictionaries, Polyglot.
I. Williams, Gareth, 1937- . II. Hunyadi, Károly.
SB613.E852D53 1987 632'.58'0947 87-575
ISBN 0-444-98969-2

Transferred to digital printing 2007

ISBN 0 444 98969 2

Joint edition published by
Elsevier Science Publishers, Amsterdam
and Akadémiai Kiadó, Budapest, Hungary

Introduction

In 1972, the Joint Panel on the Evaluation of Herbicides of the European Weed Research Council and the European Plant Protection Organisation proposed that lists of common names of weeds in the different languages of Europe should be produced. 'Elsevier's Dictionary of Weeds of Western Europe' was published in 1982, and now this companion volume for the common names and importance of weeds in the countries of Eastern Europe completes the initial stage of the work by what is now the Working Group on Education and Training of the European Weed Research Society.

We will be continuing to bring our work up to date as necessary to help in the task of passing information from scientists to growers in a form which is readily understood. Our aim has been to promote the use of common names as being more intelligible to the general public than Linnean names but to reduce as far as possible the ambiguity and confusion which is caused by one species having more than one common name in a particular language, or by one common name referring to several species. Some countries have produced official or semi-official lists of common names of weeds. Other common names have been the personal choice of contributors from many alternatives and may be superseded in future. Above all, weed populations are never static. Plants which are not regarded as weeds today may be of importance in a few years and will have to be included.

The production of this book has involved far more than those listed as the compilers. We are very grateful to all of those who have contributed, and ask them to forgive any errors in transcribing their efforts.

Details of other activities of the Working Groups of the European Weed Research Society may be obtained from the EWRS Secretariat, Biologische Bundesanstalt, Messeweg 11/12, D-3300 Braunschweig.

Gareth Williams
West of Scotland Agricultural College,
Ayr,
Scotland.

Károly Hunyadi
University of Agricultural Sciences,
Keszthely,
Hungary.

Notes on the Text

Linnean Names

The Linnean names of plants and the authorities for these names are those given in the Flora Europaea, edited by T. G. Tutin and others, and published by the Cambridge University Press: Vol. I (1964); Vol. II (1968); Vol. III (1972); Vol. IV (1976); Vol. V (1980). While it is impossible to give all synonyms for each species, it is hoped that sufficient have been included to allow a particular species to be identified and its description checked in national floras.

Distribution and Importance

The figure in brackets which follows each language symbol under a particular Linnean name shows the importance of the species as a weed in the area in which that language is spoken.

The categories of importance are:
0 — Of no importance as a weed.
1 — Only of minor importance.
2 — Only important in a few situations, although it may be widespread as a minor weed species.
3 — An important competitive weed occurring in many crops and many situations.
- — No information available on the weed status of the species

Languages

Al — Albanian; *Bg* — Bulgarian; *Cs* — Czech; *De* — German; *En* — English; *Gr* — Greek; *Hu* — Hungarian (Magyar); *Pl* — Polish; *Ro* — Romanian; *Ru* — Russian; *Sh* — Serbo-Croat; *Sk* — Slovak

Contributors

C. Chirilă, Romania
H. Domańska, Poland
N. A. Fetvadjieva, Bulgaria
C. N. Giannopolitis, Greece
V. Grinius, Hungary
B. Hundozi, Yugoslavia
K. Hurle, Federal Republic of Germany
T. Kaspirova, U.S.S.R.
V. Kohout, Czechoslovakia
M. Kojić, Yugoslavia
F. Krasniqi, Yugoslavia
M. Milroy, U.K.
J. Pichler, Austria
M. Reth, Federal Republic of Germany
T. Šarić, Yugoslavia
N. Şarpe, Romania
A. Skender, Yugoslavia
M. de O. Tollemache, Austria

BASIC TABLE

A

* ***Abutilon avicennae*** Gaertn. → *Abutilon theophrasti*

1 ***Abutilon theophrasti*** Med.
Al (–) këmacës; pambuk i egër
Bg (2) просфорник
Cs (0) mračňák Theophrastův
De (–) Lindenblättrige Schönmalve
En (–) velvetleaf
Gr (2) αγριομκαμπακιά
Hu (2) selyemmályva
Ro (–) teişor
Ru (1) канатник Теофраста
Sh (1) Teofrastova lipica
Sk (–) mračňák Theophrastov

2 ***Acanthus balcanicus*** Heyw. & Richardson
Ru (1) акант балканский
Sh (1) balkanska matruna

3 ***Acanthus spinosus*** L.
Al (–) dashter gjembake
En (–) spiny bear's-breech
Ru (1) акант колючий
Gr (1) ακούρανος
Sk (–) paznechtík ostnitý

* ***Acetosa pratensis*** Mill. → *Rumex acetosa*

* ***Acetosella vulgaris*** (Koch) Fourr. → *Rumex acetosella*

4 ***Achillea* spp.** L.
Bg (1) равнец
De (–) Schafgarbe
Gr (1) αχίλλεα
Ru (1) тысячелистник
Sk (–) mačúch

5 ***Achillea biebersteinii*** Afan.
Ru (1) тысячелистник Биберштейна
Sk (–) mačúch biebersteinov

6 ***Achillea collina*** Beck
De (–) Hügel-Schafgarbe
En (–) mountain yarrow
Sh (1) brdski sporiš
Sk (–) mačúch kopcový

7 ***Achillea millefolium*** L.
Al (–) barpezmi me njëmijë gjethe
Bg (1) бял равнец
Cs (3) řebříček obecný
De (–) Gemeine Schafgarbe
En (–) yarrow
Hu (3) közönséges cickafark
Pl (1) krwawnik pospolity
Ro (1) coada şoricelului
Ru (1) тысячелистник обыкновенный
Sh (2) obićni sporiš; obični stolisnik
Sk (–) myší chvost obyčajný

8 ***Achillea nobilis*** L.
Al (–) barpezmi i fisëm
Bg (0) благороден равнец
Cs (0) řebříček sličný
De (–) Edle Schafgarbe
En (–) graceful yarrow
Pl (1) krwawnik szlachetny
Ru (0) тысячелистник благородный
Sk (–) mačúch spanilý

9 ***Achillea ptarmica*** L.
Al (–) bar teshtitje
Cs (1) řebříček bertrám
De (–) Sumpf-Schafgarbe
En (–) sneezewort
Pl (1) krwawnik kichawiec
Ro (0) rotoţele albe
Ru (0) тысячелистник птармика
Sk (–) bertram obyčajný

10 *Acinos arvensis* (Lam.) Dandy
Al (–) trumza si borzilok
Bg (1) обикновено тиризливче
Cs (0) marulka pamětník
De (–) Stein-Bergminze
En (–) basil-thyme
Hu (1) ugari pereszlény
Pl (1) czyścica drobnokwiatowa
Ro (0) ismuşoară de cîmp
Ru (0) щебрушка полевая
Sh (1) poljska marulja
Sk (–) pamätník obyčajný

11 *Acinos rotundifolius* Pers.
Ru (1) щебрушка круглолистная
Sk (–) pamätník okruhlolistý

12 *Aconitum napellus* L.
Al (–) bar dhelpre; lulë vathi
Bg (0) реловидна санакитка
Cs (2) oměj šalamúnek
De (–) Blauer Eisenhut
En (–) monkshood
Pl (1) tojad mocny
Ro (1) trei cumnate supărate
Ru (1) аконит синий
Sk (–) prilbica modrá

13 *Aconitum variegatum* L.
Bg (0) синя санакитка
Cs (0) oměj různobarvý
De (–) Bunter Eisenhut
En (–) branched monkshood
Hu (1) karcsú sisakvirág
Pl (1) tojad dzióbaty
Ru (1) аконит пёстрый
Sk (–) prilbica pestrá

14 *Acorus calamus* L.
Al (–) prese
Bg (0) аир
Cs (2) puškvorec obecný
De (–) Kalmus
En (–) sweet-flag
Gr (1) ἀκορος
Hu (0) kálmos
Pl (1) tatarak zwyczajny
Ro (0) obligeană
Ru (–) аир обыкновенный
Sk (–) puškvorec obyčajný

15 *Acroptilon repens* (L.) DC.
En (–) Russian knapweed
Ru (1) горчак ползучий
Sk (–) nevädza plazivá

16 *Actaea spicata* L.
Bg (1) ресник
Cs (0) samorostlík klasnatý
De (–) Ähriges Christophskraut
En (–) baneberry
Pl (1) czerniec gronkowy
Ro (0) orbalţ
Sk (–) samorastlík klasovitý

17 *Adenophora lilifolia* (L.) Ledeb.
Cs (0) zvonovec liliolistý
En (–) lilyleaf ladybell
Hu (0) csengettyűvirág
Pl (1) dzwonecznik wonny
Sk (–) zvonovec ľaliolistý

18 *Adenostyles alpina* (L.) Bluff & Fingerh.
De (–) Grüner Alpendost
En (–) alpine plantain
Gr (1) αδενόστυλος
Ro (0) ciucuraş
Sk (–) mačucha alpínska

19 *Adonis* spp. L.
De (–) Teufelsauge
En (–) pheasant's-eye
Gr (1) ἀδωνης
Ru (0) адонис
Sk (–) hlaváčik

20 ***Adonis aestivalis*** L.
Al (–) adonisi i verës; sy thëllezë
Bg (2) гороцвят
Cs (2) hlaváček letní
De (–) Sommer-Teufelsauge
En (–) summer pheasant's-eye
Hu (1) vetési hérics
Pl (1) miłek letni
Ro (2) cocoşei de cîmp
Ru (1) адонис летний
Sh (1) pjegavi gorocvijet; zečji mak
Sk (–) hlaváčik letný

21 ***Adonis annua*** L.
Al (–) sydjalli
Bg (–) есенен гороцвят
Cs (0) hlaváček roční
De (–) Herbst-Teufelsauge
En (–) common pheasant's-eye
Ru (1) адонис осенний
Sh (1) jednogodišnji gorocvijet; jesenji gorocvet
Sk (–) hlaváčik ročný

22 ***Adonis flammea*** Jacq.
Al (–) syfazani
Bg (2) огнецветен гороцвят
Cs (0) hlaváček plamenný
De (–) Brennendes Teufelsauge
En (–) flame adonis
Hu (1) lángszínű hérics
Pl (1) miłek szkarłatny
Ro (1) ruscuţă
Ru (1) адонис пламенный
Sh (1) plameni gorocvet; žarki gorocvijet
Sk (–) hlaváčik plamenný

* ***Adonis miniatus*** Jacq. → *Adonis aestivalis*

* ***Adonis phoeniceus*** Frit. → *Adonis aestivalis*

23 ***Adonis vernalis*** L.
Al (–) këmbë gomari
Bg (0) слетикоса
Cs (1) hlaváček jarní
De (–) Frühlings-Teufelsauge
En (–) yellow adonis
Hu (1) tavaszi hérics
Pl (1) miłek wiosenny
Ro (1) ruscuţă primăvăratică
Ru (1) адонис весенний
Sh (1) ljuskarasti gorocvijet; prolećni gorocvet
Sk (–) hlaváčik jarný

24 ***Adoxa moschatellina*** L.
Al (–) adoksë muskate
Bg (0) обикновена мошковица
Cs (0) pižmovka mošusová
De (–) Moschuskraut
En (–) moschatel
Gr (1) άδοξα
Hu (0) pézsmaboglár
Pl (1) piżmaczek wiosenny
Ro (0) frăguţă
Ru (1) адокса мускусная
Sk (–) pižmovka obyčajná

25 ***Aegilops* spp.** L.
Bg (1) осилка
De (–) Walch
En (–) goatgrass
Gr (1) αγριόσταρο
Ru (1) эгилопс
Sk (–) mnohoštet

26 ***Aegilops cylindrica*** Host
Bg (1) цилиндрична осилка
Cs (0) mnohoštět válcovitý
De (–) Walzenförmiger Walch
En (–) jointed goatgrass
Hu (1) kecskebúza
Ro (0) ciucure
Ru (1) эгилопс цилиндрический

Sh (1) valjkasta ostika
Sk (–) mnohoštet valcovitý

27 *Aegilops geniculata* Roth
Al (–) halmuca vezake
Bg (1) тройна осилка
Cs (0) mnohoštět vejčitý
Ru (1) эгилопс коленчатый
Sk (–) mnohoštet vajcovitý

* ***Aegilops ovata*** L. → *Aegilops geniculata*

28 *Aegilops tauschii* Coss.
Ru (1) эгилопс Тауша
Sk (–) mnohoštet tauschov

29 *Aegilops triuncialis* L.
Al (–) halmuca trigisht e gjatë
Bg (0) триунциев егилопс
De (–) Ziegenbart-Walch
En (–) hard goatgrass
Ru (1) эгилопс трёхдюймовый
Sk (–) mnohoštet trojpalcový

30 *Aegopodium podagraria* L.
Al (–) bar shtergjia; këmbëdhia
Bg (0) седмолист
Cs (3) bršlice kozí noha
De (–) Gewöhnlicher Geißfuß
En (–) ground-elder
Gr (1) αιγοπόδιο
Hu (1) podagrafű
Pl (2) podagrycznik rzeczpospolity
Ro (1) piciorul caprei
Ru (2) сныть обыкновенная
Sh (1) obični sedmolist
Sk (–) kozia noha hostcová

31 *Aeluropus littoralis* (Gouan) Parl.
Ru (1) прибрежница прибрежная
Sk (–) reznačka pobrežná

32 *Aethusa cynapium* L.
Al (–) stërpuja
Bg (0) див магданоз
Cs (3) tetlucha kozí pysk
De (–) Gemeine Hundspetersilie
En (–) fool's parsley
Gr (1) αγριομαϊντανός
Hu (2) ádáz
Pl (1) blekot pospolity
Ro (1) pătrunjelul cîinelui
Ru (1) кокорыш обыкновенный
Sh (1) mali peršunak; obicna kukutica
Sk (–) tetlucha kozí pysk

* ***Agrimonia asiatica*** Juz. → *Agrimonia eupatoria*

33 *Agrimonia eupatoria* L.
Al (–) rodhëza
Bg (0) обикновен камшик
Cs (1) řepík lékařský
De (–) Gewöhnlicher Odermennig
En (–) agrimony
Gr (1) αγριμόνια
Hu (1) közönséges párlófű
Pl (1) rzepik pospolity
Ro (1) turiţă mare
Ru (1) репейничек обыкновенный
Sh (1) obični petrovac
Sk (–) repík lekársky

34 *Agropyron cristatum* (L.) Gaertn.
Bg (0) житняк
Cs (1) pýr hřebenitý
De (–) Kamm-Quecke
En (–) crested couch-grass
Gr (1) αγροπυρο
Hu (0) taréjos búzafű
Pl (1) perz grzebieniasty
Ru (1) житняк гребенчатый
Sk (–) žitňak hrebenitý

* ***Agropyron glaucum*** Roem. & Schultes → *Elymus hispidus*

* ***Agropyron intermedium*** (Host) Beauv. → *Elymus hispidus*

* ***Agropyron litorale*** auct. → *Elymus hispidus*

* ***Agropyron repens*** (L.) Beauv. → *Elymus repens*

35 ***Agrostemma githago*** L.
Al (–) konkoli; gunevgjitka
Bg (1) къклица
Cs (0) koukol polní
De (–) Korn-Rade
En (–) corncockle
Gr (2) γόγγολη
Hu (2) konkoly
Pl (1) kąkol polny
Ro (2) neghină
Ru (1) куколь обыкновенный
Sh (1) njivski kukolj; usjevni kukolj
Sk (–) kúkoľ poľný

36 ***Agrostis*** **spp.** L.
Bg (–) полевица
De (–) Straußgras
En (–) bentgrass
Gr (1) άγρωστη
Ru (1) полевица
Sk (–) psinček

* ***Agrostis alba*** auct. → *Agrostis stolonifera*

37 ***Agrostis canina*** L.
Al (–) barimza e qenit
Bg (0) кучешка полевица
Cs (0) psineček psí
De (–) Sumpf-Straußgras
En (–) brown bent
Pl (1) mietlica psia
Ro (0) bucăţel
Ru (0) полевица собачья
Sh (1) pasja rosulja; ščetinasta rosulja
Sk (–) psinček psí

38 ***Agrostis capillaris*** L.
Al (–) barimza e rëndomtë
Bg (0) обикновена полевица
Cs (1) psineček tenký
De (–) Rotes Straußgras
En (–) common bent
Hu (1) cérnatippan
Pl (3) mietlica pospolita
Ro (0) păiuş
Ru (1) полевица волосовидная
Sh (1) obična rosulja
Sk (–) psinček obyčajný

* ***Agrostis interrupta*** L → *Apera interrupta*

39 ***Agrostis gigantea*** Roth
Cs (2) psineček výběžkatý obrovský
De (–) Großes Straußgras
En (–) black bent
Ro (1) iarba cîmpului
Ru (0) полевица гигантская
Sk (–) psinček vybežkatý obrovský

* ***Agrostis spica-venti*** L. → *Apera spica-venti*

40 ***Agrostis stolonifera*** L.
Al (–) barimza e bardhë
Bg (1) бяла полевица
Cs (3) psineček výběžkatý pravý
De (–) Weißes Straußgras
En (–) creeping bent
Hu (1) fehér tippan; tarackos tippan
Pl (3) mietlica biaława
Ro (3) iarba cîmpului
Ru (1) полевица белая
Sh (1) bela rosulja; bijela rosulja
Sk (–) psinček vybežkatý pravý

* ***Agrostis tenuis*** Sibth. → *Agrostis capillaris*

* ***Agrostis verticillata*** Vill. → *Polypogon viridis*

41 *Ailanthus altissima* (Mill.) Swin.
Al (–) aringa; arra e Indisë
Cs (0) pajasan cizí
De (–) Chinesischer Götterbaum
En (–) tree-of-heaven
Gr (1) βρωμόδεντρο
Hu (0) bálványfa
Pl (0) bożodrzew gruczołowaty
Ro (0) cenuşar
Ru (0) айлант высочайший
Sk (–) pajaseň žľaznatý

* ***Aira capillaris*** Host → *Aira elegantissima*

* ***Aira cespitosa*** L. → *Deschampsia cespitosa*

* ***Aira elegans*** Willd. → *Aira elegantissima*

42 *Aira elegantissima* Schur
Al (–) veri në trajtë flokësh
Bg (1) брол
Cs (0) ovsíček lepý
En (–) silver hair-grass
Ru (0) айра изящная
Sh (1) obična metlača; vlasasta metlača
Sk (–) ovsienka pekná

43 *Ajuga chamaepitys* (L.) Schreb.
Al (–) ajuga si pishë e vogël
Bg (1) жълто срещиче
Cs (2) zběhovec yva
De (–) Gelber Günsel
En (–) ground-pine
Hu (3) kalinca ínfű
Pl (1) dąbrówka żółtokwiatowa
Ro (1) tămîiţă de cîmp
Ru (1) живучка низенькая
Sh (2) zuta ivica
Sk (–) zbehovec trojklaný

* ***Ajuga chia*** Schreb. → *Ajuga chamaepitys*

44 *Ajuga genevensis* L.
Al (–) ajuga e Gjenevës
Bg (1) обикновено срещиче
Cs (1) zběhovec ženevský
De (–) Genfer-Günsel
En (–) blue bugle
Hu (1) közönséges ínfű
Pl (1) dąbrówka kosmata
Ro (0) suliman
Ru (1) живучка женевская
Sh (1) trocjepna ivica; ženevska ivica
Sk (–) zbehovec lesný

45 *Ajuga pyramidalis* L.
Al (–) ajuga piramidale
Bg (0) пирамидално срещиче
Cs (1) zběhovec jehlancovitý
De (–) Pyramiden-Günsel
En (–) pyramidal bugle
Pl (1) dąbrówka piramidalna
Ru (0) живучка пирамидальная
Sk (–) zbehovec ihlancovitý

46 *Ajuga reptans* L.
Al (–) ajuga zvarranike
Bg (1) пълзящо срещиче
Cs (3) zběhovec plazivý
De (–) Kriechender Günsel
En (–) bugle
Hu (0) indás ínfű
Pl (2) dąbrówka rozłogowa
Ro (0) vineţică
Ru (2) живучка ползучая
Sh (2) puzava ivica
Sk (–) zbehovec plazivý

* ***Alcea biennis*** Winterl. → *Alcea pallida*

47 *Alcea pallida* (Willd.) Waldst. & Kit.
Bg (0) бледна ружа
Cs (0) proskurník bledý
De (–) Blasse Stockrose
En (–) pale hollyhock
Hu (1) halvány ziliz

Ru (0) проскурняк бледный
Sh (1) bledi slezac; blijedi bijeli sljez
Sk (–) ibiš bledý

48 *Alcea rosea* L.
Al (–) mollohire ngjyrë trëndafili
Bg (0) градинска ружа
Cs (1) proskurník topolovka
De (–) Stockrose
En (–) garden hollyhock
Gr (1) δεντρομολόχα
Hu (0) mályvarózsa
Ro (0) nalbă de grădină
Ru (0) проскурняк розовый
Sk (–) ibiš ružový

* ***Alchemilla aphanes*** L. → *Aphanes arvensis*

* ***Alchemilla arvensis*** (L.) Scop. → *Aphanes arvensis*

* ***Alchemilla pratensis*** auct. → *Alchemilla xanthochlora*

* ***Alchemilla vulgaris*** auct. → *Alchemilla xanthochlora*

49 *Alchemilla xanthochlora* Rothm.
Bg (0) обикновено щапиче
Cs (2) kontryhel obecný
De (–) Gemeiner Frauenmantel
En (–) lady's-mantle
Hu (0) közönséges palástfű
Pl (1) przywrotnik żółtawo-zielony
Ro (0) creţişoare
Ru (1) манжетка обыкновенная
Sh (1) obična vrkuta
Sk (–) alchemilka lúčna

* ***Alectorolophus glaber*** (Lam.) Beck → *Rhinanthus angustifolius*

* ***Alectorolophus hirsutus*** (Lam.) All. → *Rhinanthus alectorolophus*

* ***Alectorolophus major*** Reichenb. → *Rhinanthus angustifolius*

* ***Alectorolophus minor*** (L.) Wim. & Grab. → *Rhinanthus minor*

* ***Alectorolophus rumelicus*** (Velen.) Bord → *Rhinanthus rumelicus*

50 *Alhagi pseudalhagi* (Bieb.) Desv.
Ru (1) верблюжья колючка обыкновенная

51 *Alisma gramineum* Lej.
Cs (0) žabnik trávovitý
De (–) Grasblättriger Froschlöffel
En (–) ribbon-leaved water-plantain
Hu (1) úszó hídőr
Ru (0) частуха злаколистная
Sk (–) žabník trávovitý

52 *Alisma lanceolatum* With.
Ru (1) частуха ланцетолистная
Sk (–) žabník kopijovitý

53 *Alisma plantago-aquatica* L.
Al (–) këlkëza e ujit
Bg (1) жабина лъжица
Cs (1) žabník jitrocelový
De (–) Gemeiner Froschlöffel
En (–) common water-plantain
Gr (1) κεντάνευρο νερού
Hu (1) vízi hídőr
Pl (0) babka wodna
Ro (2) limbariţă
Ru (1) частуха болотная
Sh (1) obična žabočunka; pršljenasti žabočun
Sk (–) žabník skorocelový

54 *Alkanna tinctoria* (L.) Taus.
Al (–) alkana ngjyruese
Bg (0) багрилна айважива
Cs (0) kamejník barvířský

De (–) Alkannawurzel
En (–) dyer's alkanet
Gr (1) βαφόριζα
Ru (–) алкана красильная
Sk (–) alkana farbiarska

* ***Alliaria officinalis*** Andrz. → *Alliaria petiolata*

55 *Alliaria petiolata* (Bieb.) Cav. & Gran.
Al (–) barëhudhëre mjekësore
Bg (0) лечебна лъжичина
Cs (1) česnáček lékařský
De (–) Gemeine Knoblauchsrauke
En (0) garlic mustard
Gr (1) αλλιάρα
Hu (1) kányazsombor
Pl (1) czosnaczek pospolity
Ro (0) usturoiţă
Ru (0) чесночник черешковый
Sh (2) obična češnjača
Sk (–) rancesnak lekársky

56 *Allium ampeloprasum* L.
Al (–) hudhër vreshtash; purrigjarpëri
Bg (1) лозов лук
Cs (0) česnek pór; pór
De (–) Acker-Knoblauch
En (–) wild leek
Hu (0) nyári hagyma
Pl (0) por dziki
Ro (0) ceapă de vară
Ru (0) лук виноградный
Sk (–) pór obyčajný

57 *Allium angulosum* L.
Ru (1) лук угловатый
Sk (–) cesnak hranatý

58 *Allium atropurpureum* Waldst. & Kit.
Bg (1) тъмночервен лук
Hu (1) fekete hagyma
Ru (0) лук тёмно-красный
Sk (–) cesnak čiernonachový

59 *Allium atroviolaceum* Boiss.
Cs (–) česnek černofialový
Ru (1) лук тёмно-фиолетовый
Sk (–) cesnak čiernofialový

60 *Allium caeruleum* Pallas
Ru (1) лук голубой

61 *Allium decipiens* Fisch.
Ru (1) лук обманчивый

* ***Allium laxiflorum*** Taus. → *Allium vineale*

62 *Allium neapolitanum* Cyr.
En (–) Naples garlic
Gr (1) αγριοκρέμμυδο
Ru (–) лук неаполитанский

63 *Allium oleraceum* L.
Al (–) hudhra e egër
Bg (1) градинарски лук
Cs (0) česnek domácí
De (–) Kohl-Lauch
En (–) field garlic
Hu (0) érdes hagyma
Pl (1) czosnek zielonawy
Ro (0) ai sălbatic
Ru (0) лук огородный
Sk (–) cesnak domáci

* ***Allium rotundum*** L. → *Allium scorodoprasum* ssp. *rotundum*

64 *Allium saxatile* Bieb.
Al (–) hudhra e shëmbinjve
Bg (1) скален лук
En (–) stone leek
Ru (0) чеснок каменный
Sk (–) cesnak skalný

65 *Allium scorodoprasum* L. **ssp. *rotundum*** (L.) Stearn
Al (–) qepa rruzullore
Bg (1) кръгъл лук

Cs (0) česnek kulovitý
De (–) Runder Lauch
En (–) round-headed sand leek
Hu (1) ereszes hagyma
Pl (–) czosnek kulisty
Ro (1) pur
Ru (0) лук круглый
Sk (–) cesnak guľovitý

66 *Allium scorodoprasum* L. **ssp. *scorodoprasum***
Bg (1) врански лук
Cs (0) česnek ořešec
En (–) sand leek
Hu (1) kígyó hagyma
Pl (1) czosnek wężowy
Ro (1) aiul şarpelui
Ru (1) чеснок ореховый
Sk (–) cesnak orešec; orešec

67 *Allium scorodoprasum* L. **ssp. *waldsteinii*** (Don) Stearn
Ru (1) лук причесночный

68 *Allium sphaerocephalon* L.
Al (–) hudhra kokërruzullore
Bg (0) кръглоглавест лук
Cs (1) česnek kulatohlavý
De (–) Kugel-Lauch
En (–) round-headed leek
Hu (1) bunkós hagyma
Pl (0) czosnek główkowaty
Ru (1) лук круглоголовчатый
Sk (–) cesnak guľatohlavý

69 *Allium ursinum* L.
Al (–) hudhra e ariut
Bg (0) мечи лук
Cs (1) česnek medvědí
De (–) Bären-Lauch
En (–) ramsons
Hu (0) medvehagyma
Pl (2) czosnek niedźwiedzi
Ro (0) leurdă
Ru (1) лук медвежий
Sh (1) medvedji luk
Sk (–) cesnak medvedí

70 *Allium vineale* L.
Al (–) hudhera e vreshtave
Bg (1) лозев лук
Cs (2) česnek viniční
De (–) Weinberg-Lauch
En (–) wild onion
Gr (2) αγριοκρέμμυδο
Hu (1) bajuszos hagyma
Pl (1) czosnek winnicowy
Ru (1) лук полевой
Sh (1) divlji luk; pupčasti luk
Sk (–) cesnak poľný

* ***Allium waldsteinii*** Don → *Allium scorodoprasum* ssp. *waldsteinii*

71 *Alopecurus aequalis* Sobol.
Bg (0) равна класица
Cs (1) psárka plavá
De (–) Gelbrotes Fuchsschwanzgras
En (–) orange foxtail
Hu (1) mocsári ecsetpázsit
Pl (–) wyczyniec czerwonożółty
Ru (1) лисохвост равный
Sk (–) psiarka plavá

* ***Alopecurus agrestis*** L. → *Alopecurus myosuroides*

72 *Alopecurus geniculatus* L.
Al (–) bishtdhelpra gjunjake
Bg (0) високопланинска ломвица
Cs (1) psárka kolénkatá
De (–) Knick-Fuchsschwanzgras
En (–) marsh foxtail
Hu (1) gombos ecsetpázsit
Pl (3) wyczyniec kolankowy
Ru (1) лисохвост коленчатый
Sh (1) koljenčasti repak; sivi repak
Sk (–) psiarka kolienkatá

73 *Alopecurus myosuroides* Huds.
Al (–) bishtdhelpra e arave; pungaca si bisht miu
Bg (2) полска лисича опашка
Cs (2) psárka polní
De (–) Acker-Fuchsschwanzgras
En (–) black-grass
Gr (3) αλεποκουρά
Hu (2) parlagi ecsetpázsit
Pl (1) wyczyniec polny
Ru (2) лисохвост мышехвостный
Sh (2) mišji repak
Sk (–) psiarka roľná

74 *Alopecurus pratensis* L.
Al (–) bishtdhelpra e livadheve
Bg (1) ливадна лисича опашка
Cs (2) psárka luční
De (–) Wiesen-Fuchsschwanzgras
En (–) meadow foxtail
Hu (1) réti ecsetpázsit
Pl (1) wyczyniec łąkowy
Ro (0) coada vulpii
Ru (1) лисохвост луговой
Sk (–) psiarka lúčna

75 *Alopecurus utriculatus* Solan.
Al (–) bishtdhelpra me kacekth; micunthi
Bg (1) мехуреста лисича опашка
De (–) Aufgeblasener Fuchsschwanzgras
En (–) bladder foxtail
Sk (–) psiarka trsťovitá

* ***Alsine rubra*** Cran. → *Spergularia rubra*

* ***Alsine serpyllifolia*** Cran. → *Arenaria serpyllifolia*

76 *Althaea cannabina* L.
Al (–) mullanjadhe kërpe
Bg (1) коноповидна ружа
Cs (0) proskurník konopný
Hu (1) kender ziliz
Ru (0) алтей коноплевый
Sk (–) ibiš konopný

77 *Althaea hirsuta* L.
Bg (1) космата ружа
Cs (0) proskurník chlupatý
De (–) Rauher Eibisch
En (–) rough marsh-mallow
Gr (1) αλθέα
Hu (1) borzas ziliz
Ru (0) алтей жестковолосый
Sk (–) ibiš chlpatý

78 *Althaea officinalis* L.
Al (–) mollohire; takra
Bg (0) лечебна ружа
Cs (0) proskurník lékařský
De (–) Echter Eibisch
En (–) common marsh-mallow
Hu (1) orvosi ziliz
Pl (1) prawoślaz lekarski
Ro (1) nalbă mare
Ru (0) алтей лекарственный
Sh (1) beli slezac; ljekoviti bijeli sljez
Sk (–) ibiš lekársky

* ***Althaea pallida*** Willd. → *Alcea pallida*

* ***Althaea rosea*** (L.) Cav. → *Alcea rosea*

79 *Alyssum* spp. L.
Bg (1) игловръх
De (–) Steinkresse
Gr (1) άλυσσο
Ru (1) бурачок
Sk (–) tarica

80 *Alyssum alyssoides* (L.) L.
Al (–) serica si sericë
Bg (1) чашковиден игловръх
Cs (0) tařice kališní
De (–) Kelch-Steinkresse

En (–) small alyssum
Hu (1) közönséges ternye
Pl (1) smagliczka kielichowata
Ro (1) albiţă
Ru (1) бурачок чашечковидный
Sh (1) obična žumenica
Sk (–) tarica kališná

* ***Alyssum calycinum*** L. → *Alyssum alyssoides*

* ***Alyssum campestre*** auct. → *Alyssum minus*

81 *Alyssum desertorum* Stapf
Bg (1) пустинен игловръх
Cs (0) tařice pustinná
De (–) Wüsten-Steinkresse
Hu (0) pusztai ternye
Pl (0) smagliczka drobna
Ro (1) bărbişoară
Ru (0) бурачок пустынный
Sh (1) kamenjarka žumenica
Sk (–) tarica pustinná

82 *Alyssum hirsutum* Bieb.
En (–) hairy alyssum
Ru (0) бурачок шершавый
Sh (1) dlakava žumenica
Sk (–) tarica chlpatá

* ***Alyssum incanum*** L. → *Berteroa incana*

* ***Alyssum minimum*** auct. → *Alyssum desertorum*

83 *Alyssum minus* (L.) Rothm.
Bg (0) дребен игловръх
En (–) lesser alyssum
Ru (0) бурачок меньший
Sh (1) poljska žumenica
Sk (–) tarica menšia

84 *Alyssum minutum* Schlecht.
Bg (1) малък игловръх
Ru (0) бурачок маленький
Sk (–) tarica drobná

85 *Alyssum murale* Waldst. & Kit.
Al (–) serica e murit
Bg (0) стенен игловръх
Cs (0) tařice zední
Ru (0) бурачок стенной
Sh (1) zidna žumenica
Sk (–) tarica múrová

* ***Alyssum paniculatum*** Willd. → *Neslia paniculata* ssp. *thracica*

86 *Alyssum strigosum* Ban. & Solan.
Bg (1) полски игловръх
Ru (0) бурачок полевой
Sk (–) tarica poľná

87 *Amaranthus albus* L.
Al (–) nena e bardhë
Bg (2) бял щир
Cs (2) laskavec bílý
De (–) Weißer Fuchsschwanz
En (–) white pigweed
Gr (3) άσπρο βλήτο
Hu (3) fehér disznóparéj
Pl (0) szarłat biały
Ro (3) ştir alb
Ru (2) щирица белая
Sh (2) beli štir; bijeli štir
Sk (–) láskavec biely

88 *Amaranthus blitoides* Wats.
Bg (1) гърлицевиден щир
De (–) Westamerikanischer Fuchs-schwanz
En (–) prostrate pigweed
Gr (3) κλαγιαστό βλήτο
Hu (3) henye disznóparéj
Ro (3) ştir tîrîtor

Ru (2) щирица жминдовидная
Sh (2) bljutavi štir
Sk (–) láskavec blitovitý

* ***Amaranthus blitum*** L. → *Amaranthus lividus*

89 *Amaranthus cruentus* L.
Al (–) nena e hapur
Cs (–) laskavec zvrhlý latnatý
De (–) Ausgebreiteter Fuchsschwanz
Hu (2) terpedt disznóparéj
Ru (1) щирица багряная
Sk (–) láskavec rozkladitý

90 *Amaranthus deflexus* L.
Al (–) nena e përkulur
Bg (1) наклонен щир
Cs (1) laskavec skloněný
Hu (1) vöröslő disznóparéj
Ru (1) щирица согнутая
Sk (–) láskavec sklonený

91 *Amaranthus graecizans* L.
Al (–) nena e pyllit
Bg (0) горски щир
Cs (–) laskavec úzkolistý
De (–) Schmalblättriger Fuchsschwanz
En (–) spreading pigweed
Hu (1) cigány disznóparéj
Ro (1) ştir prost
Ru (1) щирица узколистная
Sk (–) láskavec úzkolistý

92 *Amaranthus hybridus* L.
Al (–) nena hibride
Bg (0) хибриден щир
Cs (1) laskavec zvrhlý
De (–) Grünähriger Fuchsschwanz
En (–) green amaranth
Hu (3) karcsú disznóparéj
Pl (2) szarłat prosty
Ro (3) ştir hibrid
Ru (2) щирица метельчатая
Sk (–) láskavec bastardný

93 *Amaranthus lividus* L.
Al (–) nena e rëndomtë
Bg (0) светъл щир
Cs (2) laskavec blít
De (–) Grünlicher Fuchsschwanz
En (–) livid amaranth
Gr (2) ζωηρό βλήτο
Hu (2) zöld disznóparéj
Pl (1) szarłat tępolistny
Ro (2) ştir verde
Ru (2) щирица синеватая
Sk (–) láskavec nabelastý

* ***Amaranthus patulus*** Bertol. → *Amaranthus cruentus*

94 *Amaranthus retroflexus* L.
Al (–) nena e përveshur
Bg (3) обикновен щир
Cs (3) laskavec ohnutý
De (–) Zurückgekrümmter Fuchsschwanz
En (–) common amaranth
Gr (3) τραχύ βλήτο
Hu (3) szőrös disznóparéj
Pl (3) szarłat szorstki
Ro (3) ştir sălbatic
Ru (3) щирица запрокинутая
Sh (3) obični štir
Sk (–) láskavec ohnutý

* ***Amaranthus spicatus*** L. → *Amaranthus retroflexus*

* ***Amaranthus sylvestris*** Vill. → *Amaranthus graecizans*

95 *Ambrosia artemisiifolia* L.
Cs (1) ambrosie vyvýšená
De (–) Beifußblattrige Ambrosie
En (–) common ragweed

Hu (3) parlagfű
Ru (2) амброзия полыннолистная
Sh (2) pelenasta ambrozija;
pelinolisni limundžik
Sk (–) ambrózia vyššia

* **Ambrosia elatior** L. → *Ambrosia artemisiifolia*

96 Ambrosia trifida L.
De (–) Dreilappige Ambrosie
En (–) giant ragweed
Ru (2) амброзия трёхраздельная
Sk (–) ambrózia trojdielna

97 Ammannia baccifera L.
Ru (1) аммания ягодоносная

98 Ammannia verticillata (Ard.) Lam.
Ru (1) аммания мутовчатая

* **Ammannia viridis** Hornem. → *Ammannia baccifera*

99 Ammi majus L.
Cs (0) moračina větší
De (–) Große Knorpelmöhre
En (–) bullwort
Gr (1) ασπροκέφαλος
Sk (–) parasca väčšia

100 Amorpha fruticosa L.
Cs (–) netvařec křovitý
De (–) Bastard-Indigo
En (–) false indigo
Ru (–) аморфа кустарниковая
Sh (1) obični bagremac
Sk (–) beztvarec krovitý

101 Anacamptis pyramidalis (L.) Richard
Al (–) anakampta piramidale
Cs (0) rudohlávek jehlancovitý
De (–) Pyramiden-Hundswurz
En (–) pyramidal orchid
Hu (0) vitézvirág
Pl (1) koślaczek stożkowaty
Ro (0) bujor
Ru (0) анакамптис пирамидальный
Sk (–) červenohlav ihlancovitý

102 Anacyclus spp. L.
De (–) Bertramwurzel
Gr (1) ανάκυκλος
Ru (–) анациклус
Sk (–) bertramka

103 Anagallis arvensis L.
Al (–) barcapoja; rruthe e fushave
Bg (1) обикновено огнивче
Cs (3) drchnička rolní
De (–) Roter Gauchheil
En (–) scarlet pimpernel
Gr (2) αναγαλλίδα
Hu (3) mezei tikszem
Pl (3) kurzyślad polny
Ro (2) scînteiuţă
Ru (3) очный цвет полевой
Sh (1) crvena vidovčica; poljska krika
Sk (–) drchnička roľná ohnivá

* **Anagallis caerulea** Schreb. → *Anagallis foemina*

104 Anagallis foemina Mill.
Al (–) barcapoja femër
Bg (1) синьо огнивче
Cs (2) drchnička rolní modrá
De (–) Blauer Gauchheil
En (–) blue pimpernel
Gr (1) αναγαλλίδα
Hu (1) kék tikszem
Pl (2) kurzyślad błękitny
Ro (2) ochişor albastru
Ru (2) очный цвет синий
Sh (1) modra krika; plava vidovčica
Sk (–) drchnička roľná modrá

105 ***Anagallis minima*** (L.) Krause
Cs (0) drobýšek nejmenší
De (–) Acker-Kleinling
En (–) chaffweed
Hu (1) centike
Pl (0) niedośpiałek maleńki
Ro (1) ochişor
Ru (1) низмянка малая
Sk (–) drobček najmensí

* ***Anagallis parviflora*** Hoffm. & Link → *Anagallis arvensis*

106 ***Anagyris foetida*** L.
En (–) purging trefoil
Gr (1) βρωμολιγαριά
Sk (–) osika smradľavá

107 ***Anchusa* spp.** L.
De (–) Ochsenzunge
Gr (1) αγχούζα
Ru (0) анхуза

108 ***Anchusa arvensis*** (L.) Bieb.
Al (–) zgjebëza
Bg (1) полско паче гнездо
Cs (2) prlina rolní
De (–) Wolfsauge
En (–) bugloss
Hu (1) farkasszem
Pl (1) farbownik polny
Ro (2) ochiul lupului
Ru (1) анхуза полевая
Sh (2) poljski volujak
Sk (–) přhlica roľná

109 ***Anchusa azurea*** Mill.
Al (–) gjuhëlopa e kaltër
Cs (1) pilát italský
De (–) Italienische Ochsenzunge
En (–) large blue-alkanet
Hu (1) olasz atracél
Ro (2) miruţă mare
Ru (0) анхуза итальянская
Sh (1) talijanski volujak
Sk (–) smohla modrá

110 ***Anchusa barrelieri*** (All.) Vitm.
Cs (–) pilát Barrelierův
Ru (0) анхуза Баррелье
Sh (1) mali volujak

* ***Anchusa italica*** Retz. → *Anchusa azurea*

* ***Anchusa microcalyx*** Vis. → *Anchusa officinalis*

111 ***Anchusa ochroleuca*** Bieb.
Bg (0) жълтеникаво паче гнездо
Hu (0) vajszínű atracél
Ro (1) miruţă galbenă
Ru (0) анхуза жёлто-белая
Sk (–) smohla žltkavá

112 ***Anchusa officinalis*** L.
Al (–) gjuhëlopa mjekësore
Bg (1) челебно паче гнездо
Cs (1) pilát lékařský
De (–) Gemeine Ochsenzunge
En (–) common alkanet
Hu (2) orvosi atracél
Pl (1) farbownik lekarski
Ro (1) limba boului
Ru (1) анхуза аптечная
Sh (1) ljekoviti volujak; obični volujak
Sk (–) smohla lekárska

113 ***Andrachne telephoides*** L.
Gr (1) ανδράχνη
Ru (0) андрахна телефиевидная

* ***Andropogon halepensis*** (L.) Brot. → *Sorghum halepense*

* ***Andropogon ischaemum*** L. → *Dichanthium ischaemum*

114 *Androsace elongata* L.
Ru (1) проломник удлинённый
Sk (–) pochybok predĺžený

115 *Androsace maxima* L.
Bg (1) едър оклоп
Cs (0) pochybek největší
De (–) Kelch-Mannsschild
En (–) greater rock-jasmine
Hu (1) nagy gombafű
Ru (1) проломник большой
Sh (1) štitkasta mokračica
Sk (–) pochybok najväčší

116 *Androsace septentrionalis* L.
Ru (1) проломник северный
Sk (–) pochybok severný

117 *Anemone* spp. L.
Bg (0) съсънка
De (–) Windröschen
Gr (1) ανεμώνη
Ru (1) анемона
Sk (–) veternica

118 *Anemone nemorosa* L.
Al (–) anemonë pyjesh
Bg (0) бяла съсънка
Cs (1) sasanka hajní
De (–) Busch-Windröschen
En (–) wood anemone
Hu (0) berki peónia
Pl (1) zawilec gajowy
Ro (0) floarea paştilor
Ru (1) анемона дубравная
Sh (1) glatka šumarica; šumska breberina
Sk (–) veternica hájna

* ***Anemone pratensis*** L. → *Pulsatilla pratensis*

119 *Angelica archangelica* L.
Cs (2) andelika lekarska
De (–) Echte Engelwurz
En (–) garden angelica
Hu (0) orvosi angyalgyökér
Ro (0) angelică
Ru (0) дягиль аптечный
Sh (1) ljekovita andjelika; obična andjelika
Sk (–) angelika lekárska

120 *Angelica sylvestris* L.
Al (–) angjelika e pyllit
Bg (0) горска пищялка
Cs (1) děhel lesní
De (–) Wald-Engelwurz
En (–) wild angelica
Hu (1) erdei angyalgyökér
Pl (1) dzięgiel leśny
Ru (1) дудник лесной
Sh (1) divlja andjelika
Sk (–) angelika lesná

* ***Anisantha sterilis*** (L.) Nev. → *Bromus sterilis*

* ***Anisantha tectorum*** (L.) Nev. → *Bromus tectorum*

121 *Antennaria dioica* (L.) Gaertn.
Al (–) antenaria dioike
Bg (0) витошки еделвайс
Cs (1) kociánek dvoudomý
De (–) Gemeines Katzenpfötchen
En (–) mountain everlasting
Gr (1) αντεννάρια
Hu (0) macskatalp
Pl (1) ukwap dwupienny
Ro (0) parpian
Ru (0) антеннария двудомная
Sh (1) dvodoma bubica; dvodomna bubica
Sk (–) plešivec dvojdomý

122 *Anthemis* spp. L.
Bg (1) подрумче
De (–) Hundskamille

Gr (3) ανθεμίδα
Ru (2) пупавка
Sk (–) ruman

* **Anthemis agrestis** Wallr. → *Anthemis arvensis*

123 *Anthemis altissima* L.
Al (–) syviçja maraç
Bg (1) високо подрумче
De (–) Hohe Hundskamille
En (–) tall chamomile
Ru (1) пупавка высочайшая
Sh (1) kamiličasti prstenak; osovni jarmen
Sk (–) ruman najvyšší

124 *Anthemis arvensis* L.
Al (–) syviçja e arës; luleqeni
Bg (3) полско подрумче
Cs (2) rmen rolní
De (–) Acker-Hundskamille
En (–) corn chamomile
Hu (3) parlagi pipitér
Pl (3) rumian polny
Ro (3) romaniţă de cîmp
Ru (3) пупавка полевая
Sh (2) poljski jarmen; poljski prstenak
Sk (–) ruman roľný

125 *Anthemis austriaca* Jacq.
Bg (2) австрийско подрумче
Cs (2) rmen rakouský
De (–) Österreichische Hundskamille
En (–) Austrian chamomile
Hu (3) szöszös pipitér
Pl (1) rumian austriacki
Ru (1) пупавка австрийская
Sh (1) austrijski prstenak
Sk (–) ruman rakúsky

* **Anthemis cota** auct. → *Anthemis altissima*

* **Anthemis cotiformis** Velen. → *Anthemis austriaca*

126 *Anthemis cotula* L.
Al (–) syviçja me erë të keqe
Bg (1) смрадливо подрумче
Cs (2) rmen smradlavý
De (–) Stinkende Hundskamille
En (–) stinking mayweed
Hu (2) büdös pipitér
Pl (2) rumian psi
Ro (2) romaniţă puturoasă
Ru (2) пупавка собачья
Sh (1) titričasti prstenak
Sk (–) ruman smradľavý

127 *Anthemis ruthenica* Bieb.
Bg (1) руско подрумче
Cs (0) rmen rusínský
De (–) Ruthenische Hundskamille
Hu (1) homoki pipitér
Pl (1) rumian ruski
Ru (1) пупавка русская
Sh (1) trepavičasti prstenak
Sk (–) ruman rusísnky

* **Anthemis subtinctoria** Dobrocz. → *Anthemis tinctoria*

128 *Anthemis tinctoria* L.
Al (–) syviçja ngjyruese
Bg (1) багрилно подрумче
Cs (0) rmen barvířský
De (–) Färber-Hundskamille
En (–) dyer's chamomile
Hu (1) festő pipitér
Pl (1) rumian żółty
Ro (1) floare de perină
Ru (1) пупавка красильная
Sh (1) žuti prstenak; žutocvjetni jarmen
Sk (–) rumanček farbiarsky

129 *Anthericum liliago* L.
Al (–) mjekërdhiza me lule lila
Bg (0) неразклонен вянечник
Cs (0) bělozářka liliovitá
De (–) Astlose Graslilie
En (–) Saint Bernard's lily
Hu (0) füstös homokliliom
Pl (1) pajęcznica liliowata
Ru (1) антерикум обыкновенный
Sk (–) jagavka ľaliovitá

130 *Anthericum ramosum* L.
Al (–) mjekërdhiza e degëzuar
Bg (0) разклонен венечник
Cs (0) bělozářka větevnatá
De (–) Ästige Graslilie
Hu (0) ágas homokliliom
Pl (1) pajęcznica gałęzista
Ro (0) liliuţă
Ru (0) антерикум ветвистый
Sh (1) granata visika
Sk (–) jagavka vetvistá

131 *Anthoxanthum aristatum* Boiss.
Cs (0) tomka Puelova
De (–) Begranntes Ruchgras
En (–) annual vernal-grass
Pl (2) tomka óscista
Ru (1) колосок остистый
Sk (–) tomka osinatá

132 *Anthoxanthum odoratum* L.
Al (–) pesta erëmirzë
Bg (0) ароматна миризливка
Cs (2) tomka vonná
De (–) Wohlriechendes Ruchgras
En (–) sweet vernal-grass
Hu (0) borjúpázsit
Pl (1) tomka wonna
Ro (0) viţelar
Ru (1) колосок душистый
Sk (–) tomka voňavá

* ***Anthoxanthum puelii*** Lec. & Lam. → *Anthoxanthum aristatum*

133 *Anthriscus* spp. Pers.
Bg (0) азмацук
De (–) Kerbel
Gr (1) ἀνθρισκος
Ru (0) купырь
Sk (–) trebuľka

134 *Anthriscus caucalis* Bieb.
Al (–) antriskusi i rëndomtë
Cs (1) kerblík obyčejný
De (–) Hunds-Kerbel
En (–) bur chervil
Hu (1) borzas turbolya
Pl (1) trybula pospolita
Sk (–) trebuľka obyčajná

135 *Anthriscus cerefolium* (L.) Hoffm.
Cs (2) kerblík třebule
De (–) Garten-Kerbel
En (–) garden chervil
Hu (1) zamatos turbolya
Pl (1) trybula ogrodowa
Ro (0) asmăţui
Ru (1) купырь садовый
Sh (–) obična krasuljica
Sk (–) trebuľka pravá

136 *Anthriscus nitida* (Wahlen.) Gar.
Cs (0) kerblík lesklý
De (–) Glänzender Kerbel
En (–) glossy cow-parsley
Hu (0) hamvas turbolya
Pl (1) trybula lśniąca
Ro (0) asmăţui de munte
Ru (0) купырь блестящий
Sk (–) trebuľka lesklá

* ***Anthriscus sicula*** Vis. → *Anthriscus sylvestris*

137 *Anthriscus sylvestris* (L.) Hoffm.
Al (–) antriskusi i pyllit
Bg (0) горски азмацук
Cs (3) kerblík lesní

De (–) Wiesen-Kerbel
En (–) cow-parsley
Hu (1) erdei turbolya
Pl (1) trybula leśna
Ro (0) haşmaciucă
Ru (1) купырь лесной
Sh (1) velika krasuljica
Sk (–) trebuľka lesná

* **Anthriscus vulgaris** Pers. → *Anthriscus caucalis*

138 Anthyllis spp. L.
De (–) Wundklee
Gr (1) ανθυλλίδα
Ru (0) язвенник
Sk (–) úročník

* **Anthyllis alpestris** Heget. → *Anthyllis vulneraria*

139 Anthyllis vulneraria L.
Al (–) antilida shërojcë
Bg (0) раноцелебна раменка
Cs (1) úročník bolhoj
De (–) Gemeiner Wundklee
En (–) kidney-vetch
Hu (0) réti nyúlhere
Pl (1) przelot pospolity
Ro (0) vătămătoare
Ru (1) язвенник Линнея
Sh (1) obični belodun; obični bjelodun
Sk (–) bôľhoj lekársky

* **Anthyllis weldeniana** Reichenb. → *Anthyllis vulneraria*

140 Antirrhinum majus L.
Al (–) gojasllani i madhë
Bg (0) лъбска муцунка
Cs (1) hledík větší
De (–) Großes Löwenmaul
En (–) common snapdragon
Gr (1) σκυλάκι
Hu (0) kerti oroszlánszáj
Ro (0) gura leului
Ru (0) львиный зев
Sk (–) škľabinec väčší

* **Antirrhinum orontium** L. → *Misopates orontium*

141 Apera interrupta (L.) Beauv.
Cs (0) chundelka přetrhovaná
De (–) Unterbrochener Windhalm
En (–) dense silky-bent
Hu (1) rongyos széltippan
Ru (1) мятлица прерывистая
Sk (–) metlička pretrhovaná

142 Apera spica-venti (L.) Beauv.
Al (–) apera kallierë
Bg (2) ветрушка
Cs (3) chundelka metlice
De (–) Gemeiner Windhalm
En (–) loose silky-bent
Gr (1) ανεμόχορτο
Hu (3) nagy széltippan
Pl (3) miotła zbożowa
Ro (3) iarba vîntului
Ru (3) мятлица обыкновенная
Sh (2) obična slakoperka; obični stršac
Sk (–) metlička obyčajná

143 Aphanes arvensis L.
Al (–) këmbë luani
Bg (1) полско шапиче
Cs (1) nepatrnec rolní
De (–) Acker-Sinau
En (–) parsley-piert
Hu (3) nagy ugarpalástfű
Pl (2) skrytek polny
Ro (1) creţişoare
Ru (2) манжетка полевая
Sh (2) njivski virak; poljska vrkuta
Sk (–) drobnobyľ roľná

144 ***Aquilegia vulgaris*** L.
Al (–) akilega e zakonshme; kolumbina
Bg (0) обикновена кандилка
Cs (0) orlíček planý
En (–) columbine
Pl (1) orlik pospolity
Ro (0) căldăruşă
Ru (0) водосбор обыкновенный
Sk (–) orlíček pospolitý

145 ***Arabidopsis pumila*** (Steph.) Busch.
Ru (1) резушка низкая
Sk (–) hydozelíček trpasličí

146 ***Arabidopsis thaliana*** (L.) Heynh.
Al (–) arabidopsisi i Talies
Bg (1) Талев арабидопсис
Cs (1) huseníček rolní
De (–) Acker-Schmalwand
En (–) thale-cress
Gr (1) αραβίδοψη
Hu (1) lúdfű
Pl (2) rzodkiewnik pospolity
Ro (2) gîscariţă
Ru (2) резушка Таля
Sh (1) obični uročnjak
Sk (–) hydozelíček roľný

147 ***Arabis* spp.** L.
Bg (0) гъшарка
De (–) Gänsekresse
Gr (1) αραβίδα
Ru (1) резуха
Sk (–) hydozel

* ***Arabis arenosa*** (L.) Scop. → *Cardaminopsis arenosa*

* ***Arabis auriculata*** DC. → *Arabis recta*

148 ***Arabis glabra*** (L.) Bernh.
En (–) tower mustard
Ro (0) turicel
Ru (1) башеница голая
Sh (1) glatka gušarka
Sk (–) hydozel lysý

149 ***Arabis hirsuta*** (L.) Scop.
Al (–) arabza qimeashpër
Bg (0) косматa гъшарка
Cs (0) huseník chlupatý
De (–) Rauhe Gänsekresse
En (–) hairy rock-cress
Hu (0) borzas ikravirág
Pl (1) gąsiówka szorstkowłosista
Ro (0) gîscariţă
Ru (1) резуха шершавая
Sh (1) dlakava gušarka
Sk (–) hydozel chlpatý

150 ***Arabis pendula*** L.
Ru (1) резуха повислая

151 ***Arabis recta*** Vill.
Bg (0) права гъшарка
Cs (0) huseník ouškatý
De (–) Öhrchen-Gänsekresse
En (–) annual rock-cress
Hu (1) füles ikravirág
Pl (0) gąsiówka uszkowata
Ro (0) gîscariţă
Ru (0) резуха прямая
Sk (–) hydozel ušatý

* ***Arabis thaliana*** L. → *Arabidopsis thaliana*

152 ***Arabis verna*** (L.) R Br.
Al (–) arabza pranverore
En (–) spring rock-cress
Ru (1) резуха весенняя
Sh (1) prolećna gušarka; proljetna gušarka
Sk (–) hydozel jarný

153 ***Arctium lappa*** L.
Al (–) rrodhja
Bg (1) обикновен репей

Cs (3) lopuch větší
De (–) Große Klette
En (–) greater burdock
Hu (2) nagy bojtorján
Pl (2) łopian większy
Ro (1) brustulan
Ru (2) лопух репейник
Sh (1) obični čičak
Sk (–) lopúch obyčajný

154 ***Arctium minus*** Bernh.
Al (–) rrodhja e vogël
Bg (0) малък репей
Cs (0) lopuch menší
De (–) Kleine Klette
En (–) lesser burdock
Hu (1) kis bojtorján
Pl (2) łopian mniejszy
Ru (1) лопух малый
Sh (1) crveni čičak; zelenkasti čičak
Sk (–) lopúch menší

155 ***Arctium tomentosum*** Mill.
Bg (0) мъхнат репей
Cs (3) lopuch plstnatý
De (–) Filzige Klette
En (–) woolly burdock
Hu (1) pókhálós bojtorján
Pl (1) łopian pajęczynowaty
Ro (1) brusture
Ru (1) лопух войлочный
Sh (1) pusteni čičak
Sk (–) lopúch plstnatý

156 ***Arctostaphylos uva-ursi*** (L.) Spreng.
Al (–) rrush arushe
Bg (0) мечо грозде
Cs (–) medvědice léčivá
De (–) Immergrüne Bärentraube
En (–) bearberry
Pl (1) mącznica lekarska
Ro (0) strugurii ursului
Ru (0) толокнянка медвежья
Sk (–) medvedica lekárska

157 ***Arenaria* spp.** L.
Bg (–) пясъчница
De (–) Sandkraut
En (–) sandwort
Gr (1) αμμόφυτο
Ru (1) песчанка
Sk (–) piesočnica

158 ***Arenaria ciliata*** L.
Cs (0) písečnice brvitá
De (–) Bewimpertes Sandkraut
En (–) hairy sandwort
Pl (1) piaskowiec orzęsiony
Sk (–) piesočnica riasinková

* ***Arenaria rubra*** L. → *Spergularia rubra*

159 ***Arenaria serpyllifolia*** L.
Al (–) arenaria me gjethe lisre
Bg (1) полегнала пясъчница
Cs (2) písečnice douškolistá
De (–) Quendelblättriges Sandkraut
En (–) thyme-leaved sandwort
Hu (3) kakukkhomokhúr
Pl (2) piaskowiec macierzankowy
Ro (2) studeniţă
Ru (2) песчанка тимьянолистная
Sh (2) obična pešćanka; obična pjeskarica
Sk (–) piesočnica duškolistá

160 ***Argusia sibirica*** (L.) Dandy
Ru (1) аргузия сибирская

161 ***Arisarum vulgare*** Targ.
De (–) Kappenzehrwurz
En (–) friar's cowl
Gr (1) λυχνοδρακοντιά
Ru (0) аризарум обыкновенный

162 ***Aristolochia clematitis*** L.
Al (–) kolumaqe; petriku kulpër
Bg (2) обикновена вълча ябълка
Cs (1) podražec křovištní

De (–) Gemeine Osterluzei
En (–) birthwort
Gr (1) αριστολόχια
Hu (2) közönséges farkasalma
Pl (0) kokornak powojnikowy
Ro (3) cucurbețică
Ru (1) кирказон ломоносовидный
Sh (2) obična kokotinja; vučja stopa
Sk (–) vlkovec obyčajný

163 ***Armeria rumelica*** Boiss.
Bg (1) румелийска армерия
Sk (–) ťravnička rumelská

164 ***Armoracia rusticana*** Gaertn., Meyer & Scherb.
Bg (0) хрян
Cs (2) křen selský
De (–) Meerrettich
En (–) horse-radish
Hu (1) közönséges torma
Pl (2) chrzan pospolity
Ro (1) hrean
Ru (1) хрен деревенский
Sh (1) divlji hren; seoski hren
Sk (–) chren dedinský

165 ***Arnica montana*** L.
Al (–) duhan mali
Cs (1) prha arnika
De (–) Arnika
En (–) arnica
Hu (0) árnika
Pl (1) arnika górska
Ro (0) arnică
Ru (0) арника горная
Sk (–) hornica

166 ***Arnoseris minima*** (L.) Schweig. & Koer.
Cs (0) písečnatka nejmenší
De (–) Lämmersalat
En (–) lamb's succory
Pl (1) chłodek drobny
Ru (0) арнозерис малый
Sk (–) jahňací šalát

167 ***Arrhenatherum elatius*** (L.) Beauv.
Al (–) tërshëra e egër; arenatera e lartë
Bg (0) френцузки райграс
Cs (2) ovsík vyvýšený
De (–) Glatthafer
En (–) false oat-grass
Hu (0) francia perje
Pl (2) owsik wyniosły; rajgras wyniosły
Ro (0) ovăscior
Ru (1) райграс высокий
Sk (–) ovsík vyvýšený

168 ***Artemisia*** **spp.** L.
Bg (1) пелин
De (–) Beifuß
En (–) wormwood
Gr (1) αρτεμίσια
Ru (2) полынь
Sk (–) palina

169 ***Artemisia abrotanum*** L.
Al (–) bar kamfuri
Cs (1) pelyněk brotan
De (–) Eberrraute
En (–) southernwood
Hu (0) istenfa
Pl (0) boże drzewko
Ro (0) lemnul Domnului
Ru (0) полынь лечебная
Sk (–) palina božie drievko

170 ***Artemisia absinthium*** L.
Al (–) pelini
Bg (1) обикновен пелин
Cs (1) pelyněk pravý
De (–) Wermut
En (–) common wormwood
Gr (1) αψιθιά
Hu (1) fehér üröm
Pl (1) bylica piołun

Ro (1) pelin alb
Ru (1) полынь горькая
Sh (1) gorki pelin; obični pelen
Sk (–) palina pravá

171 *Artemisia annua* L.
Bg (1) едногодишен пелин
Cs (0) pelyněk roční
De (–) Einjähriger Beifuß
En (–) annual wormwood
Hu (1) egynyári üröm
Pl (1) bylica roczna
Ro (1) năfurică
Ru (1) полынь однолетняя
Sh (1) jednogodišnji pelin
Sk (–) palina ročná

172 *Artemisia austriaca* Jacq.
Bg (0) австрийски пелин
Cs (0) pelyněk rakouský
De (–) Österreichischer Beifuß
En (–) Austrian wormwood
Hu (1) selymes üröm
Pl (1) bylica austriacka
Ro (1) pelinițã
Ru (1) полынь австрийская
Sk (–) palina rakúska

173 *Artemisia campestris* L.
Al (–) pelini i fushës
Bg (0) полски пелин
Cs (0) pelyněk ladní
De (–) Feld-Beifuß
En (–) field wormwood
Hu (1) mezei üröm
Pl (1) bylica polna
Ro (0) pelin nemirositor
Ru (1) полынь полевая
Sh (–) poljski pelin
Sk (–) palina poľná pravá

* ***Artemisia matricarioides*** Less. → *Chamomilla suaveolens*

* ***Artemisia paniculata*** Lam. → *Artemisia abrotanum*

174 *Artemisia scoparia* Waldst. & Kit.
Bg (1) метловиден пелин
Cs (1) pelyněk metlovitý
De (–) Besen-Beifuß
Hu (1) seprő üröm
Pl (1) bylica miotłowa
Ro (0) pelin de mături
Ru (1) полынь метельчатая
Sk (–) palina metlovitá

175 *Artemisia siversiana* (Ehrh.) Willd.
Ru (1) полынь Сиверса
Sk (–) palina sieversova

176 *Artemisia vulgaris* L.
Al (–) pelini i egër
Bg (1) див пелин
Cs (3) pelyněk černobýl
De (–) Gemeiner Beifuß
En (–) mugwort
Hu (3) fehér üröm
Pl (2) bylica pospolita
Ro (1) pelinarițã
Ru (2) полынь обыкновенная
Sh (2) crni pelin; obični pelin
Sk (–) palina obyčajná

177 *Arum italicum* Mill.
Al (–) kallomoç gjarpni; këlkaza
Bg (0) италиански змиярник
Cs (0) aron italský
De (–) Aronstab
En (–) Italian arum
Gr (2) κοινή δρακοντιά
Hu (0) olasz kontyvirág
Ru (0) аронник итальянский
Sk (–) aron taliansky

178 *Arum maculatum* L.
Al (–) bar plagësh
Bg (0) петнист змиярник

Cs (1) aron plamatý
De (–) Gefleckter Aronstab
En (–) cuckoo-pint
Gr (2) στικτή δρακοντιά
Hu (0) foltos kontyvirág
Pl (1) obrazki plamiste
Ro (0) rodul pămîntului
Ru (1) аронник пятнистый
Sk (–) aron škvrnitý

179 *Aruncus dioicus* (Walt.) Fern.
Cs (1) udatna lesní
De (–) Wald-Geißbart
En (–) goat's-beard
Hu (0) tündérfürt
Pl (1) parzydło leśne
Ro (0) barba popei
Ru (0) волжанка
Sk (–) udatka lesná

180 *Arundo donax* L.
Al (–) kallami
Bg (0) италианска тръстика
De (–) Pfahlrohr
En (–) giant reed
Gr (3) καλάμι
Ro (0) trestie italiană
Ru (1) арундо тростниковый
Sk (–) trsť obrovská

181 *Asarum europaeum* L.
Al (–) kopedani
Bg (0) европейски копитник
Cs (1) kopytník evropský
De (–) Haselwurz
En (–) European ginger
Hu (0) kapotnyak
Pl (1) kopytnik pospolity
Ro (0) pochivnic
Ru (0) копытень европейский
Sk (–) kopytník európsky

182 *Asclepias syriaca* L.
Al (–) kërpi i Sirisë
Cs (1) klejicha vatočník
De (–) Seidenpflanze
En (–) common milkweed
Hu (2) selyemkóró
Pl (0) trojeść amerykańska
Ro (0) ceara albinei
Ru (0) ваточник сирийский
Sh (1) cigansko perje; pustenolisna svilenica
Sk (–) glejovka sýrska

183 *Asparagus* spp. L.
Bg (–) аспержа
De (–) Spargel
En (–) asparagus
Gr (1) αγριοσπαράγγι
Ru (0) спаржа
Sk (–) špargľa

184 *Asparagus officinalis* L.
Al (–) shpargulli i rëndomtë
Bg (1) лечебна зайча сянка
Cs (1) chřest lékařský
De (–) Spargel
En (–) wild asparagus
Hu (1) nyúlárnyék; spárga
Pl (1) szparag lekarski
Ro (0) sparanghel
Ru (0) спаржа лекарственная
Sk (–) špargľa lekárska

185 *Asperugo procumbens* L.
Bg (1) разперен остред
Cs (2) ostrolist poléhavý
De (–) Scharfkraut
En (–) madwort
Gr (1) ασπέρουγκος
Hu (1) magiszák
Pl (1) lepczyca rozesłana
Ro (1) lipicioasă
Ru (1) асперуга простёртая
Sh (1) obični prilep; puzava broćanica
Sk (–) ostrolist poliehavý

186 ***Asperula* spp.** L.
Bg (1) лазаркиня
De (–) Meier
En (–) woodruff
Gr (1) ασπέρουλα
Ru (0) ясменник
Sk (–) marinka

187 ***Asperula arvensis*** L.
Bg (1) полска лазаркиня
Cs (1) mařinka rolní
De (–) Acker-Meier
En (–) blue woodruff
Hu (1) vetési müge
Ro (1) lipitoare
Ru (1) ясменник полевой
Sh (1) njivska lazarkinja
Sk (–) marinka roľná

* ***Asperula ciliata*** Roch. → *Asperula tinctoria*

188 ***Asperula cynanchica*** L.
Cs (0) mařinka psí
De (–) Hügel-Meier
En (–) squinancywort
Hu (0) ebfojtó müge
Pl (1) marzanka pogórkowata
Ru (0) ясменник розоватый
Sk (–) marinka psia

189 ***Asperula tinctoria*** L.
Bg (0) ресничеста лазаркиня
Cs (0) mařinka barvířská
De (–) Färber-Meier
En (–) dyer's woodruff
Hu (0) festő müge
Pl (1) marzanka barwierska
Ro (0) lipitoare de vopsit
Ru (0) ясменник красильный
Sk (–) marinka farbiarska

190 ***Asphodelus aestivus*** Brot.
Al (–) badhra frutvogël
En (–) asphodel
Gr (2) ασφόδελος
Ru (0) асфоделус летний
Sk (–) kopýtko letné

* ***Asphodelus microcarpus*** Viv. → *Asphodelus aestivus*

191 ***Asplenium trichomanes*** L.
Al (–) fierguri trikoman
Bg (0) влакнесто изтравниче
Cs (1) sleziník červený
De (–) Brauner Streifenfarn
En (–) maidenhair spleenwort
Hu (0) aranyos fodorka
Pl (1) zanokcica skalna
Ro (0) straşnic
Ru (0) костенец волосовидный
Sh (1) krugolisna slezenica; okrugla sleznica
Sk (–) slezinník červený

192 ***Aster amellus*** L.
Bg (0) обикновено димитровче
Cs (0) hvězdnice chlumní
De (–) Berg-Aster
En (–) summer Michaelmas-daisy
Hu (0) csillag őszirózsa
Pl (1) aster gawędka
Ro (1) steliţă vînătă
Ru (1) астра итальянская
Sk (–) pleskáč horský

193 ***Aster lanceolatus*** Willd.
Cs (0) hvězdnice kopinatá
De (–) Lanzettblättrige Aster
Hu (0) lándzsás őszirózsa
Pl (1) aster lancetowaty
Ru (1) астра ланцетовидная
Sk (–) pleskáč kopijovitý

194 ***Aster novi-belgii*** L.
Cs (0) hvězdnice novobelgická
De (–) Neubelgische Aster

En (–) common Michaelmas-daisy
Hu (0) sötétlila őszirózsa
Pl (1) aster wirginijski
Ru (0) астра виргинская
Sk (–) pleskáč virginský

* **Aster tradescantia** auct. → *Aster lanceolatus*

195 Asteriscus spp. Mill.
Gr (1) αστερίσκος
Ru (0) астерискус

* **Asteriscus spinosus** (L.) Schultz → *Pallenis spinosa*

196 Astragalus spp. L.
Bg (0) клин
De (–) Tragant
En (–) milkvetch
Gr (1) αστράγαλος
Ru (1) астрагал
Sk (–) kozinec

197 Astragalus austriacus Jacq.
Bg (0) австрийски клин
Cs (1) kozinec rakouský
De (–) Österreichischer Tragant
En (–) Austrian milkvetch
Hu (0) kisvirágú csüdfű
Pl (0) traganek austriacki
Ru (0) астрагал австрийский
Sk (–) kozinec rakúsky

198 Astragalus cicer L.
Al (–) qiqri i egër
Bg (0) нахутово клинавче
Cs (1) kozinec cizrnovitý
De (–) Kicher-Tragant
En (–) chickpea milkvetch
Hu (0) hólyagos csüdfű
Pl (1) traganek pęcherzykowaty
Ro (0) coşaci
Ru (0) астрагал нутовый
Sk (–) kozinec cícerovitý

199 Astragalus glycyphyllos L.
Al (–) arithja gjetheëmbël
Bg (0) сладколистно клинавче
Cs (1) kozinec sladkolistý
De (–) Süßer Tragant
En (–) common milkvetch
Hu (0) édeslevelű csüdfű
Pl (1) traganek szerokolistny
Ro (1) unghia găii
Ru (1) астрагал сладколистный
Sh (1) medika kozinac; puzavi kozlinac
Sk (–) kozinec sladkolistý

200 Astragalus onobrychis L.
Al (–) arithja esparsetë
Bg (0) еспарзетов клин
Cs (1) kozinec vičencovitý
De (–) Esparsetten-Tragant
En (–) sainfoin milkvetch
Hu (0) zászlós csüdfű
Pl (1) traganek długokwiatowy
Ro (0) coşaci
Ru (1) астрагал эспарцетовый
Sh (1) tankolisni kozlinac
Sk (–) kosinec ľadničkovitý

201 Astrantia major L.
Al (–) astranta e madhe
Bg (0) едра астранция
Cs (1) jarmanka větší
De (–) Große Sterndolde
En (–) tall sanicle
Hu (0) völgycsillag
Pl (1) jarzmianka większa
Ro (0) ştevie de munte
Ru (0) астранция крупная
Sk (–) jarmanka väčšia

202 Athyrium filix-femina (L.) Roth
Al (–) atiriumi fier femër
Bg (0) женска папрат
Cs (1) papradka samice
De (–) Wald-Frauenfarn
En (–) lady-fern

Hu (0) hölgypáfrány
Pl (1) wietlica samcza
Ro (0) spinarea lupului
Ru (0) кочедыжник женский
Sh (1) ženska bujadika
Sk (–) papradka ženská lesná

203 ***Atriplex* spp.** L.
Bg (1) лобода
De (–) Melde
Gr (1) ατρίπλια
Ru (2) лебеда
Sk (–) loboda

* ***Atriplex acuminatum*** Waldst. & Kit. → *Atriplex nitens*

* ***Atriplex angustifolium*** Gaudin → *Atriplex patula*

204 ***Atriplex hastata*** L.
Al (–) laboti si shigjetë
Bg (1) копиевидна лобода
Cs (–) lebeda hrálovitá
De (–) Spießblättrige Melde
En (–) spear-leaved orache
Hu (1) dárdás laboda
Pl (1) łoboda oszczepowata
Ru (1) лебеда копьевидная
Sh (1) kopljasta loboda
Sk (–) loboda strelovitá

205 ***Atriplex hortensis*** L.
Al (–) laboti i kopshtit
Bg (0) градинска лобода
Cs (1) lebeda zahradní
De (–) Garten-Melde
En (–) garden orache
Hu (1) kerti laboda
Pl (1) łoboda ogrodowa
Ro (0) lobodă de grădină
Ru (1) лебеда садовая
Sh (2) baštenska loboda
Sk (–) loboda záhradná

* ***Atriplex latifolium*** Wahlen. → *Atriplex hastata*

206 ***Atriplex nitens*** Schk.
Bg (0) лъскава лобода
Cs (3) lebeda lesklá
De (–) Glänzende Melde
En (–) glossy-leaved orache
Hu (1) fényes laboda
Pl (2) łoboda błyszcząca
Ro (1) lobodă de drumuri
Ru (1) лебеда лоснящаяся
Sk (–) loboda leskla

207 ***Atriplex oblongifolia*** Waldst. & Kit.
Bg (0) продълговата лобода
Cs (2) lebeda podlouhlolistá
De (–) Langblättrige Melde
En (–) long-leaved orache
Hu (1) hosszúlevelű laboda
Pl (1) łoboda wąskolistna
Ru (1) лебеда продолголистная
Sk (–) loboda dlholistá

208 ***Atriplex patula*** L.
Bg (1) разстлана лобода
Cs (3) lebeda rozkladitá
De (–) Gemeine Melde
En (–) common orache
Hu (2) terebélyes laboda
Pl (2) łoboda rozłożysta
Ru (2) лебеда раскидистая
Sh (2) divlja loboda; obična loboda
Sk (–) loboda rozložitá

* ***Atriplex polysperma*** Cran. → *Chenopodium polyspermum*

209 ***Atriplex rosea*** L.
Al (–) laboti ngjyrë trëndafili
Bg (1) розова лобода
Cs (0) lebeda růžová
De (–) Rosen-Melde
Hu (1) fehér laboda

Pl (2) łoboda gwiazdowata
Ru (1) лебеда розовая
Sk (–) loboda ružová

210 *Atriplex sphaeromorpha* Iljin
Ru (1) лебеда шарообразная

211 *Atriplex tatarica* L.
Al (–) laboti i Tatarisë
Bg (1) татарска лобода
Cs (1) lebeda tatarská
De (–) Tatarische Melde
Hu (2) tatár laboda
Pl (1) łoboda szara
Ro (3) lobodă sălbatică
Ru (2) лебеда татарская
Sk (–) loboda tatárska

* ***Atriplex triangularis*** Willd. → *Atriplex patula*

212 *Atropa bella-donna* L.
Al (–) helmarina
Bg (0) старо биле
Cs (1) rulík zlomocný
De (–) Tollkirsche
En (–) deadly nightshade
Gr (1) μπελλαντόνα
Hu (0) nadragulya
Pl (1) pokrzyk wilczajagoda
Ro (0) mătrăgună
Ru (0) красавка белладонна
Sh (1) gorsko velebilje
Sk (–) ľuľkovec zlomočný

213 *Avena barbata* Pott
Al (–) tërshëra mjekëroshe
Bg (0) брадат овес
De (–) Barthafer
En (–) slender oat
Gr (2) μικρή αγριοβρώμη
Ru (1) овёс бородатый
Sk (–) ovos bradatý

* ***Avena elatior*** L. → *Arrhenatherum elatius*

214 *Avena fatua* L.
Al (–) tërshëra e egër
Bg (2) лив овес
Cs (3) oves hluchý
De (–) Flughafer
En (–) spring wild-oat
Hu (3) héla zab
Pl (3) owies głuchy
Ro (3) odos
Ru (3) овёс пустой; овсюг пустой
Sh (2) divlja zob; divlji ovas
Sk (–) ovos hluchý

* ***Avena ludoviciana*** Durieu → *Avena sterilis* ssp. *ludoviciana*

* ***Avena nigra*** Wallr. → *Avena fatua*

* ***Avena pratensis*** L. → *Avenula pratensis*

* ***Avena pubescens*** Huds. → *Avenula pubescens*

215 *Avena sativa* L.
Al (–) tërshëra
Bg (0) обикновен овес
Cs (1) oves setý
De (–) Saat-Hafer
En (–) common oat
Hu (0) abrak zab
Pl (1) owies zwyczajny
Ro (0) ovăz
Ru (1) овёс посевной
Sk (–) ovos siaty

216 *Avena sterilis* L. **spp. *ludoviciana*** (Durieu) Nyman
Bg (2) Людовиков овес
Cs (0) oves hluchý ozimý

De (–) Winter-Flughafer
En (–) winter wild-oat
Ro (1) obsigă
Ru (1) овёс Людовика
Sh (1) Ludovikov ovas; Ludovikova zob
Sk (–) ovsík ozimný

217 ***Avena sterilis*** L. **ssp.** ***sterilis***
Al (–) tërshëra shterpë
Cs (–) oves jalový
De (–) Tauber Hafer
En (–) sterile oat
Gr (3) μεγάλη αγριοβρώμη
Ru (1) овёс стерильный
Sk (–) ovos jalový

218 ***Avena strigosa*** Schreb.
Al (–) tërshëra e ranishteve
Cs (0) oves hubený
De (–) Sand-Hafer
En (–) bristle oat
Pl (1) owies szorstki
Ru (0) овёс щетинистый
Sk (–) ovos ježatý

* ***Avena trichophylla*** Koch → *Avena sterilis* ssp. *sterilis*

219 ***Avenula pratensis*** (L.) Dumort.
Cs (0) oves lučni
De (–) Rauher Wiesenhafer
En (–) meadow oat-grass
Pl (1) owsica łąkowa
Ru (0) овсец луговой
Sk (–) ovsica lúčna

220 ***Avenula pubescens*** (Huds.) Dumort.
Cs (0) oves pýřitý
De (–) Flaumiger Wiesenhafer
En (–) downy oat-grass
Pl (1) owsica omszona
Ro (0) ovăscior
Ru (0) овсец опушенный
Sk (–) ovsica páperistá

221 ***Axyris amaranthoides*** L.
De (–) Amaranthähnliches Axyris
En (–) Russian pigweed
Ru (1) аксирис обыкновенный
Sk (–) sivovlasec láskavcovitý

B

* **Baldingera arundinacea** (L.) Dumort. → *Phalaris arundinacea*

* **Ballota foetida** Lam. → *Ballota nigra*

* **Barbarea arcuata** (Opiz) Reichenb. → *Barbarea vulgaris*

222 **Ballota nigra** L.
Al (–) balota e zezë
Bg (1) черно кандилниче
Cs (2) měrnice černá
De (–) Stinkandorn
En (–) black horehound
Hu (2) fekete peszterce
Pl (1) mierznica czarna
Ro (1) cătuşe
Ru (1) белокудренник чёрный
Sh (1) modri tetrljan; obični crnoglavac
Sk (–) mernica čierna

223 **Barbarea vulgaris** R Br.
Al (–) barbarea e zakonshme
Bg (1) обикновена злина
Cs (2) barborka obecná
De (–) Echtes Barbenkraut
En (–) winter-cress
Hu (1) közönséges borbálafű
Pl (2) gorczycznik pospolity
Ro (1) bărbuşoară
Ru (2) сурепица обыкновенная
Sh (1) obična repnica
Sk (–) barborka obyčajná

224 **Bassia hyssopifolia** (Pallas) Volk.
En (–) hyssop-leaved seablite
Ru (1) бассия иссополистная
Sk (–) trnoplodník hyzopolistý

225 **Beckmannia eruciformis** (L.) Host
Al (–) bekmania
Bg (1) биволско просо
Cs (0) housenkovec roketový
De (–) Raupenähre
En (–) Japanese sloughgrass
Gr (1) μπεκμάνια
Hu (1) hernyópázsit
Pl (2) bekmania robaczkowata
Ro (0) ierboi de bahnă
Ru (1) бекманния обыкновенная
Sh (1) obična bekmanija
Sk (–) beckmannia zdurelá

226 **Beckmannia syzigachne** (Steud.) Fern.
Ru (1) бекманния восточная

227 **Bellis spp.** L.
De (–) Gänseblümchen
Ru (1) маргаритка
Gr (1) μπέλλα

228 **Bellis perennis** L.
Al (–) luledhensh; luleshqerra
Bg (0) паричка
Cs (2) sedmikráska chudobka
De (–) Gemeines Gänseblümchen
En (–) daisy
Hu (1) százszorszép
Pl (2) stokrotka łąkowa
Ro (0) părăluţă
Ru (0) маргаритка многолетняя
Sh (1) obični krasuljak; trajnocvjetna tratinčica
Sk (–) sedmokráska obyčajná

229 **Berberis cretica** L.
Gr (1) ξαγκαθιά
Ru (0) барбарис критский
Sk (–) dráč krétsky

230 **Berberis vulgaris** L.
Al (–) mëlqinja e rëndomtë
Bg (0) обикновен кисел трън

Cs (2) dřišťál obecný
De (–) Gemeiner Sauerdorn
En (–) barberry
Gr (1) ξαγκαθιά
Hu (0) sóskafa
Pl (1) berberys zwyczajny
Ro (1) dracilă
Ru (0) барбарис обыкновенный
Sh (1) obična šimširika
Sk (–) dráč obyčajný

231 **Berteroa incana** (L.) DC.
Al (–) berteroja e përhimët
Bg (1) сива турия
Cs (1) šedivka šedivá
De (–) Gemeine Graukresse
En (–) hoary alison
Hu (1) hamuka
Pl (1) pyleniec pospolity
Ro (1) ciucuşoară
Ru (1) икотник седой
Sh (2) obična sivka; obični sivac
Sk (–) šedivka sivá

232 **Berteroa mutabilis** (Vent.) DC.
Bg (0) изменчива турия
Gr (1) μπερτερόα
Ru (0) икотник изменчивый
Sk (–) šedivka premenlivá

233 **Berula erecta** (Huds.) Cov.
Al (–) siumi gjethengushtë
Cs (0) potočník vzpřímený
De (–) Aufrechter Merk
En (–) lesser water-parsnip
Hu (1) mérges békakorsó
Pl (1) potocznik wąskolistny
Ro (0) cosiţel
Ru (1) берула прямостоячая
Sh (1) obični bobur; uskolisni hrenak
Sk (–) berla úzkolistá

234 **Beta vulgaris** L.
Al (–) panxhar i rëndomtë; pazi
Bg (0) обикновено цвекло
Cs (0) řepa burák
De (–) Wilde Runkelrübe
En (–) sea beet
Hu (0) répa
Pl (1) burak zwyczajny
Ro (0) sfeclă
Ru (0) свёкла обыкновенная
Sk (–) repa obyčajná

* **Betonica annua** L. → *Stachys annua*

* **Betonica officinalis** L. → *Stachys officinalis*

* **Betonica serotina** Host → *Stachys officinalis*

235 **Betula pubescens** Ehrh.
Al (–) mështekna pushëlore
Cs (0) bříza pýřitá
De (–) Moor-Birke
En (–) common birch
Hu (0) szőrös nyírfa
Pl (1) brzoza omszona
Ro (–) mesteacăn pufos
Ru (0) берёза пушистая
Sk (–) breza pýrita pravá

236 **Bidens bipinnata** L.
De (–) Fiederblättriger Zweizahn
Ru (1) череда дваждыперистая

237 **Bidens cernua** L.
Bg (1) наведен бутрак
Cs (0) dvouzubec nící
De (–) Nickender Zweizahn
En (–) nodding bur-marigold
Hu (1) bókoló farkasfog
Pl (1) uczep uwisły
Ro (1) iarbă roşie
Ru (1) череда поникшая
Sk (–) dvojzubec prievislý

238 *Bidens radiata* Thuill.
Ru (1) череда лучистая
Sk (–) dvojzubec skvelý

239 *Bidens tripartita* L.
Al (–) dydhëmbëshja tripjesëshe
Bg (1) тpиделeн бутрак
Cs (3) dvouzubec trojdílný
De (–) Dreiteiliger Zweizahn
En (–) trifid bur-marigold
Hu (2) subás farkasfog
Pl (2) uczep trójlistkowy
Ro (2) dentiţă
Ru (2) череда трёхраздельная
Sh (2) obični dvozub; trodjelni dvozub
Sk (–) dvojzubec trojdielny

* ***Bifora floscula*** Bieb. → *Bifora testiculata*

240 *Bifora radians* Bieb.
Al (–) bifora rrezake
Bg (2) колендро
Cs (1) štěničník paprskující
De (–) Strahlender Hohlsame
Gr (2) μπιφόρα
Hu (2) poloskafű
Pl (2) sparzyca promienista
Ro (2) iarbă puturoasă
Ru (2) бифора лучистая
Sh (1) obična smrdulja
Sk (–) dvojguľka lúčovitá

241 *Bifora testiculata* (L.) Roth
Bg (1) яйцевидно колендро
De (–) Kleiner Koriander
En (–) small coriander
Hu (2) poloskafű
Ru (1) бифора яйцевидная
Sk (–) dvojguľka vreckatá

242 *Bilderdykia convolvulus* (L.) Dumort.
Al (–) bilderdykia dredhje
Bg (2) повитиновидно пиперче
Cs (3) pohanka svlačcovitá
De (–) Winden-Knöterich
En (–) black-bindweed
Gr (2) αναρριχώμενο πολύγωνο
Hu (3) ugari szulákpohánka
Pl (3) rdest powojowy
Ro (3) hrişcă urcătoare
Ru (3) горец вьюнковый
Sh (2) njivski vijušac
Sk (–) pohanka opletavá

243 *Bilderdykia dumetorum* (L.) Dumort.
Al (–) bilderdykia shkurretash
Cs (0) pohanka křovištní
De (–) Hecken-Knöterich
En (–) copse-bindweed
Hu (1) erdei szulákpohánka
Pl (2) rdest zaroślowy
Ro (0) hrişcă deasă
Ru (1) горец кустарниковый
Sh (1) veliki vijušac
Sk (–) pohanka krovisková

* ***Bistorta major*** Gray → *Polygonum bistorta*

244 *Blackstonia perfoliata* (L.) Huds.
Cs (0) žlutnice prorostlá
En (–) yellow-wort
Hu (1) gyíkpohár
Ru (0) блакстония пронзённолистная
Sk (–) zarcanka prerastaná pravá

* ***Blackstonia serotina*** (Koch) Beck → *Blackstonia perfoliata*

245 *Blechnum spicant* (L.) Roth
Cs (0) žebrovice různolistá
De (–) Wald-Rippenfarn
En (–) hard fern
Hu (0) bordapáfrány
Pl (1) podrzeń żebrowiec
Ro (0) scăriţă muntelui

Ru (0) дербянка колосистая
Sk (–) rebrovka rôžnolistá

246 ***Blysmus compressus*** (L.) Panz.
Al (–) blisma e rrasur
Bg (0) плосък блисмус
Cs (0) skřípinka smáčknutá
De (–) Zusammengedrücktes Quellried
En (–) flat-sedge
Hu (0) kétsoros káka
Pl (1) ostrzew spłaszczowy
Ru (0) блисмус сжатый
Sk (–) škripinka stisnutá

* ***Bolboschoenus maritimus*** (L.) Palla → *Scirpus maritimus*

* ***Bolboschoenus popovii*** Egorova → *Scirpus maritimus*

247 ***Bombycilaena erecta*** (L.) Smolj.
Bg (1) изправен микропус
De (–) Falzblume
En (–) micropus
Hu (1) topagyom
Ru (1) бомбицилена прямостоящая
Sh (1) uspravna malica
Sk (–) plsťovka rovná

248 ***Borago officinalis*** L.
Al (–) shaja mjekësore
Bg (–) пореч
Cs (1) brutnák lékařský
De (–) Boretsch, Borretsch
En (–) borage
Gr (1) βορατσίνο
Pl (1) ogórecznik lekarski
Ro (0) limba mielului
Ru (0) бурачник лекарственный
Sk (–) borák lekársky

* ***Bothriochloa ischaemum*** (L.) Keng → *Dichanthium ischaemum*

249 ***Brachiaria eruciformis*** (Sibth. & Sm.) Gris.
Gr (1) βραχιάρια
Ru (1) ветвянка гусеницевидная
Sk (–) ježatka rakytová

250 ***Brachypodium*** **spp.** Beauv.
Bg (0) късонот
De (–) Zwenke
Gr (1) βραχυπόδιο
Ru (0) коротконожка
Sk (–) mrvica

251 ***Brachypodium pinnatum*** (L.) Beauv.
Al (–) rudithi pendor
Bg (0) перест късоног
Cs (2) válečka prápořitá
De (–) Fieder-Zwenke
En (–) tor-grass
Hu (0) tollas szálkaperje
Pl (1) kłosownica pierzasta
Ro (0) obsigă
Ru (0) коротконожка перистая
Sh (1) perasta goštica
Sk (–) mrvica perovitá

252 ***Brachypodium sylvaticum*** (Huds.) Beauv.
Al (–) rudithi i pyjeve
Bg (0) горски късоног
Cs (0) válečka lesní
De (–) Wald-Zwenke
En (–) false brome
Hu (0) erdei szálkaperje
Pl (1) kłosownica leśna
Ru (1) коротконожка лесная
Sh (1) tanana goštica; šumska pasjača
Sk (–) mrvica lesná

* ***Brassica arvensis*** L. → *Moricandia arvensis*

* ***Brassica campestris*** L. → *Brassica rapa* ssp. *sylvestris*

253 *Brassica elongata* Ehrh.
Cs (–) brukev prodloužená
De (–) Langrispiger Kohl
Ru (1) рогачка меловая
Sh (1) renolika repica
Sk (–) kapusta predĺžená

* ***Brassica eruca*** L. → *Eruca vesicaria*

254 *Brassica juncea* (L.) Czern.
De (–) Ruten-Kohl
En (–) Japanese mustard
Ru (1) горчица сарептская
Sh (1) njivska repica
Sk (–) horčica sareptská

* ***Brassica kaber*** (DC.) Wheeler → *Sinapis arvensis*

* ***Brassica muralis*** Huds. → *Diplotaxis muralis*

255 *Brassica nigra* (L.) Koch
Al (–) lakra e zezë
Bg (1) черен синап
Cs (2) brukev černohořčice
De (–) Senf-Kohl
En (–) black mustard
Hu (1) fekete mustár
Pl (1) kapusta czarna
Ro (2) muştar negru
Ru (1) горчица чёрная
Sh (2) crna repica; crni kupus
Sk (–) horčica čierna

* ***Brassica orientalis*** L. → *Conringia orientalis*

256 *Brassica rapa* L. **ssp. *sylvestris*** (L.) Janch.
Al (–) repa
Bg (1) полска репица
Cs (0) brukev řepák
De (–) Rübsen
En (–) wild turnip
Hu (2) vadkáposzta
Pl (1) kapusta polna
Ro (3) rapiţă
Ru (2) турнепс
Sh (2) poljska repica
Sk (–) repák

257 *Briza maxima* L.
Al (–) briza e madhe
Bg (0) едра сълзица
De (–) Großes Zittergras
En (–) great quaking-grass
Gr (1) μεγάλο σκουλαρίκι
Ru (0) трясунка большая
Sk (–) traslica veľká

258 *Briza media* L.
Al (–) bari i kulaçit; briza e mesme
Bg (0) средна сълзица
Cs (2) třeslice prostřední
De (–) Mittleres Zittergras
En (–) perennial quaking-grass
Gr (1) μεσαίο σκουλαρίκι
Hu (0) rezgőpázsit
Pl (1) drżączka średnia
Ro (0) tremurătoare
Ru (1) трясунка средняя
Sh (1) obična treslica
Sk (–) traslica prostredná

259 *Briza minor* L.
Al (–) briza e vogël
De (–) Kleines Zittergras
En (–) lesser quaking-grass
Gr (1) μικρό σκουλαρίκι
Ru (0) трясунка меньшая
Sk (–) traslica menšia

260 *Bromus* spp. L.
Bg (1) овсига
De (–) Trespe
En (–) brome grass

Gr (2) βρόμος
Ru (3) костёр
Sk (–) stoklas

* ***Bromus affinis*** Dumort. → *Bromus hordeaceus* ssp. *hordeaceus*

261 *Bromus arvensis* L.
Al (–) bilba e fushave; bromusi i arave
Bg (1) полска овсига
Cs (1) sveřep rolní
De (–) Acker-Trespe
En (–) field brome
Hu (1) mezei rozsnok
Pl (1) stokłosa polna
Ru (1) костёр полевой
Sh (2) poljska klasača
Sk (–) stoklas roľný

262 *Bromus commutatus* Schrad.
Al (–) bromusi i ndryshueshëm
Bg (1) променлива овсига
Cs (0) sveřep pomíchaný
De (–) Verwechselte Trespe
En (–) meadow brome
Hu (1) bókoló rozsnok
Pl (1) stokłosa zmieniona
Ru (1) костёр переменчивый
Sk (–) stoklas lúčny pravý

263 *Bromus erectus* Huds.
Al (–) bromusi i përpjetë; bilba e ngritur
Bg (0) изправена овсига
Cs (2) sveřep vzpřímený
De (–) Aufrechte Trespe
En (–) upright brome
Hu (0) sudár rozsnok
Pl (1) stokłosa prosta
Ro (0) obsigă aristată
Ru (0) костёр прямой
Sk (–) stoklas vzpriamený

264 *Bromus hordeaceus* L. **ssp. *hordeaceus***
Al (–) bilba e butë; bromusi butlosh
Bg (1) мека овсига
Cs (2) sveřep měkký
De (–) Weiche Trespe
En (–) soft brome
Hu (1) puha rozsnok
Pl (1) stokłosa miękka
Ru (1) костёр мягкий
Sh (2) maljava klasača; meki ovsik
Sk (–) stoklas mäkký

265 *Bromus inermis* Leys.
Al (–) bilba pa hala
Bg (0) безосилеста овсига
Cs (1) sveřep bezbranný
De (–) Unbegrannte Trespe
En (–) Hungarian brome
Hu (0) árva rozsnok
Pl (1) stokłosa bezostna
Ro (0) tîrsacă
Ru (1) костёр безостый
Sk (–) stoklas bezbranný

266 *Bromus japonicus* Thunb.
Al (–) bilba e Japonisë
Cs (0) sveřep japonský
De (–) Japanische Trespe
En (–) Japanese brome
Hu (1) parlagi rozsnok
Ru (1) костёр японский
Sk (–) stoklas japonský

267 *Bromus lanceolatus* Roth
Cs (0) sveřep velkoklasý
En (–) lanceolate brome
Pl (1) stokłosa długoością
Ru (1) костёр крупноколосковый
Sk (–) stoklas kopijovitý

* ***Bromus macrostachys*** Desf. → *Bromus lanceolatus*

* ***Bromus mollis*** L. → *Bromus hordeaceus* ssp. *hordeaceus*

* ***Bromus multiflorus*** Weigel → *Bromus arvensis*

268 ***Bromus racemosus*** L.
Al (–) bromusi me vile
Bg (0) гроздовидна овсига
Cs (0) sveřep hroznatý
De (–) Trauben-Trespe
En (–) smooth brome
Hu (0) fürtös rozsnok
Pl (1) stokłosa groniasta
Ru (0) костёр кистевидный
Sh (1) grozdasti ovsik; livadska klasača
Sk (–) stoklas strapcovitý

269 ***Bromus ramosus*** Huds.
Al (–) bromusi i degëzuar
Bg (0) клонеста овсига
Cs (0) sveřep větevnatý
De (–) Wald-Trespe
En (–) hairy brome
Hu (0) ágas rozsnok
Pl (1) stokłosa gałęzista
Ru (0) костёр ветвистый
Sk (–) stoklas vetvistý

270 ***Bromus secalinus*** L.
Al (–) bromusi si thekër
Bg (1) овсига ружовидная
Cs (1) sveřep stoklasa
De (–) Roggen-Trespe
En (–) rye brome
Hu (2) gabona rozsnok
Pl (2) stokłosa żytnia
Ru (2) костёр ржаной
Sk (–) stoklas obilný

271 ***Bromus squarrosus*** L.
Al (–) bromusi i ashpër
Bg (0) разперена овсига
Cs (0) sveřep kostrbatý
De (–) Sperrige Trespe
En (–) rough brome
Hu (1) berzedt rozsnok
Pl (1) stokłosa łuskowata
Ru (1) костёр растопыренный
Sk (–) stoklas kostrbatý

272 ***Bromus sterilis*** L.
Al (–) bromusi shterpë
Bg (1) стерилна овсига
Cs (1) sveřep jalový
De (–) Taube Trespe
En (–) barren brome
Hu (1) meddő rozsnok
Pl (1) stokłosa płona
Ru (1) костёр бесплодный
Sh (2) dvocvjetni ovsik; oštra klašaca
Sk (–) stoklas jalový

273 ***Bromus tectorum*** L.
Al (–) bromusi i pullazeve
Bg (1) покривна овсига
Cs (2) sveřep střešní
De (–) Dach-Trespe
En (–) downy brome
Hu (1) fedél rozsnok
Pl (1) stokłosa dachowa
Ru (1) костёр кровельный
Sh (1) crvena klasača; klimavi ovsik
Sk (–) stoklas múrový

* ***Brunella vulgaris*** auct. → *Prunella vulgaris*

274 ***Bryonia alba*** L.
Bg (0) бяла лива тиква
Cs (1) posed bílý
De (–) Weiße Zaunrübe
En (–) white bryony
Hu (1) fekete földitök
Pl (1) przestęp biały
Ro (0) mutătoare
Ru (0) переступень белый

Sh (1) bijeli bljuštac; obični bljuštac
Sk (–) posed biely

275 *Bryonia cretica* L.
Al (–) stërkungulli
Bg (0) критска дива тиква
Cs (0) posed dvoudomý
De (–) Zweihäusige Zaunrübe
En (–) common bryony
Gr (1) αγριοκολοκύθα
Hu (1) piros földitök
Pl (1) przestęp dwupienny
Ru (1) переступень двудомный
Sh (1) dvodomi bljuštac
Sk (–) posed dvojdomý

* ***Bryonia dioica*** Jacq. → *Bryonia cretica*

276 *Bufonia tenuifolia* L.
De (–) Rispige Büffonie
Sh (1) obična bufonija

277 *Buglossoides arvensis* (L.) Johnst.
Al (–) kokërruja e arave
Bg (1) полско врабчово семе
Cs (2) kamejka rolní
De (–) Acker-Steinsame
En (–) field gromwell
Hu (3) mezei gyöngyköles
Pl (3) nawrot polny
Ro (2) mărgeluşă
Ru (2) буглосойдес полевой
Sh (1) poljska biserka; poljsko vrapseme
Sk (–) kamienka roľná

278 *Buglossoides purpurocaerulea* (L.) Johnst.
Cs (–) kamejka modronachová
De (–) Blauroter Steinsame
Ru (0) эгонихон фиолетово-голубой
Sh (1) rumeno vrapseme
Sk (–) kamienka modronachová

279 *Bunias erucago* L.
Al (–) brokra
Bg (0) свърбига
Cs (0) rukevník roketovitý
De (–) Senfblättriges Zackenschötchen
En (–) crested bunias
Gr (1) βουνιάδα
Pl (1) rukiewnik właściwy
Ro (1) hreniţă
Ru (1) свербига полевая
Sh (1) obični repušac
Sk (–) roripovník roketovitý

* ***Bunias macroptera*** Reichenb. → *Bunias erucago*

280 *Bunias orientalis* L.
Bg (1) източна свърбига
Cs (1) rukevník východní
De (–) Morgenländisches Zackenschötchen
En (–) warty cabbage
Hu (1) szümcső
Pl (2) rukiewnik wschodni
Ro (1) brăbin
Ru (1) свербига восточная
Sh (1) istočni repušac
Sk (–) roripovník východný

281 *Bunium bulbocastanum* L.
Al (–) gështenja e tokës
Cs (2) bulvuška hlíznatá
De (–) Gewöhnliche Erdknolle
En (–) great pignut
Sk (–) buľka hľuznatá

282 *Bupleurum* spp. L.
Bg (1) урока
De (–) Hasenohr
Gr (1) βούπλευρο
Ru (0) володушка
Sk (–) prerastlík

283 *Bupleurum longifolium* L.
Al (–) brinjëkau gjethegjatë
Cs (0) prorostlík dlouholistý
De (–) Langblättriges Hasenohr
Hu (0) hosszúlevelű buvákfű
Pl (1) przewiercień długolistny
Ro (0) salată
Ru (0) володушка долголистовая
Sk (–) prerastlík dlholistý

284 *Bupleurum rotundifolium* L.
Al (–) brinjëkau me gjethe të rrumbullakët
Bg (1) кръглолистна урока
Cs (1) prorostlík okrouhlolistý
De (–) Durchwachsenes Hasenohr
En (–) thorow-wax
Hu (1) kereklevelű buvákfű
Pl (1) przewiercień okrągłolistny
Ro (1) urechea iepurelui
Ru (1) володушка круглолистная
Sh (1) krugolisni zvinčac; obična propašljika
Sk (–) prerastlík okrúhlolistý

285 *Bupleurum tenuissimum* L.
Al (–) brinjëkau shumë i hollë
Bg (1) най-тънка урока
Cs (0) prorostlík nejtenší
En (–) slender hare's-ear
Hu (1) sziki buvákfű
Pl (1) przewiercień cienki
Ru (1) володушка тончайшая
Sk (–) prerastlík najtenší

286 *Butomus umbellatus* L.
Al (–) bliqini; hudhra ujëse
Bg (1) водолюба
Cs (1) šmel okoličnatý
De (–) Doldige Schwanenblume
En (–) flowering rush
Gr (1) βούτομο
Hu (1) virágkáka
Pl (1) łączeń baldaszkowy
Ro (1) crin de baltă
Ru (1) сусак зонтичный
Sh (1) štitasti srčak; štitkasti vodoljub
Sk (–) šmel okolíkatý

C

287 ***Calamagrostis* spp.** Adans.
Bg (0) вейник
De (–) Reitgras
Gr (1) καλαμάγρωστη
Ru (1) вейник
Sk (–) smlz

288 ***Calamagrostis epigejos*** (L.) Roth
Al (–) kallami i mbitokë
Bg (0) приземен вейник
Cs (2) třtina křovištní
De (–) Land-Reitgras
En (–) wood small-reed
Hu (2) siska nádtippan
Pl (2) trzcinnik piaskowy
Ro (1) trestie de cîmp
Ru (1) вейник наземный
Sh (2) obična sašuljica
Sk (–) smlz kroviskový

289 ***Calamagrostis pseudophragmites*** (Haller) Koeler
Bg (0) тръстикоподобен вейник
Cs (1) třtina pobřežní
De (–) Ufer-Reitgras
Hu (0) parti nádtippan
Pl (1) trzcinnik szuwarowy
Ro (0) trestioară
Ru (1) вейник ложнотростниковый
Sk (–) smlz pobrežný

290 ***Calamagrostis varia*** (Schrad.) Host
Cs (0) třtina pestrá
De (–) Berg-Reitgras
En (–) variegated small-reed
Hu (0) tarka nádtippan
Pl (1) trzcinnik pstry
Ro (0) stuh de baltă
Ru (0) вейник изменчивый
Sk (–) smlz pestrý

291 ***Calamintha* spp.** Mill.
Cs (1) marulka
De (–) Bergminze
Gr (1) καλαμίθρα
Ru (1) душевка
Sk (–) marulka

292 ***Calamintha grandiflora*** (L.) Moen.
Cs (1) marulka velkokvětá
De (–) Buchenwaldmelisse
En (–) large-flowered calamint
Ru (1) душевка крупноцветная
Sk (–) marulka veľkokvetá

293 ***Calamintha sylvatica*** Bromf.
Cs (1) marulka lesní
De (–) Echte Bergminze
En (–) wood calamint
Ro (0) izma pădurilor
Ru (0) душевка мятолистная
Sk (–) marulka lesná pravá

* ***Calamintha vulgaris*** (L.) Halá. → *Clinopodium vulgare*

* ***Calcitrapa solstitialis*** Lam. → *Centaurea solstitialis*

294 ***Calendula arvensis*** L.
De (–) Acker-Ringelblume
En (–) field marigold
Gr (1) καλεντούλα
Ru (1) календула полевая
Sk (–) nechtík roľný

295 ***Calendula officinalis*** L.
Al (–) kalendula mjekësore
Bg (0) лечебен невен
Cs (1) měsíček lékařský
De (–) Gemeine Ringelblume
En (–) pot marigold
Hu (0) körömvirág
Pl (0) nagietek lekarski

Ro (0) filimică
Ru (0) календула лекарственная
Sk (–) nechtik lekársky

* **Calepina corvini** (All.) Desv. → *Calepina irregularis*

296 Calepina irregularis (Asso) Thell.
Al (–) kalepina e çrregullt
Bg (1) кривица
De (–) Unregelmäßiger Wendich
En (–) white ball-mustard
Gr (1) καλεπίνα
Hu (1) matyó
Ru (1) калепина неравномерная
Sh (1) naborana gorušika; obična kalepina
Sk (–) katranička rôznolistá

297 Callitriche palustris L.
Al (–) ylli i ujit
Cs (0) hvězdoš jarní
De (–) Sumpf-Wasserstern
En (–) marsh water-starwort
Hu (0) mocsárhúr
Pl (1) rzęśl wiosenna
Ro (0) drenţe
Ru (1) болотник болотный
Sk (–) hviezdoš jarný

* **Callitriche verna** L. → *Callitriche palustris*

298 Calluna vulgaris (L.) Hull
Al (–) brushtulli
Bg (0) обикновена калуна
Cs (2) vřes obecný
De (–) Heidekraut
En (–) heather
Hu (0) csarab
Ro (0) iarbă neagră
Ru (1) вереск обыкновенный
Sh (1) obični vres; obični vrijes
Sk (–) vres metlovitý

299 Caltha palustris L.
Al (–) lëpushtra e kënetës
Bg (0) блатняк
Cs (2) blatouch bahenní
De (–) Sumpf-Dotterblume
En (–) marsh-marigold
Gr (1) κάλθα
Hu (1) mocsári gólyahír
Pl (1) kaczyniec; knieć błotna
Ro (0) calcea calului
Ru (1) калужница болотная
Sh (1) močvarna kaljužnica; veliki žuti žabljak
Sk (–) záružlie močiarne

300 Calystegia sepium (L.) R Br.
Al (–) gështalla e gardhave
Bg (1) ограден ладолеж
Cs (2) opletník plotní
De (–) Gemeine Zaunwinde
En (–) hedge bindweed
Gr (1) μεγάλη κεριπλοκάδα
Hu (2) sövényszulák
Pl (1) kielisznik zaroślowy
Ro (2) cupa vacii
Ru (2) повой заборный
Sh (3) obični ladolež; veliki slak
Sk (–) povoja plotná

* **Camelina albiflora** Kots. → *Camelina rumelica*

301 Camelina alyssum (Mill.) Thell.
Bg (1) ленова камелина
Cs (1) lnička tařicovitá
De (–) Gezähnter Leindotter
En (–) flax gold-of-pleasure
Hu (1) duzzadt gomborka
Pl (2) lnicznik właściwy
Ro (1) oul inului
Ru (1) рыжик льновый
Sk (–) ľaničník taricovitý

* **Camelina austriaca** Pers. → *Rorippa austriaca*

* ***Camelina dentata*** (Willd.) Pers. → *Camelina alyssum*

* ***Camelina foetida*** (Schk.) Fries → *Camelina alyssum*

* ***Camelina glabrata*** (DC.) Frit. → *Camelina sativa*

* ***Camelina linicola*** Schim. & Spen. → *Camelina alyssum*

302 ***Camelina macrocarpa*** Wier.
Bg (1) ленен ленак
De (–) Großsamiger Leindotter
Hu (1) sárga gomborka
Ru (1) рыжик крупноплодный
Sk (–) ľaničník veľkoplodý

303 ***Camelina microcarpa*** Andrz.
Bg (1) дребноплоден ленак
Cs (2) lnička drobnoplodá
De (–) Kleinfrüchtiger Leindotter
En (–) hairy gold-of-pleasure
Gr (1) καμελίνα
Hu (3) vetési gomborka
Pl (1) lnicznik drobnoowocowy
Ro (2) lubiţ
Ru (2) рыжик мелкоплодный
Sk (–) ľaničník maloplodý

304 ***Camelina rumelica*** Velen.
Bg (1) румелийски ленак
Cs (–) lnička rumelská
De (–) Bulgarischer Leindotter
Gr (1) καμελίνα
Hu (1) fehér gomborka
Ru (1) рыжик румелийский
Sk (–) ľaničník rumelský

305 ***Camelina sativa*** (L.) Cran.
Bg (1) възгол ленак
Cs (0) lnička setá
De (–) Saat-Leindotter
En (–) gold-of-pleasure
Hu (1) sárga gomborka
Pl (1) lnicznik jary
Ro (0) lubiţ
Ru (1) рыжик озимый
Sh (1) obični lanik; usjevni podlanak
Sk (–) ľaničník siaty

* ***Camelina sylvestris*** Wallr. → *Camelina microcarpa*

306 ***Campanula* spp.** L.
Bg (–) звънче
De (–) Glockenblume
En (–) bellflower
Gr (1) καμπανούλα
Ru (1) колокольчик
Sk (–) zvonček

307 ***Campanula bononiensis*** L.
Al (–) lule këmbane e Bolonjës
Bg (0) болонско звънче
Cs (0) zvonek boloňský
De (–) Filzige Glockenblume
Hu (0) olasz harangvirág
Pl (–) dzwonek boloński
Ro (1) clopoţei
Ru (0) колокольчик болонский
Sk (–) zvonček boloňský

* ***Campanula glaucophylla*** Schloss. & Vukot. → *Campanula glomerata*

308 ***Campanula glomerata*** L.
Al (–) lule këmbane lëmshore
Bg (1) главесто звънче
Cs (0) zvonek klubkatý
De (–) Geknäuelte Glockenblume
En (–) clustered bellflower
Hu (0) csomós harangvirág
Pl (–) dzwonek skupiony
Ro (0) ciucure
Ru (0) колокольчик сборный
Sh (1) gronjasti zvončić
Sk (–) zvonček klbový

* **Campanula hybrida** (L.) Gér. → *Legousia hybrida*

309 **Campanula lingulata** Waldst. & Kit.
Al (–) lule këmbane në trajtë gjuhe
Bg (1) езичесто звънче
Ru (0) колокольчик язычковый
Sk (–) zvonček jazyčný

310 **Campanula macrostachya** Waldst. & Kit.
Bg (1) класовидно звънче
Cs (0) zvonek velkoklasý
Hu (0) hosszúfüzérű harangvirág
Ro (0) clopoţei
Ru (0) колокольчик крупноколосый
Sk (–) zvonček veľkoklasatý

311 **Campanula patula** L.
Al (–) lule këmbane e hapur
Cs (1) zvonek rozkladitý
De (–) Wiesen-Glockenblume
En (–) spreading bellflower
Hu (0) terebélyes harangvirág
Pl (1) dzwonek rozpierzchły
Ro (0) cupa oii
Ru (1) колокольчик раскидистый
Sh (1) široki zvončić; tobolčasti zvončić
Sk (–) zvonček konáristý

312 **Campanula rapunculoides** L.
Bg (0) репоподобно звънче
Cs (3) zvonek řepkovitý
De (–) Acker-Glockenblume
En (–) creeping bellflower
Hu (1) kánya harangvirág
Pl (1) dzwonek jednostronny
Ru (1) колокольчик репчатовидный
Sh (1) jednostrani zvončić
Sk (–) zvonček repkovitý

313 **Campanula rapunculus** L.
Al (–) lule këmbane fitemë
Bg (0) реповидно звънче
Cs (1) zvonek řepka
De (–) Rapunzel-Glockenblume
En (–) rampion bellflower
Hu (0) rapons harangvirág
Pl (0) dzwonek rapunkuł
Ro (0) clopoţei
Ru (0) колокольчик рапунцель
Sh (1) repasti zvončić
Sk (–) zvonček repka

314 **Campanula rotundifolia** L.
Al (–) lule këmbane gjetherrumbullakët
Bg (0) кръглолистно звънче
Cs (1) zvonek okrouhlolistý
De (–) Rundblättrige Glockenblume
En (–) harebell
Hu (0) kereklevelű harangvirág
Pl (1) dzwonek okrągłolistny
Ro (0) albăstrică
Ru (0) колокольчик круглолистный
Sk (–) zvonček okrúhlolistý

315 **Campanula scutellata** Gris.
Sk (1) zvonček štítovitý

316 **Campanula sibirica** L.
Bg (0) сибирско звънче
Cs (1) zvonek sibiřský
De (–) Sibirische Glockenblume
En (–) Siberian bellflower
Hu (0) pongyola harangvirág
Pl (1) dzwonek syberyjski
Ru (0) колокольчик сибирский
Sk (–) zvonček sibírsky

* **Campanula speculum-veneris** L. → *Legousia speculum-veneris*

317 **Cannabis ruderalis** Janisch.
De (–) Wild-Hanf
Hu (3) kender
Pl (1) konopie dzikie
Ro (2) cînepă sălbatică
Ru (1) конопля сорная
Sk (–) konope rumištné

318 *Cannabis sativa* L.
Al (–) kërpi
Bg (1) посевен коноп
Cs (0) konopě setá
De (–) Gewöhnlicher Hanf
En (–) hemp
Hu (3) kender
Pl (1) konopie siewne
Ro (0) cînepă
Ru (1) конопля посевная
Sk (–) konope siate

319 *Capparis* spp. L.
De (–) Kaper
Gr (1) κάππαρη
Ru (0) каперс
Sk (–) kaprica

* ***Capriola dactylon*** Kuntze → *Cynodon dactylon*

320 *Capsella bursa-pastoris* (L.) Med.
Al (–) shtrapëra; torba e bariut
Bg (2) овчарска торбичка
Cs (3) kokoška pastuší tobolka
De (–) Gemeines Hirtentäschelkraut
En (–) shepherd's-purse
Gr (3) χαψέλλα
Hu (3) pásztortáska
Pl (3) tasznik pospolity
Ro (3) traista ciobanului
Ru (3) пастушья сумка обыкновенная
Sh (1) obična rusomača
Sk (–) pastierska kapsička obyčajná

321 *Capsella rubella* Reut.
Cs (2) kokoška červenavá
De (–) Rötliches Hirtentäschelkraut
En (–) pink shepherd's-purse
Ru (0) пастушья сумка красноватая
Sk (–) pastierska kapsička červenastá

322 *Cardamine* spp. L.
De (–) Schaumkraut
En (–) bittercress
Gr (1) καρδαμίνα
Ru (0) сердечник
Sk (–) žerušnica

323 *Cardamine amara* L.
Bg (0) горчива горва
Cs (1) řeřišnice hořká
De (–) Bitteres Schaumkraut
En (–) large bittercress
Hu (0) keserű kakukktorma
Pl (1) rzeżucha gorzka
Ro (0) stupitul cucului
Ru (0) сердечник горький
Sk (–) žerušnica horká

324 *Cardamine hirsuta* L.
Al (–) kardamina qimeashpër
Bg (0) космата горва
Cs (0) řeřišnice chlupatá
De (–) Behaartes Schaumkraut
En (–) hairy bittercress
Hu (0) borzas kakukktorma
Pl (1) rzeżucha włochata
Ro (1) rîjnică
Ru (0) сердечник шершавый
Sh (1) dlakava režuha
Sk (–) žerušnica chlpatá

325 *Cardamine impatiens* L.
Al (–) kardamina e padurueshme
Bg (0) нетърпелива горва
Cs (0) řeřišnice nedůtklivá
De (–) Spring-Schaumkraut
En (–) narrow-leaved bittercress
Hu (0) virágrúgó kakukktorma
Pl (1) rzeżucha niecierpkowa
Ro (0) rîjnică
Ru (0) сердечник недотрога
Sk (–) žerušnica nedotklivá

326 *Cardamine pratensis* L.
Al (–) kardamina livadhesh
Bg (0) ливадна горва

Cs (–) řeřišnice luční
De (–) Wiesen-Schaumkraut
En (–) cuckooflower
Hu (1) réti kakukktorma
Pl (1) rzeżucha łąkowa
Ro (0) stupitul cucului
Ru (1) сердечник луговой
Sh (1) livadska režuha
Sk (–) žerušnica lúčna

327 ***Cardamine resedifolia*** L.
Cs (0) řeřišnice rýtolistá
De (–) Resedablättriges Schaumkraut
Pl (1) rzeżucha rezedolistna
Sk (–) žerušnica rytolistá

328 ***Cardaminopsis arenosa*** (L.) Hayek
Bg (0) песъчлива гъшарка
Cs (1) řeřišničník písečný
De (–) Sand-Schaumkresse
En (–) sand rock-cress
Hu (0) közönséges dercevirág
Pl (1) gąsiówka piaskowa
Ro (1) frigurele
Ru (1) резуха песчаная
Sk (–) žerušničník piesočný

329 ***Cardaria draba*** (L.) Desv.
Al (–) djegësi drabë
Bg (2) горуха
Cs (3) vesnovka obecná
De (–) Gemeine Pfeilkresse
En (–) hoary cress
Gr (3) βρωμολάχανο
Hu (3) útszéli zsázsa
Pl (1) pieprzyczni przydrożny
Ro (3) urda vacii
Ru (3) клоповник крупковидный
Sh (2) obična renika; sivkasta grbica
Sk (–) vesnovka obyčajná

330 ***Carduus acanthoides*** L.
Al (–) freshkulli si gjemb
Bg (1) магарешки бодил
Cs (2) bodlák obecný
De (–) Weg-Distel
En (–) welted thistle
Hu (2) útszéli bogáncs
Pl (1) oset nastroszony
Ro (2) spin
Ru (2) чертополох колючий
Sh (1) obični stričak
Sk (–) bodliak tŕnistý

* ***Carduus arvensis*** Sm. → *Cirsium arvense*

* ***Carduus axillaris*** Gaudin → *Carduus acanthoides*

331 ***Carduus candicans*** Waldst. & Kit.
Bg (1) беловат магарешки бодил
Sk (–) bodliak bielavý

332 ***Carduus crispus*** L.
Bg (0) къдраволистен магарешки бодил
Cs (1) bodlák kadeřavý
De (–) Krause Distel
En (–) curly welted-thistle
Hu (1) fodros bogáncs
Pl (1) oset kędzierzawy
Ro (0) pălămidă
Ru (1) чертополох курчавый
Sk (–) bodliak kučeravý

333 ***Carduus hamulosus*** Ehrh.
Bg (0) кукест магарешки бодил
Cs (0) bodlák háčkatý
Hu (1) horgas bogáncs
Pl (–) oset haczykowaty
Ro (0) armirai sălbatic
Ru (0) чертополох крючочковый
Sk (–) bodliak háčkaty

334 ***Carduus nutans*** L.
Al (–) freshkulli; therë kali
Bg (0) наведен магарешки бодил

Cs (1) bodlák nící
De (–) Nickende Distel
En (–) musk thistle
Gr (2) γαϊδουράγκαθο
Hu (1) bókoló bogáncs
Pl (1) oset zwisły
Ro (2) ciulin
Ru (1) чертополох поникающий
Sh (1) crveni stričak; klimavi stričak
Sk (–) bodliak kývajúci

335 *Carduus personata* (L.) Jacq.
Bg (0) гъстоглавест магарешки бодил
Cs (0) bodlák lopuchovitý
De (–) Berg-Distel
En (–) great marsh-thistle
Pl (1) oset łopianowaty
Ru (0) чертополох замаскированный
Sk (–) bodliak lopúchovitý

* ***Carduus polyanthus*** Schreb. → *Carduus acanthoides*

336 *Carduus pycnocephalus* L.
Al (–) freshkulli kokëdendur
Bg (1) гъстоглавест магарешки бодил
De (–) Knäuelköpfige Distel
En (–) Corsican thistle
Gr (2) φιθάγκάθο
Ru (1) чертополох мелкоголовный
Sk (–) bodliak tučnolisty

337 *Carduus uncinatus* Bieb.
Ru (1) чертополох крючковатый
Sk (–) bodliak hákovitý

338 *Carex* spp. L.
Al (–) presja
Bg (1) острица
Cs (2) ostřice
De (–) Segge
En (–) sedge
Gr (2) ξιφάρα
Hu (1) sás
Pl (2) turzyce
Ro (3) rogoz
Ru (2) осока
Sh (1) oštrica; šaš
Sk (–) ostrica

339 *Carlina* spp. L.
Bg (1) решетка
De (–) Eberwurz
Gr (1) καρλίνα
Ru (1) колючник
Sk (–) krasovlas

340 *Carlina acanthifolia* L.
Al (–) ushonjëza gjethedashtre
Bg (1) безстъблена решетка
Sk (–) krasovlas voňavý

341 *Carlina acaulis* L.
Al (–) ushojza
Cs (1) pupava bezlodyžná
De (–) Stengellose Eberwurz
En (–) dwarf carline-thistle
Hu (1) szártalan bábakalács
Pl (1) dziewieciosił bezłodygowy
Ro (1) turtă
Ru (1) колючник бесстебельный
Sh (1) bezstabljični kravljak; niski kravljak
Sk (–) krasovlas bezbyľový

342 *Carlina vulgaris* L.
Al (–) ferëkali; ushojza e rëndomtë
Bg (1) обикновена решетка
Cs (1) pupava obecná
De (–) Gemeine Eberwurz
En (–) common carline-thistle
Hu (1) közönséges bábakalács
Pl (1) dziewieciosił pospolity
Ro (1) turtă
Sh (1) trepavičasti kravljak; veliki kravljak
Ru (1) колючник обыкновенный
Sk (–) krasovlas obyčajný pravý

343 ***Carthamus dentatus*** (Forsk.) Vahl
Bg (1) зъбчест аспурт
Sk (–) požlt zubatý

344 ***Carthamus glaucus*** Bieb.
Bg (0) син аспурт
Ru (1) сафлор сизый

345 ***Carthamus lanatus*** L.
Al (–) kartami leshatak
Bg (1) жълт аспурт
Cs (0) světlice vlnatá
De (–) Wolliges Spornblatt
En (–) woolly safflower
Hu (1) vadpórsáfrány
Ro (1) pintenoagă
Ru (1) сафлор шерстистый
Sh (1) obični bodalj; vunasta šafranka
Sk (–) požlt vlnatý

346 ***Carum carvi*** L.
Al (–) karaboti
Bg (0) кимион
Cs (1) kmín kořenný
De (–) Echter Kümmel
En (–) caraway
Hu (0) kömény
Pl (1) kminek zwyczajny
Ro (0) chimen
Ru (1) тмин обыкновенный
Sh (1) obični kim
Sk (–) rasca lúčna

347 ***Catabrosa aquatica*** (L.) Beauv.
Al (–) katabroza e ujit
Bg (1) водна катаброса
Cs (0) odemka vodní
De (–) Quellgras
En (–) whorl-grass
Hu (0) forrásperje
Pl (1) brodobrzanka wodna
Ro (0) lăcrămiţă
Ru (0) поручейница водяная
Sk (–) odemka vodná

* ***Catapodium rigidum*** (L.) Hub. → *Desmazeria rigida*

* ***Caucalis daucoides*** L. → *Caucalis platycarpos*

* ***Caucalis lappula*** Gran. → *Caucalis platycarpos*

* ***Caucalis latifolia*** L. → *Turgenia latifolia*

* ***Caucalis muricata*** Bisch. → *Caucalis platycarpos*

348 ***Caucalis platycarpos*** L.
Al (–) hermëthi frutgjërë
Bg (1) обикновен бабинец
Cs (1) dejvorec stroškový
De (–) Kletten-Haftdolde
En (–) small bur-parsley
Gr (2) καυκαλίδα
Hu (1) tüskés ördögbocskor
Pl (1) włóczydło polne
Ro (2) ruşinătoare
Ru (1) прицепник широкоплодный
Sh (1) kljunasta podlanica
Sk (–) dyvorec lopúšikový

349 ***Cenchrus echinatus*** L.
Cs (0) ostrokvět ježatý
De (–) Stacheliges Klettengras
En (–) bur-grass
Hu (2) átoktüske
Sk (–) ostrokvet ježatý

350 ***Centaurea* spp.** L.
Bg (2) метличина
Cs (–) chrpa
De (–) Flockenblume
Gr (1) κενταύρια
Ru (2) василёк

* ***Centaurea amara*** L. → *Centaurea jacea*

351 *Centaurea calcitrapa* L.
Al (–) kokoçeli yllor
Bg (0) бодлива метличина
Cs (0) chrpa sikavice
De (–) Stern-Flockenblume
En (–) red star-thistle
Hu (1) úti imola
Pl (1) chaber kolący
Ru (1) василёк колючеголовный
Sk (–) sekavica červená

352 *Centaurea cyanus* L.
Al (–) kokoçeli ngjyrë qielli
Bg (2) полска метличина
Cs (2) chrpa modrák
De (–) Kornblume
En (–) cornflower
Gr (2) κενταύρια
Hu (3) kék búzavirág
Pl (3) chaber bławatek
Ro (3) albăstriţă
Ru (3) василёк синий
Sh (2) modri različak; njivski različak
Sk (–) nevädza poľná

353 *Centaurea depressa* Bieb.
Ru (1) василёк сжатый

354 *Centaurea diffusa* Lam.
Bg (0) широкоразклонена метличина
Cs (0) chrpa rozvětvená
De (–) Sparrige Flockenblume
Hu (0) terpedt imola
Pl (0) chaber drobnogłówkowy
Ru (1) василёк раскидистый
Sk (–) nevädza rozložitá

355 *Centaurea iberica* Trev.
Ru (1) василёк иберийский

356 *Centaurea jacea* L.
Al (–) kokoçeli i livadhit; perenik i livadheve
Bg (1) ливадна метличина
Cs (1) chrpa luční
De (–) Gemeine Flockenblume
En (–) brown knapweed
Hu (1) réti imola
Pl (2) chaber łąkowy
Ru (2) василёк луговой
Sh (1) livadski različak
Sk (–) krupina lúčna

357 *Centaurea maculosa* Lam.
Cs (0) chrpa latnatá skvrnitá
De (–) Gescheckte Flockenblume
En (–) spotted knapweed
Hu (1) kisfészkű imola
Sk (–) čakanica škvrnitá

358 *Centaurea phrygia* L.
Al (–) kokoçeli i Frigias
Bg (0) фригийска метличина
Cs (0) chrpa parukářka
De (–) Perücken-Flockenblume
En (–) wig knapweed
Hu (0) parókaimola
Pl (1) chaber perukowy
Ru (0) василёк фригийский
Sh (1) bosanski različak
Sk (–) krupina parochniarka

* ***Centaurea pseudophrygia*** Meyer → *Centaurea phrygia*

* ***Centaurea repens*** L. → *Acroptilon repens*

359 *Centaurea rhenana* Boreau
Al (–) kokcçeli i Renit
Bg (1) петлистна метличина
Cs (2) chrpa latnatá porýnská
De (–) Rispen-Flockenblume
Hu (1) rajnai imola
Pl (1) chaber nadreński
Ru (0) василёк рейнский
Sh (1) kožuhasti različak
Sk (–) nevädza porýnska pravá

360 ***Centaurea scabiosa*** L.
Al (–) kroma
Bg (1) ресничеста метличина
Cs (1) chrpa čekánek
De (–) Skabiosen-Flockenblume
En (–) greater knapweed
Hu (1) vastövű imola
Pl (2) chaber driakiewnik
Ru (2) василёк скабиозовый
Sh (1) češljasti različak
Sk (–) cigořinek obyčajný pravý

361 ***Centaurea solstitialis*** L.
Al (–) ferrëbardha
Bg (2) слънчева метличина
Cs (0) chrpa letní
De (–) Sonnwend-Flockenblume
En (–) yellow star-thistle
Hu (1) sáfrányos imola
Pl (1) chaber wełnisty
Ro (1) scai galben
Ru (1) василёк солнечный
Sh (2) letni različak; ljetni različak
Sk (–) sekavica žltá

362 ***Centaurea stenolepis*** Kerner
Ru (0) василёк ушкочешуйчатый
Sh (1) uski različak
Sk (–) krupina úzkoperá

* ***Centaurea stoebe*** L. → *Centaurea rhenana*

363 ***Centaurium erythraea*** Rafn
Al (–) bari i ethes
Cs (0) zeměžluč obecná
De (–) Echtes Tausendgüldenkraut
En (–) common centaury
Hu (1) kis ezerjófű
Pl (2) centuria pospolita
Ro (1) fierea pămîntului
Ru (1) золототысячник обыкновенный
Sh (1) mala kičica; obična kičica
Sk (–) zemežlč horká pravá

364 ***Centaurium pulchellum*** (Swartz) Druce
Al (–) bari i ethes i bukur
Bg (1) красив кантарион
Cs (0) zeměžluč spanilá
De (–) Ästiges Tausendgüldenkraut
En (–) lesser centaury
Hu (1) csinos ezerjófű
Pl (1) centuria nadobna
Ru (1) золототысячник красивый
Sk (–) zemežlč spanilá

* ***Centaurium umbellatum*** auct. → *Centaurium pulchellum*

365 ***Centranthus*** **spp.** DC.
De (–) Spornblume
Gr (1) κέντρανθος
Ru (0) кентрантус
Sk (–) mavúň

* ***Centunculus minimus*** L. → *Anagallis minima*

366 ***Cephalaria laevigata*** (Waldst. & Kit.) Schrad.
Sh (1) glatka praskoč

367 ***Cephalaria syriaca*** (L.) Roem. & Schultes
Bg (1) сирийска звездоглавка
Cs (0) hlavatka syrská
De (–) Syrischer Schuppenkopf
Gr (1) κεφαλάρια
Ru (1) головчатка сирийская
Sk (–) hlavička sýrska

368 ***Cephalaria transylvanica*** (L.) Roem. & Schultes
Al (–) cefalaria e Transilvanisë
Bg (1) трансилванска звездоглавка
Cs (0) hlavatka sedmihradská
Gr (1) κεφαλάρια
Hu (1) fehér fejvirág

Ro (2) sipică
Ru (1) головчатка трансильванская
Sh (1) modrikasta glavatka; obična praskoč
Sk (–) hlavička sedmohradská

369 *Cerastium* spp. L.
Bg (1) рожец
De (–) Hornkraut
Gr (1) κεράστιο
Ru (2) ясколка
Sk (–) rožec

* ***Cerastium anomalum*** Waldst. & Kit. → *Cerastium dubium*

370 *Cerastium arvense* L.
Al (–) qerasti i arave
Bg (0) полски рожец
Cs (3) rožec rolní
De (–) Acker-Hornkraut
En (–) field mouse-ear
Gr (2) κεράστιο
Hu (0) parlagi madárhúr
Pl (2) rogownica polna
Ro (1) cornut
Ru (2) ясколка полевая
Sh (1) poljski tičinac
Sk (–) rožec roľný

371 *Cerastium brachypetalum* Pers.
Al (–) qerasti petëlshkurtër
Bg (1) късовенечен рожец
Cs (0) rožec krátkoplátečný
De (–) Bärtiges Hornkraut
En (–) grey mouse-ear
Hu (1) ugari madárhúr
Pl (1) rogownica drobnokwiatowa
Ru (1) ясколка крымская
Sh (1) submediteranski tičinac
Sk (–) rožec krátkolistočný

* ***Cerastium caespitosum*** Gilib. → *Cerastium fontanum*

372 *Cerastium dubium* (Bast.) Schwarz
Bg (1) отклоняващ се рожец
Cs (0) rožec pochybný
De (–) Klebriges Hornkraut
En (–) starwort mouse-ear
Hu (1) sziki madárhúr
Ru (1) ясколка уклоняющаяся
Pl (1) rogownica lepka
Sk (–) rožec odchylný

373 *Cerastium fontanum* Baumg.
Bg (0) крайизворен рожец
Cs (3) rožec obecný
De (–) Quellen-Hornkraut
En (–) common mouse-ear
Hu (2) gyepes madárhúr
Pl (2) rogownica pospolita
Ru (2) ясколка дернистая
Sh (1) krupni tičinac
Sk (–) rožec pramenistý

374 *Cerastium glomeratum* Thuill.
Al (–) qerasti lëmshor
Bg (1) кичурест рожец
Cs (1) rožec lepivý
De (–) Kleinblütiges Hornkraut
En (–) sticky mouse-ear
Hu (2) gomolyos madárhúr
Ru (2) ясколка скученноцветная
Sh (1) runjavi tičinac
Sk (–) rožec kĺbkový

* ***Cerastium glutinosum*** Fries → *Cerastium pumilum*

* ***Cerastium holosteoides*** Fries → *Cerastium fontanum*

* ***Cerastium murale*** Salisb. → *Holosteum umbellatum*

375 *Cerastium perfoliatum* L.
Ru (1) ясколка пронзённолистная
Sk (–) rožec prerastaný

376 ***Cerastium pumilum*** Curt.
Cs (0) rožec nízký bledý
De (–) Niedriges Hornkraut
En (–) dwarf mouse-ear
Hu (1) törpe madárhúr
Pl (1) rogownica drobna
Ru (1) ясколка липкая
Sk (–) rožec nízky

377 ***Cerastium semidecandrum*** L.
Bg (1) петтичинков рожец
Cs (1) rožec pětimužný
De (–) Sand-Hornkraut
En (–) little mouse-ear
Hu (1) homoki madárhúr
Pl (1) rogownica pięciopręcikowa
Ru (1) ясколка пятитычинковая
Sh (1) sitni tičinac
Sk (–) rožec pätmužný

* ***Cerastium umbellatum*** Cran. → *Holosteum umbellatum*

* ***Cerastium viscosum*** auct. → *Cerastium glomeratum*

* ***Cerastium vulgare*** Hartm. → *Cerastium fontanum*

* ***Cerastium vulgatum*** L. → *Cerastium fontanum*

378 ***Ceratocarpus arenarius*** L.
Ru (1) рогач песчаный

379 ***Ceratocephalus falcatus*** (L.) Pers.
Bg (0) сърповиден рогоглавец
Cs (0) rohohlavec srpovitý
De (–) Sichelfrüchtiges Hornköpfchen
Hu (1) sarlóboglárka
Pl (1) jaskierek
Ru (1) рогоглавник серповидный
Sk (–) rohohlavec kosákovitý

380 ***Ceratocephalus testiculatus*** (Cran.) Roth
Bg (1) прав рогоглавец
Cs (0) roholavec rovnorohý
De (–) Gradfrüchtiges Hornköpfchen
Hu (1) sarlóboglárka
Ro (1) ploşnicar
Sk (–) rohohlavec rovnorohý

381 ***Ceratophyllum* spp.** L.
Bg (–) роголистник
De (–) Hornblatt
En (–) hornwort
Gr (1) κερατόφυλλο
Ru (1) роголистник
Sk (–) rožkatec

382 ***Ceratophyllum demersum*** L.
Al (–) gjethebri gati i zhytur
Bg (–) подводен роголистник
Cs (0) růžkatec ponořený
De (–) Gemeines Hornblatt
En (–) rigid hornwort
Hu (1) érdes borzhínár
Pl (3) rogatek sztywny
Ro (1) cosor
Ru (2) роголистник погруженный
Sk (–) rožkatec ponorený

383 ***Ceratophyllum muricatum*** Cham.
Bg (1) подводен роголистник

384 ***Ceratophyllum submersum*** L.
Bg (0) потопен роголистник
Cs (0) růzkatec potopený
De (–) Zartes Hornblatt
En (–) soft hornwort
Hu (1) sima borzhínár
Pl (0) rogatek krótkoszyjkowy
Ro (1) cosor
Ru (1) роголистник полупогруженный
Sk (–) rožkatec potopený

385 ***Cerinthe* spp.** L.
Bg (0) меденик
De (–) Wachsblume
En (–) honeywort
Gr (1) κερίνθη
Ru (1) восковцветник
Sk (–) voskovka

386 ***Cerinthe major*** L.
Cs (0) voskovka větší
De (–) Große Wachsblume
En (–) common honeywort
Ro (1) pidosnic
Ru (1) восковцветник большой
Sk (–) voskovka väčšia

387 ***Cerinthe minor*** L.
Al (–) qerintha më e vogël
Bg (0) малък меденик
Cs (1) voskovka menší
De (–) Kleine Wachsblume
En (–) lesser honeywort
Hu (2) szeplőlapu
Pl (0) ośmiał mniejszy
Ro (1) pidosnic
Ru (1) восковцветник малый
Sh (1) mala visika
Sk (–) voskovka menšia

* ***Cervaria rivini*** Gaertn. → *Peucedanum carvifolia*

388 ***Ceterach officinarum*** DC.
Al (–) bargjarpëri mjekësor
Bg (0) златиста папрат
Cs (0) kyvor lékařský
De (–) Spreuschuppiger Milzfarn
En (–) rustyback fern
Hu (0) nyugati pikkelypáfrány
Pl (0) śledzionka skalna
Ro (0) unghia ciutei
Ru (0) скребница аптечная
Sh (1) obična zlatarica
Sk (–) ceterák lekársky pravý

389 ***Chaenorhinum minus*** (L.) Lange
Al (–) kenorhinumi më i vogël
Bg (1) дребна луличка
Cs (1) hledíček menší
De (–) Kleiner Orant
En (–) small toadflax
Hu (2) tátos
Pl (3) lnica mała
Ru (2) хеноринум малый
Sh (1) mala zevalica; mala zijevalica
Sk (–) škľabinček menši

390 ***Chaerophyllum* spp.** L.
Bg (–) балдаран
De (–) Kälberkropf
En (–) chervil
Gr (1) χερόφυλλο
Ru (1) бутень
Sk (–) krkoška

391 ***Chaerophyllum aromaticum*** L.
Al (–) stërpuja aromatike
Cs (2) krabilice zápašná
De (–) Gewürzhafter Kälberkropf
En (–) sweet chervil
Hu (0) fűszeres baraboly
Pl (0) świerząbek korzenny
Ro (1) antonică
Ru (1) бутень ароматный
Sk (–) krkoška zápašná

392 ***Chaerophyllum bulbosum*** L.
Al (–) stërpuja me qepujkë
Bg (0) грудков балдаран
Cs (3) krabilice hliznatá
De (–) Knolliger Kälberkropf
En (–) tuberous chervil
Hu (0) csemegebaraboly
Pl (1) świerząbek bulwiasty
Ro (1) baraboi
Ru (1) бутень луковичный
Sk (–) krkoška hľuznatá

* ***Chaerophyllum prescottii*** DC. → *Chaerophyllum bulbosum*

* **Chaerophyllum sylvestre** L. → *Anthriscus sylvestris*

393 **Chaerophyllum temulentum** L.
Bg (1) опиянявaщ шашан
Cs (3) krabilice mámivá
De (–) Hecken-Kälberkropf
En (–) rough chervil
Pl (1) świerząbek gajowy
Ru (1) бутень одуряющий
Sh (1) otrovna krabljica
Sk (–) krkoška mámivá

394 **Chamaecytisus austriacus** (L.) Link
Cs (0) čilimník rakouský
En (–) Austrian clustered-broom
Hu (0) buglyos zanót
Pl (0) szczodrzeniec austriacki
Sh (1) obični zanovet; obični zanovjet
Sk (–) zanovät rakúsky

395 **Chamaecytisus heuffelii** (Wierzb.) Rothm.
Sh (1) pešćarski zanovet; pešćarski zanovjet

396 **Chamaecytisus hirsutus** (L.) Link
Al (–) vjexhësa qimeashpër
Bg (0) космат зановец
Cs (0) čilimník chlupatý
En (–) hairy broom
Hu (0) borzas zanót
Pl (1) szczodrzeniec włochaty
Ro (0) drob de munte
Sh (1) runjavi zanovet; runjavi zanovjet
Sk (–) zanovät chlpatý

* **Chamaenerion angustifolium** (L.) Scop. → *Epilobium angustifolium*

* **Chamaeplium officinale** (L.) Wallr. → *Sisymbrium officinale*

397 **Chamaespartium sagittale** (L.) Gibbs
Al (–) gjineshtra në tratjë shigjete
Cs (0) kručinečka křídlatá
De (–) Flügel-Ginster
En (–) winged broom
Pl (1) janowiec skrzydlaty
Ro (0) grozamă
Sh (1) streličasta žutica
Sk (–) krutinôčka krídlatá

* **Chamaesyce humifusa** (Willd.) Prokh. → *Euphorbia humifusa*

398 **Chamomilla recutita** (L.) Rausch.
Al (–) kamomili
Bg (1) лайка същинска
Cs (3) heřmánek pravý
De (–) Echte Kamille
En (–) scented mayweed
Gr (3) χαμομήλι
Hu (3) orvosi szikfű; székfű
Pl (3) rumianek pospolity
Ro (1) mușețel
Ru (2) ромашка ободранная
Sk (3) prava kamilica
Sk (–) rumanček pravý

399 **Chamomilla suaveolens** (Pursh) Rydb.
Cs (3) heřmánek terčovitý
De (–) Strahllose Kamille
En (–) pineapple-weed
Hu (1) sugártalan szikfű
Pl (2) rumianek bezpromienny
Ru (2) ромашка ромашковидная
Sk (–) rumanček terčovitý

400 **Cheilanthes marantae** (L.) Domin.
En (–) hardy cloak-fern
Sh (1) serpentinski vodijerak; serpentinski voderak
Ru (0) ложнопокровница марантовая
Sk (–) šupinatka marantová

401 *Cheiranthus cheiri* L.
Al (–) lulja e Misirit
Cs (0) chejr vonný
De (–) Gold-Lack
En (–) garden wallflower
Hu (0) sárgaviola
Pl (0) lak pospolity
Ro (0) micşunele ruginite
Ru (0) лакфиоль обыкновенный
Sh (1) baštenski šeboj; žuti šeboj
Sk (–) cheir voňavý

402 *Chelidonium majus* L.
Al (–) latrapeci i madhë; tamlagjaku
Bg (1) обикновено змийско мляко
Cs (2) vlaštovičník větší
De (–) Großes Schöllkraut
En (–) greater celandine
Gr (1) χελιδονιά
Hu (1) vérehulló fecskefű
Pl (2) glistnik jaskółcze ziele
Ro (1) rostopască
Ru (1) чистотел большой
Sh (1) majska rusa; perastolisni rosopas
Sk (–) lastovičník väčší

403 *Chenopodium* spp. L.
Bg (1) куча лобода
De (–) Gänsefuß
Gr (2) λουβουδιά
Ru (3) марь
Sk (–) mrlík

404 *Chenopodium album* L.
Al (–) minueri i bardhë
Bg (3) бяла куча лобода
Cs (3) merlík bílý
De (–) Weißer Gänsefuß
En (–) fat-hen
Gr (3) λουβουδιά
Hu (3) fehér libaparéj
Pl (3) komosa biała
Ro (3) spanac sălbatic
Ru (3) марь белая
Sh (3) obična pepeljuga
Sk (–) mrlík biely

405 *Chenopodium ambrosioides* L.
Al (–) minueri si pelin deti
Cs (0) merlík vonný
De (–) Wohlriechender Gänsefuß
En (–) Indian goosefoot
Hu (1) mirha libaparéj
Pl (0) komosa piżmowa
Ro (0) tamîiţă
Ru (0) марь амброзиевидная
Sk (–) mrlík voňavý

406 *Chenopodium aristatum* L.
Ru (1) марь остистая

407 *Chenopodium bonus-henricus* L.
Al (–) pjerëza
Bg (1) планинска куча лобода
Cs (1) merlík všedobr
De (–) Guter Heinrich
En (–) good-King-Henry
Hu (1) hegyi libaparéj
Pl (2) komosa strzałkowata
Ro (1) spanacul ciobanului
Ru (1) марь Доброго Генриха
Sh (1) brašnjava pepeljuga
Sk (–) všeliek

408 *Chenopodium botrys* L.
Al (–) minueri vile rrushi
Bg (1) огниче
Cs (0) merlík hroznovitý
De (–) Klebriger Gänsefuß
En (–) Jerusalem goosefoot
Hu (1) mirigyes libaparéj
Pl (2) komosa wonna
Ro (1) peliniţă grecească
Ru (1) марь душистая
Sk (–) mrlík strapcový

409 *Chenopodium ficifolium* Sm.
Bg (0) смокволистна куча лобода

Cs (2) merlík fíkolistý
De (–) Feigenblatt-Gänsefuß
En (–) fig-leaved goosefoot
Hu (1) fügelevelű libaparéj
Pl (1) komosa jesienna
Ru (1) марь фиголистная
Sk (–) mrlík smokvoňolisty

410 *Chenopodium glaucum* L.
Bg (0) синьозелена куча лобода
Cs (1) merlík sivy
De (–) Graugrüner Gänsefuß
En (–) oak-leaved goosefoot
Hu (1) fakó libaparéj
Pl (2) komosa sina
Ru (1) марь сизая
Sh (1) polegla pepeljuga
Sk (–) mrlík dubolistý

411 *Chenopodium hybridum* L.
Bg (1) хибридна куча лобода
Cs (2) merlík zvrhlý
De (–) Bastard-Gänsefuß
En (–) maple-leaved goosefoot
Hu (3) pokolvar libaparéj
Pl (2) komosa wielkolistna
Ro (1) talpa gîştii
Ru (2) марь гибридная
Sh (1) srcolisna pepeljuga
Sk (–) mrlík bastardný

412 *Chenopodium murale* L.
Al (–) minueri muresh
Bg (0) стенна куча лобода
Cs (0) merlík zední
De (–) Mauer-Gänsefuß
En (–) nettle-leaved goosefoot
Hu (1) fali libaparéj
Pl (1) komosa murowa
Ro (0) frunză de potcă
Ru (1) марь стенная
Sh (2) mala pepeljuga
Sk (–) mrlík múrový

413 *Chenopodium opulifolium* Schrad.
Bg (0) калинолистна куча лобода
Cs (1) merlík kalinolistý
De (–) Schneeballblättriger Gänsefuß
En (–) grey goosefoot
Hu (1) terpedt libaparéj
Pl (1) komosa kalinolistna
Ru (1) марь калинолистная
Sk (–) mrlík kalinolistý

414 *Chenopodium polyspermum* L.
Bg (2) сладък бурен
Cs (3) merlík mnohosemenný
De (–) Vielsamiger Gänsefuß
En (–) many-seeded goosefoot
Hu (3) sokmagvú libaparéj
Pl (2) komosa wielonasienna
Ru (2) марь многосемянная
Sh (3) baštenska pepeljuga; mnogocvjetna pepeljuga
Sk (–) mrlík mnohosemenný

415 *Chenopodium rubrum* L.
Bg (0) червена куча лобода
Cs (1) merlík červený
De (–) Roter Gänsefuß
En (–) red goosefoot
Hu (1) vörös libaparéj
Pl (2) komosa czerwonawa
Ru (1) марь красная
Sk (–) mrlík červený

416 *Chenopodium schraderianum* Schultes
Ru (1) марь Шрадера
Sk (–) mrlík citrónový

* ***Chenopodium striatum*** (Kraš.) Murr → *Chenopodium album*

* ***Chenopodium strictum*** Roth → *Chenopodium album*

417 *Chenopodium suecicum* Murr
Cs (0) merlík švédsky
De (–) Grüner Gänsefuß
Hu (1) zöld libaparéj
Pl (2) komosa zielona
Ru (1) марь шведская
Sk (–) mrlík švédsky

418 *Chenopodium urbicum* L.
Al (–) minueri i qytetit
Bg (0) градска куча лобода
Cs (–) merlík mestský
De (–) Städte-Gänsefuß
En (–) upright goosefoot
Hu (2) falusi libaparéj
Pl (2) komosa trójkątna
Ru (2) марь городская
Sk (–) mrlík mestský

* ***Chenopodium viride*** auct. → *Chenopodium suecicum*

* ***Chenopodium vulgare*** Gold. → *Chenopodium album*

419 *Chenopodium vulvaria* L.
Al (–) nena e qelbët
Bg (1) струпейниче
Cs (1) merlík smradlavý
De (–) Stinkender Gänsefuß
En (–) stinking goosefoot
Hu (1) büdös libaparéj
Pl (0) komosa mierzliwa
Ro (0) lobodă puturoasă
Ru (1) марь зловонная
Sk (–) mrlík smradľavý

420 *Chondrilla juncea* L.
Al (–) kondrila si zunkth; xhanabarthi
Bg (1) кривец
Cs (0) radyk sítinovity
De (–) Binsen-Knorpellattich
En (–) skeletonweed
Gr (1) χονδρίλλα
Hu (2) nyúlparéj
Pl (1) chondrilla sztywna
Ro (2) răsfug
Ru (1) хондрилла ситниковидная
Sh (1) obična žutenica
Sk (–) radyk sitinovitý

421 *Chorispora tenella* (Pallas) DC.
Ru (1) хориспора нежная
Sk (–) drobnosemienka útla

422 *Chrozophora tinctoria* (L.) Juss.
Al (–) krozofora ngjyruese
Bg (1) синилна трева
De (–) Färberkroton
En (–) dyer's litmus
Gr (1) χρωζοφόρα
Ru (1) хрозофора красильная
Sh (1) bojadisarka mljěcika; obična hrozofora
Sk (–) mliečnik farbiarsky

423 *Chrysanthemum* spp. L.
De (–) Wucherblume
Gr (2) αγριομαργαρίτα
Ru (1) хризантема
Sk (–) králik

424 *Chrysanthemum coronarium* L.
Al (–) krizantema e rëndomtë
Cs (0) kopretina věncová
De (–) Goldblume
En (–) crown daisy
Gr (3) μαργαρίτα
Hu (0) koronás margitvirág
Ru (1) хризантема килеватая
Sh (1) srčana popadika
Sk (–) králik vencovitý

* ***Chrysanthemum leucanthemum*** L. → *Leucanthemum vulgare*

* ***Chrysanthemum parthenium*** (L.) Bernh. → *Tanacetum parthenium*

425 ***Chrysanthemum segetum*** L.
Cs (1) kopretina osenní
De (–) Saat-Wucherblume
En (–) corn marigold
Pl (1) złocień polny
Ru (1) хризантема посевная
Sh (1) njivska popadika; žutocvjetna ivančica
Sk (–) králik siatinovy

* ***Chrysanthemum tanacetum*** Karsch → *Tanacetum vulgare*

* ***Chrysanthemum vulgare*** (L.) Bernh. → *Tanacetum vulgare*

426 ***Cichorium intybus*** L.
Al (–) cikore; lulekore
Bg (1) синя жлъчка
Cs (2) čekanka obecná
De (–) Gemeine Wegwarte
En (–) chicory
Gr (3) ραδίκι
Hu (2) mezei katáng
Pl (1) cykoria podróżnik
Ro (1) cicoare
Ru (1) цикорий обыкновенный
Sh (1) obična cikorija; obična vodopija
Sk (–) čakanka obyčajná

427 ***Cicuta virosa*** L.
Al (–) helmëza
Bg (1) вонящ бучиниш
Cs (2) rozpuk jízlivý
De (–) Gift-Wasserschierling
En (–) cowbane
Hu (1) csomorika
Pl (1) szalej jadowity
Ro (1) cucută de apă
Ru (1) цикута ядовитая
Sh (1) smrdljiva trubeljika
Sk (–) rozpuk jedovatý

428 ***Circaea lutetiana*** L.
Bg (0) парижка чаровница
Cs (0) čarovník pařížský
De (–) Gemeines Hexenkraut
En (–) enchanter's nightshade
Hu (0) erdei varázslófű
Pl (1) czartawa pospolita
Ro (0) tilişcă
Ru (0) цирцея парижская
Sk (–) čarovník lesklý

429 ***Cirsium* spp.** Mill.
De (–) Kratzdistel
Gr (1) κίρσιο
Ru (2) бодяк
Sk (–) pichliač

430 ***Cirsium acaule*** Scop.
Cs (0) pcháč bezlodyžný
De (–) Stengellose Kratzdistel
En (–) dwarf thistle
Pl (1) ostrożeń krótkołodygowy
Ru (0) бодяк бесстебельный
Sk (–) pichliač bezbylový

431 ***Cirsium arvense*** (L.) Scop.
Al (–) cirza e arave; perenik
Bg (2) паламида
Cs (3) pcháč oset
De (–) Acker-Kratzdistel
En (–) creeping thistle
Gr (3) κίρσιο
Hu (3) mezei acat
Pl (3) ostrożeń polny
Ro (3) pălămidă
Ru (3) бодяк полевой, осот розовый
Sh (3) njivska palamida; poljski osjak
Sk (–) pichliač roľný

432 ***Cirsium brachycephalum*** Jurat.
Cs (0) pcháč krátkohlavý
Hu (1) alföldi acat
Sk (–) pichliač krátkohlavý

433 ***Cirsium canum*** (L.) All.
Bg (1) бяла паламида
Cs (1) pcháč šedý
De (–) Graue Kratzdistel
Hu (1) szürke acat
Pl (0) ostrożeń siwy
Ru (1) бодяк серый
Sh (1) bela palamida; sijedi osjak
Sk (–) pichliač šedivý

434 ***Cirsium eriophorum*** (L.) Scop.
Al (–) cirza e Vandasit
Cs (1) pcháč bělohlavý
De (–) Wollige Kratzdistel
En (–) woolly thistle
Hu (1) gyapjas acat
Pl (1) ostrożeń głowacz
Sk (–) pichliač bielohlavý

435 ***Cirsium erisithales*** (Jacq.) Scop.
Cs (1) pcháč lepkavý
De (–) Klebrige Kratzdistel
En (–) yellow melancholy-thistle
Hu (0) enyves acat
Pl (1) ostrożeń lepki
Ru (0) бодяк клейкий
Sk (–) pichliač lepkavý

436 ***Cirsium helenioides*** (L.) Hill
Bg (0) разночетинеста паламида
Cs (0) pcháč různolistý
De (–) Verschiedenblättrige Kratzdistel
En (–) common melancholy-thistle
Pl (1) ostrożeń dwubarwny
Ru (0) бодяк разнолистный
Sk (–) pichliač rôznolistý

* ***Cirsium heterophyllum*** (L.) Hill → *Cirsium helenioides*

* ***Cirsium incanum*** (Gmel.) Fisch. → *Cirsium arvense*

* ***Cirsium lanceolatum*** (L.) Scop. → *Cirsium vulgare*

437 ***Cirsium oleraceum*** (L.) Scop.
Cs (2) pcháč zelinný
De (–) Kohl-Kratzdistel
En (–) cabbage thistle
Hu (1) halvány acat
Pl (2) ostrożeń warzywny
Ro (1) nilocea
Ru (1) бодяк огородный
Sh (1) mlečasta palamida; žutkasti osjak
Sk (–) pichliač zelinný

438 ***Cirsium palustre*** (L.) Scop.
Al (–) cirza e kënetës
Cs (1) pocháč bahenní
De (–) Sumpf-Kratzdistel
En (–) marsh thistle
Hu (1) mocsári acat
Pl (1) ostrożeń błotny
Ru (1) бодяк болотный
Sh (1) močvarna palamida; močvarni osjak
Sk (–) pichliač močiarny

439 ***Cirsium rivulare*** (Jacq.) All.
Cs (0) pcháč potoční
De (–) Bach-Kratzdistel
Hu (1) csermely acat
Pl (1) ostrożeń łąkowy
Ru (1) бодяк приручейный
Sk (–) pichliač potočný

* ***Cirsium setosum*** (Willd.) Bess. → *Cirsium arvense*

440 ***Cirsium vulgare*** (Savi) Ten.
Al (–) cirza e zakonshme
Bg (1) ланцетолистна паламида
Cs (1) pcháč obecný
De (–) Gemeine Kratzdistel
En (–) spear thistle
Hu (1) lándzsás acat
Pl (2) ostrożeń lancetowaty
Ro (1) crăpuşnic
Ru (1) бодяк обыкновенный

Sh (1) lancetasta palamida
Sk (–) pichliač obyčajný

441 *Cladium mariscus* (L.) Pohl
Bg (0) режещ кладиум
Cs (0) mařice pilovitá
De (–) Sumpf-Schneide
En (–) great fen-sedge
Gr (1) κλάδιο
Hu (0) télisás
Pl (1) knieć wiechowata
Ru (0) меч-трава обыкновенная
Sk (–) marica pilovitá

442 *Clematis flammula* L.
Al (–) kulpëra e butë
Bg (0) парлив павит
Cs (0) plamének palčivý
De (–) Brennende Waldrebe
En (–) fragrant clematis
Gr (1) αγράμπελλη
Ru (0) ломонос жгучий
Sk (–) plamienok plamenný

443 *Clematis integrifolia* L.
Cs (–) plamének celolistý
De (–) Ganzblättrige Waldrebe
En (–) solitary clematis
Ru (0) ломонос цельнолистный
Sh (1) plava pavit; poljska kandilka
Sk (–) plamienok celolistý

444 *Clematis recta* L.
Bg (0) прав павит
Cs (0) plamének přímý
De (–) Aufrechte Waldrebe
Gr (1) ορθάμπελλη
Hu (1) felálló iszalag
Pl (1) powojnik prosty
Ro (0) luminoasă
Ru (0) ломонос прямой
Sk (–) plamienok priamy

445 *Clematis vitalba* L.
Al (–) kërpnaja; kulpëra e egër
Bg (0) обикновен павит
Cs (1) plamének plotní
De (–) Gemeine Waldrebe
En (–) old-man's-beard
Hu (1) közönséges iszalag
Pl (1) powojnik pnący
Ro (0) curpen de pădure
Ru (1) ломонос виноградолистный
Sk (–) plamienok plotný

446 *Clematis viticella* L.
Al (–) zgjebça
Bg (0) лозовиден павит
Cs (0) plamének révovitý
De (–) Italienische Waldrebe
Gr (1) μικράμπελλη
Hu (0) olasz iszalag
Ru (0) ломонос фиолетовый
Sk (–) plamienok révovitý

447 *Clinopodium vulgare* L.
Al (–) trumza e rëndomtë
Bg (0) черновръх
Cs (0) marulka klinopád
De (–) Wirbeldost
En (–) wild basil
Gr (1) κοψόχορτο
Hu (1) borsfű
Pl (0) czyścica storzyszek
Ro (1) apărătoare
Ru (1) пахучка обыкновенная
Sh (1) obična marulja
Sk (–) marulka

448 *Cnicus benedictus* L.
Bg (1) пресечка
Cs (0) benedikt lékařský
De (–) Benediktenkraut
En (–) blessed thistle
Gr (1) καλάγκαθο
Hu (0) áldott bárcs

Ro (0) schinel
Ru (1) кникус благословенный
Sh (1) blaženi čakalj
Sk (–) Benedikt lekársky

* **Cochlearia armoracia** L. → *Armoracia rusticana*

449 **Colchicum autumnale** L.
Al (–) lulë vjeshte; xhërrokulli vjeshtor
Bg (0) обикновен мразовец
Cs (0) ocún jesenní
De (–) Herbst-Zeitlose
En (–) autumn crocus
Gr (1) σπασόχορτο
Hu (1) őszi kikerics
Pl (1) zimowit jesienny
Ro (2) brînduşă de toamnă
Ru (1) безвременник осенний
Sh (1) jesenski mrazovac; obični mrazovac
Sk (–) jesienka obyčajná

450 **Comandra elegans** (Roch.) Reichenb.
Sh (1) obična komandră

451 **Commelina communis** L.
De (–) Gemeine Commeline
En (–) day-flower
Ru (1) коммелина обыкновенная
Sk (–) križatka obyčajna

452 **Conium maculatum** L.
Al (–) kakuda; kukuta e njollosur
Bg (1) бучиниш
Cs (1) bolehlav plamatý
De (–) Gefleckter Schierling
En (–) hemlock
Gr (1) κώνειο
Hu (2) bürök
Pl (2) szczwół plamisty
Ro (1) cucută
Ru (2) болиголов крапчатый
Sh (1) pegava kukuta; velika kukuta
Sk (–) bolehlav škvrnitý

453 **Conringia austriaca** (Jacq.) Sweet
Bg (0) австрийска конрингия
Cs (0) hořinka rakouská
De (–) Österreichischer Ackerkohl
Hu (0) osztrák nyilasfű
Sh (1) obična gušćarica
Ru (0) конрингия австрийская
Sk (–) horinka rakúska

454 **Conringia orientalis** (L.) Dumort.
Bg (1) източна целебница
Cs (2) hořinka východní
De (–) Weißer Ackerkohl
En (–) hare's-ear mustard
Gr (1) κονριγκία
Hu (1) keleti nyilasfű
Pl (1) pszonacznik wschodni
Ru (1) конрингия восточная
Sh (1) orientalna gušćarica
Sk (–) horinka východná

* **Consolida ajacis** auct. → *Consolida ambigua*

455 **Consolida ambigua** (L.) Ball & Heyw.
Bg (0) аяксова ралица
Cs (0) ostrožka znamenaná
De (–) Garten-Rittersporn
En (–) larkspur
Gr (1) καπουτσίνος
Pl (1) ostróżka ogrodowa
Ro (0) surguci
Ru (1) живокость аяксова
Sk (–) ostrôžka obyčajná

* **Consolida divaricata** (Ledeb.) Schröd. → *Consolida regalis*

456 **Consolida orientalis** (Gay) Schröd.
Bg (2) източна ралица
Cs (0) ostrožka východní

De (–) Orientalischer Rittersporn
En (–) eastern larkspur
Hu (3) keleti szarkaláb
Ru (3) живокость восточная
Sh (1) orijentalni žarovnjak
Sk (–) ostrôžka východná

457 *Consolida regalis* Gray
Al (–) gjuhënusja e fortë
Bg (2) полска ралица
Cs (2) ostrozka stračka
De (–) Feld-Rittersporn
En (–) forking larkspur
Gr (1) καπουτσίνος
Hu (3) mezei szarkaláb
Pl (3) ostróżka polna
Ro (3) nemţişor de cîmp
Ru (3) живокость полевая
Sh (1) obični žarovnjak
Sk (–) ostrôžka poľná

* ***Consolida segetum*** (Lam.) Gray → *Consolida regalis*

458 *Convolvulus* spp. L.
Bg (1) повитица
De (–) Winde
En (–) bindweed
Gr (2) περιπλοκάδα
Ru (3) вьюнок
Sk (–) pupenec

459 *Convolvulus arvensis* L.
Al (–) dredhja e arës
Bg (2) полска повитица
Cs (3) svlačec rolní
De (–) Acker-Winde
En (–) field bindweed
Gr (3) περιπλοκάδα
Hu (3) apró szulák; folyondárszulák
Pl (3) powój polny
Ro (3) volbură
Ru (3) вьюнок полевой
Sh (3) njivski poponac; obični slak
Sk (–) pupenec roľný

460 *Convolvulus betonicifolius* Mill.
Al (–) dredhja leshatake
Bg (1) влакнеста повитица
Ru (1) вьюнок бетониколистный
Sk (–) pupenec bukvicolistý

* ***Convolvulus hirsutus*** Bieb. → *Convolvulus betonicifolius*

* ***Convolvulus minor*** Gilib. → *Convolvulus arvensis*

* ***Convolvulus sepium*** L. → *Calystegia sepium*

461 *Conyza bonariensis* (L.) Cronq.
En (–) hairy fleabane
Ru (1) многолепестник бонский
Sk (–) turan kučeravý

462 *Conyza canadensis* (L.) Cronq.
Al (–) erigeroni i Kanadës
Bg (2) злолетница
Cs (3) turan kanadský
De (–) Kanadisches Berufkraut
En (–) Canadian fleabane
Hu (3) betyárkóró
Pl (3) przymiotno kanadyjskie
Ro (2) bătrîniş
Ru (2) мелколепестник канадский
Sh (1) kanadska hudoletnica; kanadska hudoljetnica
Sk (–) turan kanadský

463 *Coriandrum sativum* L.
De (–) Garten-Koriander
En (–) coriander
Ru (1) кориандр посевной
Sh (1) baštenski korijander
Sk (–) kyšnec záhradný

464 *Corispermum declinatum* Stephan
Ru (1) верблюдка повислая

465 ***Corispermum hyssopifolium*** L.
De (–) Großblättriger Wanzensame
Ru (1) верблюдка иссополистная
Sh (1) isopolisna steničnica; isopolisna stjeničnica

466 ***Corispermum nitidum*** Kit.
De (–) Glanzender Wanzensame
Ru (1) верблюдка лоснящаяся
Sh (1) sjajna steničnica; sjajna stjeničnica
Sk (–) plostičosemä lesklé

467 ***Coronilla*** **spp.** L.
Bg (1) зайчина
De (–) Kronwicke
Gr (1) κορονίλλα
Ru (1) вязель
Sk (–) ranostaj

468 ***Coronilla vaginalis*** Lam.
Cs (0) čičorka pochvatá
De (–) Scheiden-Kronwicke
En (–) small scorpion-vetch
Hu (0) terpedt koronafürt
Pl (1) cieciorka pochewkowata
Sk (–) ranostaj pošvatý

469 ***Coronilla varia*** L.
Al (–) milëza e ndryshme
Bg (1) пъстроцветна зайчина
Cs (2) čičorka pestrá
De (–) Bunte Kronwicke
En (–) crownvetch
Hu (2) tarka koronafürt
Pl (2) cieciorka pstra
Ro (1) coronişte
Ru (1) вязель пёстрый
Sh (1) obična ajčica; šareni grašak
Sk (–) ranostaj pestrý

470 ***Coronopus squamatus*** (Forsk.) Asch.
Cs (0) vranožka šupinatá
De (–) Niederliegender Krähenfuß
En (–) swine-cress
Hu (1) varjúláb
Ro (1) talpa stîncii
Ru (1) воронья лапа простёртая
Sh (1) ljuskava odra; polegla stazarica
Sk (–) vrania nôžka šupinatá

471 ***Corydalis*** **spp.** Vent.
Bg (1) лисичина
De (–) Lerchensporn
Gr (1) κορυδαλίδα
Ru (1) хохлятка
Sk (–) dymica

472 ***Corydalis bulbosa*** (L.) DC.
Al (–) koridali bulboz
Bg (0) куха лисичина
Cs (1) dymnivka dutá
De (–) Hohler Lerchensporn
En (–) purple corydalis
Hu (0) odvas keltike
Pl (1) kokorycz pusta
Ro (0) brebenel
Ru (1) хохлятка полая
Sk (–) dymica dutá

* ***Corydalis cava*** (L.) Schweig. & Koer.
→ *Corydalis bulbosa*

473 ***Corydalis solida*** (L.) Swartz
Bg (0) плътна лисичина
Cs (0) dymnivka plná
De (–) Gefingerter Lerchensporn
En (–) bulbous corydalis
Hu (0) ujjas keltike
Pl (1) kokorycz pełna
Ro (0) brebenel
Ru (1) хохлятка плотная
Sk (–) dymica plná

474 ***Crambe tataria*** Sebeők
Bg (0) татарско морско зеле
Cs (0) katrán tatarský
En (–) Tartarian sea-kale
Hu (0) tátorján

Pl (0) modrak tatarski
Ro (0) tîrtan
Ru (0) катран татарский
Sh (1) obični krambe
Sk (–) katrán tatársky

475 ***Crataegus monogyna*** Jacq.
Cs (–) hloh jednoblizný
De (–) Eingriffeliger Weißdorn
En (–) hawthorn
Ru (1) боярышник однопестичный
Sh (1) obični glog
Sk (–) hloh jednočnelkový

476 ***Crepis* spp.** L.
Bg (1) дрипавка
De (–) Pippau
En (–) hawk's-beard
Gr (2) πικραλίδα
Ru (2) скерда
Sk (–) škarda

477 ***Crepis biennis*** L.
Al (–) shmanga dyvjeçare
Bg (1) двегодишна дрипавка
Cs (2) škarda dvouletá
De (–) Wiesen-Pippau
En (–) rough hawk's-beard
Hu (1) réti zörgőfű
Pl (1) pępawa dwuletnia
Ru (2) скерда двулетняя
Sh (2) crna čekinjuša; dvogodišnji dimak
Sk (–) škarda dvojročná

* ***Crepis blavii*** (Asch.) Stadl. → *Crepis pannonica*

478 ***Crepis capillaris*** (L.) Wallr.
Al (–) shmanga e hollë si qime floku
Cs (0) škarda vláskovitá
De (–) Grüner Pippau
En (–) smooth hawk's-beard
Hu (2) vékony zörgőfű
Pl (2) pępawa zielona
Ru (2) скерда волосовидная
Sh (1) nitasti dimak
Sk (–) škarda vlásočnatá

479 ***Crepis foetida*** L.
Al (–) shmanga me erë të keqe
Bg (1) смрадлива дрипавка
Cs (2) škarda smrdutá
De (–) Stinkender Pippau
En (–) stinking hawk's-beard
Hu (2) büdös zörgőfű
Pl (1) pępawa makolistna
Ro (1) gălbenuş
Ru (1) скерда вонючая
Sh (1) vonjavi dimak
Sk (–) škarda smradľavá pravá

480 ***Crepis mollis*** (Jacq.) Asch.
Bg (1) нежна дрипавка
Cs (0) škarda měkká
De (–) Weicher Pippau
En (–) northern hawk's-beard
Pl (1) pępawa miękka
Ru (0) скерда мягковолосистая
Sk (–) škarda mäkká

481 ***Crepis paludosa*** (L.) Moen.
Bg (1) блатна дрипавка
Cs (0) škarda bažinná
De (–) Sumpf-Pippau
En (–) marsh hawk's-beard
Hu (0) mocsári zörgőfű
Pl (1) pępawa błotna
Ru (0) скерда болотная
Sk (–) škarda močiarna

482 ***Crepis pannonica*** (Jacq.) Koch
Bg (1) панонска дрипавка
Cs (0) škarda panonská
Hu (0) magyar zörgőfű
Pl (0) pępawa sztywna
Ru (0) скерда венгерская
Sk (–) škarda panónska

483 *Crepis pulchra* L.
Al (–) shmanga e bukur
Bg (1) красива дрипавка
Cs (0) škarda pěkná
De (–) Glanz-Pippau
Hu (1) szép zörgőfű
Ru (1) скерда красивая
Sk (–) škarda pekná

* ***Crepis rhoeadifolia*** Bieb. → *Crepis foetida*

484 *Crepis setosa* Haller
Al (–) shmanga me qime të mëndafshta
Bg (1) четинеста дрипавка
Cs (0) škarda štětinatá
De (–) Borsten-Pippau
En (–) bristly hawk's-beard
Hu (1) sertés zörgőfű
Pl (0) pępawa szczeciniasta
Ro (1) iarba găinilor
Ru (1) скерда щетинистая
Sh (1) čekinjasta čekinjuša; čekinjasti dimak
Sk (–) škarda štetinatá

* ***Crepis taraxacifolia*** Thuill. → *Crepis vesicaria*

485 *Crepis tectorum* L.
Cs (0) škarda střešní
De (–) Dach-Pippau
En (–) succory hawk's-beard
Hu (2) hamvas zörgőfű
Pl (2) pępawa dachowa
Ru (2) скерда кровельная
Sh (1) uskolisni dimak
Sk (–) škarda strešná

486 *Crepis vesicaria* L.
De (–) Blasen-Pippau
En (–) beaked hawk's-beard
Hu (0) pitypanglevelű zörgőfű
Sh (1) obični dimak
Sk (–) škarda odhryznutá

487 *Crepis zacintha* (L.) Bab.
Al (–) zacinta me lythe
Bg (1) брадавичеста зацинта
De (–) Hocker-Lattich

488 *Crocus* spp. L.
Bg (0) минзухар
De (–) Safran
Gr (1) κρόκος
Ru (0) шафран
Sk (–) krokus

489 *Crocus biflorus* Mill.
Bg (0) двуцветен минзухар
De (–) Zweiblütiger Safran
Pl (0) krokus dwukwiatowy; szafran dwukwiatowy
Ru (0) шафран двуцветковый
Sh (1) dvocvetni šafran; dvocvjetni šafran
Sk (–) krokus dvojkvetý

490 *Crucianella graeca* Boiss.
Ru (0) крестовница греческая
Sh (1) grčki broćevac
Sk (–) brotinka grécka

491 *Cruciata glabra* (L.) Ehrend.
De (–) Kahles Kreuzlabkraut
En (–) slender crosswort
Hu (0) tavaszi keresztfű
Ru (0) круциата гладкая
Sh (1) žuta broćika
Sk (–) lipkavec lysý

492 *Cruciata pedemontana* (Bella.) Ehrend.
Bg (0) пиемонтско еньовче
Cs (0) svízel piemontský
Hu (0) apró keresztfű
Pl (1) przytulia krzyżowa
Ru (0) круциата пьемонтская
Sk (–) lipkavec piemontský

493 *Crupina vulgaris* Cass.
Cs (–) chrpinka obecná

De (–) Gemeiner Schlupfsame
Ru (0) крупина обыкновенная
Sh (1) obični nosan
Sk (–) krupinka obyčajná

494 *Crypsis aculeata* (L.) Aiton
Al (–) kripsi me gjemba
Bg (0) бодлива скритокласица
Cs (0) skrytěnka bodlinatá
De (–) Dorngras
Hu (1) bajuszpázsit
Ro (0) ghimpariţă
Ru (0) скрытница колючковая
Sh (1) bodljikava trnica
Sk (–) skrytka ostnatá

495 *Crypsis alopecuroides* (Pill. & Mitt.) Schrad.
Al (–) kripsi si pungacë
Bg (1) класицеподобен крипсис
Cs (0) bahenka psárkovitá
De (–) Fuchsschwanzartiges Sumpfgras
Hu (1) karcsú bajuszfű
Ru (0) скрытница лисохвостовидная
Sk (–) bahienka psiarkovitá

496 *Crypsis schoenoides* (L.) Lam.
Ru (1) скрытница камышевидная
Sk (–) bahienka šašinovitá

497 *Cucubalus baccifer* L.
De (–) Beeren-Taubenkropf
Ru (0) волдырник ягодный
Sh (1) obična gušavica
Sk (–) nadutica bobuľnatá

498 *Cuscuta* spp. L.
Bg (1) кукувича прежда
De (–) Seide
En (–) dodder
Gr (3) κουσκούτα
Ru (1) повилика
Sk (–) kukučina

499 *Cuscuta approximata* Bab.
Ru (1) повилика сближенная
Sh (1) prepletena vilina kosica

500 *Cuscuta australis* R Br.
Cs (–) kokotice jižní
De (–) Südliche Seide
Ru (0) повилика южная
Sh (1) južna vilina kosica
Sk (–) kukučina južná

501 *Cuscuta campestris* Yunck.
Cs (2) kokotice ladní
De (–) Nordamerikanische Seide
En (–) field dodder
Hu (2) nagy aranka
Pl (1) kanianka polna
Ro (3) cuscută mare
Ru (2) повилика полевая
Sh (1) poljska vilina kosica
Sk (–) kukučina poľná

* ***Cuscuta cesattiana*** Bertol. → *Cuscuta australis*

* ***Cuscuta densiflora*** Soy. → *Cuscuta epilinum*

502 *Cuscuta epilinum* Weihe
Al (–) kuskuta e lirit
Bg (1) ленена кукувича прежда
Cs (1) kokotice hubilen
De (–) Flachs-Seide
En (–) flax dodder
Hu (2) lenfojtó aranka
Ru (2) повилика льняная
Sh (1) lanova vilina kosica
Sk (–) kukučina ľanová

503 *Cuscuta epithymum* (L.) L.
Al (–) kuskuta e tërfilit
Bg (2) обикновена кукувича прежда
Cs (3) kokotice jetelová
De (–) Quendel-Seide

En (–) common dodder
Hu (2) kis aranka
Pl (2) kanianka koniczynowa
Ru (2) повилика клеверная
Sh (2) djetelinska vilina kosica; obična vilina kosica
Sk (–) kukučina ďatelinová

504 ***Cuscuta europaea*** L.
Al (–) kuskuta e evropës; teli i dreqit
Bg (1) европеиска кукувича прежда
Cs (2) kokotice evropská
De (–) Europäische Seide
En (–) greater dodder
Hu (2) közönséges aranka
Pl (2) kanianka pospolita
Ro (3) torţel
Ru (2) повилика европейская
Sh (2) evropska vilina kosica
Sk (–) kukučina európska

505 ***Cuscuta gronovii*** Willd.
Ru (1) повилика Грановия
Sk (–) kukučina gronov

* ***Cuscuta indica*** (Englem.) Petrov → *Cuscuta pellucida*

506 ***Cuscuta lupuliformis*** Krock.
Cs (0) kokotice chmelová
De (–) Pappel-Seide
En (–) hop dodder
Hu (1) komlóaranka
Pl (1) kanianka wielka
Ru (1) повилика хмелевидная
Sh (1) vrbova vilina kosica
Sk (–) kukučina chmeľová

507 ***Cuscuta monogyna*** Vahl
Ru (1) повилика однопестичная
Sh (1) žućkasta vilina kosica
Sk (–) kukučina jednočnelková

508 ***Cuscuta pellucida*** Butkov
Ru (1) повилика индийская

509 ***Cuscuta pentagona*** Eng.
Bg (2) едрозърнеста кукувича прежда
Hu (2) nagy aranka
Ru (1) повилика пятигранная
Sh (1) ruderalna vilina kosica
Sk (–) kukučina päťhranná

510 ***Cuscuta planiflora*** Ten.
Ru (1) повилика коротколстолбиковая
Sk (–) kukučina plochokvetá

511 ***Cuscuta suaveolens*** Ser.
Al (–) kuskuta amerikane
Cs (0) kokotice libovonná
De (–) Chilenische Seide
En (–) Chile dodder
Pl (1) kanianka wonna
Ru (1) повилика душистая
Sh (1) lucerkina vilina kosica
Sk (–) kukučina voňavá

* ***Cuscuta tinei*** Insenga → *Cuscuta australis*

* ***Cuscuta trifolii*** Bab. → *Cuscuta epithymum*

* ***Cyanus arvensis*** Moen. → *Centaurea cyanus*

* ***Cyanus diffusa*** Lam. → *Centaurea diffusa*

* ***Cyanus scabiosa*** Presl. → *Centaurea scabiosa*

* ***Cyanus segetum*** Lam. → *Centaurea cyanus*

* ***Cyanus stoebe*** L. → *Centaurea rhenana*

* ***Cyclachaena xanthifolia*** (Nutt.) Fresen. → *Iva xanthifolia*

* ***Cymbopogon hirtus*** (L.) Thom. → *Hyparrhenia hirta*

512 *Cynanchum acutum* L.
Bg (1) увивен устрел
En (–) stranglewort
Gr (1) κύναγχο
Ro (1) plechiuşcă
Ru (1) цинанхум острый
Sk (–) luskáč ostrolistý

* ***Cynanchum vincetoxicum*** (L.) Pers. → *Vincetoxicum hirundinaria*

513 *Cynodon dactylon* (L.) Pers.
Al (–) krisja
Bg (2) троскот
Cs (2) troskut prstnatý
De (–) Hundszahngras
En (–) Bermuda grass
Gr (3) αγριάδα
Hu (3) csillagpázsit
Pl (0) psi ząb
Ro (3) pir gros
Ru (3) свинорой пальчатый
Sh (3) obična zubača
Sk (–) prstnatec obyčajný

514 *Cynoglossum* spp. L.
Bg (0) нишинец
De (–) Hundszunge
En (–) hound's-tongue
Gr (1) κυνόγλωσσα
Ru (0) чернокорень
Sk (–) psí jazyk

515 *Cynoglossum creticum* Mill.
Al (–) gjuhëqeni i Kretës
Bg (0) пъстър нишинец
De (–) Kreta-Hundszunge
En (–) blue hound's-tongue
Ru (0) чернокорень критский
Sk (–) psí jazyk pestrý

516 *Cynoglossum officinale* L.
Al (–) gjuhëqeni mjekësor
Bg (0) планински нишинец
Cs (0) užanka lékařská
De (–) Gemeine Hundszunge
En (–) common hound's-tongue
Hu (0) közönséges ebnyelvűfű
Pl (1) ostrzeń pospolity
Ro (1) limba cîinelui
Ru (1) чернокорень лекарственный
Sh (1) jednostruki jezik pasji; obični mišinac
Sk (–) psí jazyk lekársky

517 *Cynosurus cristatus* L.
Al (–) bishtqeni me kreshtë; grumi i egër
Bg (1) гребенест сеноклас
Cs (1) poháňka hřebenitá
De (–) Wiesen-Kammgras
En (–) crested dogstail
Hu (0) taréjos cincor
Pl (1) grzebienica pospolita
Ro (0) pieptănariță
Ru (1) гребенник обыкновенный
Sk (–) hrebienka obyčajná

518 *Cynosurus echinatus* L.
Al (–) bishtqeni me gjemba
Bg (1) четинест сеноклас
Cs (0) poháňka ježatá
De (–) Stacheliges Kammgras
En (–) rough dogstail
Gr (1) κυνόσουρος
Hu (0) tüskés cincor
Ru (1) гребенник колючий
Sk (–) hrebienka ježatá

519 *Cyperus difformis* L.
Al (–) truska dyformash
Bg (1) разновидна кисела трева
De (–) Verschiedengestaltige Zyper
En (–) small-flowered nutsedge
Hu (1) rizspalka

Ro (2) părul porcului de baltă
Ru (2) сыть разнородная
Sk (–) šachorec nepravidelný

520 *Cyperus esculentus* L.
Bg (1) чуфа
De (–) Erdmandel
En (–) yellow nutsedge
Gr (2) κίτρινη κύπερη
Ru (1) сыть съедобная
Sk (–) šachorník jedlý

521 *Cyperus flavescens* L.
Bg (1) жълтеникав шивар
Cs (0) šáchorek žlutavý
De (–) Gelbliches Zypergras
Hu (1) sárgás palka
Pl (2) cibora żółta
Ro (0) căprişor
Sk (–) šachorec žltavý

522 *Cyperus flavidus* Retz.
Bg (1) кълбовиден циперус
Ru (1) сыть желтоватая
Sk (–) šachorec nažltlý

523 *Cyperus fuscus* L.
Al (–) truska e zeshkët
Bg (1) кафяв шавар
Cs (0) šáchor hnědý
De (–) Braunes Zypergras
En (–) brown galingale
Hu (1) barna palka
Pl (2) cibora brunatna
Ro (1) căprişor oacheş
Ru (1) сыть бурая
Sh (1) smedjasti šilj
Sk (–) šachor tmavý

524 *Cyperus glaber* L.
Bg (1) гол шавар
Ru (1) сыть гладкая
Sk (–) šachorec lysý

* ***Cyperus globosus*** All. → *Cyperus flavidus*

525 *Cyperus glomeratus* L.
Bg (1) топчест шавар
De (–) Knäuel-Zypergras
En (–) round-headed nutsedge
Hu (1) csomós palka
Ro (1) căprişor bălan
Ru (1) сыть скученная
Sk (–) šachorník kĺbkatý

526 *Cyperus longus* L.
Al (–) shtiza
Bg (1) дълъг шавар
De (–) Langes Zypergras
En (–) common galingale
Ru (1) сыть длинная
Sk (–) šachorník dlhý pravý

* ***Cyperus monti*** L. → *Cyperus serotinus*

527 *Cyperus rotundus* L.
Al (–) truska
Bg (2) клубенообразуваща кисела трева
De (–) Nußgras
En (–) purple nutsedge
Gr (3) κορφυρή κύπερη
Ru (2) сыть круглая
Sh (1) gomoljasta peronjika; gomoljasti šilj
Sk (–) šachor hľuznatý

528 *Cyperus serotinus* Rottb.
Bg (0) късен циперус
De (–) Spätblühendes Zypergras
En (–) mountain nutsedge
Ru (0) сытничек поздний
Sh (1) crvenomrki šilj
Sk (–) šachor neskorý

* ***Cytisus austriacus*** L. → *Chamaecytisus hirsutus*

* ***Cytisus nigricans*** L. → *Lembotropis nigricans*

529 ***Cytisus procumbens*** (Waldst. & Kit.) Spreng.
Al (–) vjexhësa e shtririë
Bg (0) полегнал зановец
Cs (1) čilimník poléhavý
Hu (0) sziklai zanót
Sk (–) kručinkovec poliehavý

530 ***Cytisus scoparius*** (L.) Link
Al (–) gjineshtër për fshesë
Cs (2) jenovec metlatý
De (–) Gemeiner Besenginster
En (–) common broom
Hu (0) seprő zanót
Pl (1) żarnowiec miotlasty
Ro (0) drob
Ru (0) жарновец метельчатый
Sk (–) prútnatec metlovitý

D

531 *Dactylis glomerata* L.
Al (–) telishi
Bg (1) ежова главица
Cs (2) srha říznačka
De (–) Knaulgras
En (–) cocksfoot
Gr (2) δακτυλίδα
Hu (0) csomós ebír
Pl (3) kupkówka pospolita
Ro (0) golomăţ
Ru (2) ежа сборная
Sk (–) reznačka laločnatá

* ***Dactylis hispanica*** Roth → *Dactylis glomerata*

532 *Dactylorhiza maculata* (L.) Soó
De (–) Geflecktes Knabenkraut
En (–) heath spotted-orchid
Hu (0) foltos ujjaskosbor
Ro (0) mîna Maicii Domnului
Ru (0) пальчатокореннник пятнистый
Sh (1) pegava kaćunka; pjegava kaćunka
Sk (–) prstovka škvrnitá

533 *Dactylorhiza majalis* (Reichenb.) Hunt & Summ.
Al (–) salep gjethegjërë
Cs (0) vstavač májový
De (0) Breitblättriges Knabenkraut
En (–) broad-leaved marsh-orchid
Pl (1) storczyk szerokolistny
Ru (0) пульчатокоренник майский
Sk (–) prstovka širokolistá

534 *Danthonia decumbens* (L.) DC.
Cs (0) trojzubec poléhavý
De (–) Dreizahn
En (–) heath-grass
Gr (1) ντανθόνια
Ru (0) трёхзубка распростёртая
Sk (–) trojzub položený

535 *Daphne cneorum* L.
Al (–) holokuqi i shkurtër
Bg (0) червена вълча ягода
Cs (0) lýkovec vonný
De (–) Heideröschen
Hu (0) henye boroszlán
Pl (1) wawrzynek główkowy
Ro (0) tulichină
Ru (0) волчеягодник боровик
Sk (–) lykovec voňavý

536 *Dasypyrum villosum* (L.) Cand.
Al (–) gruri i thneglave
Bg (1) власата класатича
Cs (0) kosmatka huňatá
De (–) Haarweizen
Gr (2) τριχοκρίθαρο
Hu (1) Haynald füve
Ru (1) дазипирум мохнатый
Sh (1) vunasta pšenica
Sk (–) kosmač huňatý

537 *Datura stramonium* L.
Al (–) tatulla
Bg (1) татул
Cs (3) durman obecný
De (–) Stechapfel
En (–) thorn-apple
Gr (2) τάτουλας
Hu (3) maszlag
Pl (2) bieluń dziędzierzawa
Ro (1) ciumăfaie
Ru (2) дурман обыкновенный
Sh (1) obična tatula
Sk (–) durman obyčajný

538 *Daucus carota* L.
Al (–) karota e egër
Bg (1) див морков
Cs (3) mrkev obecná
De (–) Wilde Möhre

En (–) wild carrot
Gr (2) αγριοκαρώτο
Hu (3) vadmurok
Pl (2) marchew zwyczajna
Ro (2) morcov sălbatic
Ru (2) морковь дикая
Sh (2) divlja mrkva
Sk (–) mrkva obyčajná

* **Delphinium consolida** L. →
Consolida regalis

* **Delphinium orientale** Gay →
Consolida orientalis

* **Delphinium paniculatum** Host →
Consolida regalis

* **Delphinium segetum** Lam. →
Consolida regalis

539 Deschampsia cespitosa (L.) Beauv.
Al (–) deshampsia tufore; thekçi
Bg (1) туфеста пластица
Cs (3) metlice trsnatá
De (–) Rasen-Schmiele
En (–) tufted hair-grass
Hu (1) gyepes sédbúza
Pl (2) śmiałek darniowy
Ro (1) tîrsă
Ru (2) луговик дернистый
Sk (–) metlica trsnatá

540 Deschampsia flexosa (L.) Trin.
Al (–) deshampsia e epshme
Bg (0) вълновидна пластица
Cs (2) metlice křivolaká
De (–) Geschlängelte Schmiele
En (–) wavy hair-grass
Hu (0) erdei sédbúza
Pl (2) śmiałek pogięty
Ro (0) păiuş
Ru (2) луговик извилистый
Sk (–) metlica krivolaká

541 Descurainia sophia (L.) Webb
Bg (1) обикновена войничица
Cs (3) úhorník mnohodílný
De (–) Gemeine Besenrauke
En (–) flixweed
Hu (3) sebforrasztó zsombor
Pl (2) stulicha psia
Ro (3) voinicică
Ru (2) дескурения София
Sh (1) rasperjani oranj
Sk (–) úhorník liečivý

542 Desmazeria rigida (L.) Tutin
En (–) hard meadow-grass
Gr (1) σκληροπόα
Sk (–) tvrdovka tuhá

543 Dianthus spp. L.
Bg (0) карамфил
De (–) Nelke
En (–) pink
Gr (1) δίανθος
Ru (1) гвоздика
Sk (–) klinček

544 Dianthus armeria L.
Al (–) karafili Armeria; karafil i egër
Bg (0) армериявиден карамфил
Cs (0) hvozdík svazčitý
De (–) Büschel-Nelke
En (–) Deptford pink
Hu (0) szeplős szegfű
Pl (2) goździk kosmaty
Ru (0) гвоздика армериевидная
Sh (1) divlji karanfil
Sk (–) klinček sväzčitý

545 Dianthus carthusianorum L.
Al (–) karafili kartusianorum
Bg (0) картузиански карамфил
Cs (1) hvozdík kartouzek
De (–) Karthäuser-Nelke
Hu (0) barátszegfű
Pl (1) goździk kartuzek

Ro (0) garoafă
Ru (0) гвоздика картузианская
Sk (–) klinček kartuziánsky

546 *Dianthus deltoides* L.
Al (–) karafili deltoid
Bg (0) делтовиден карамфил
Cs (0) hvozdík kropenatý
De (–) Heide-Nelke
En (–) maiden pink
Hu (0) mezei szegfű
Pl (1) goździk kropkowany
Ru (0) гвоздика травянка
Sh (1) šareni karanfil
Sk (–) klinček slzičky Panny Márie

* ***Dianthus sanguineus*** Vis. → *Dianthus carthusianorum*

547 *Dianthus superbus* L.
Bg (0) великолепен карамфил
Cs (0) hvozdík pyšný
De (–) Pracht-Nelke
En (–) large pink
Hu (0) buglyos szegfű
Pl (1) goździk pyszny
Ro (0) garoafă de munte
Ru (0) гвоздика пышная
Sk (–) klinček pyšný

548 *Dichanthium ischaemum* (L.) Rob.
Al (–) fëngu
Bg (0) бяла садина
Cs (1) vousatka prstnatá
En (–) beardgrass
Hu (0) fenyérfű
Pl (1) palczatka kosmata
Ro (0) bărboasă
Ru (0) бородач кровеостанавливающий
Sh (1) obična belešina; obična vlaska
Sk (–) fúzatka prstnatá

549 *Dictamnus albus* L.
Al (–) bari i uzos
Bg (–) бял росен
Cs (1) třemdava bílá
De (–) Weißer Diptam
En (–) burning bush
Hu (0) nagy ezerjófű
Pl (0) dyptam jesionolistny
Ro (0) frăsinel
Ru (0) ясенец белый
Sk (–) tremdava biela

550 *Digitalis* spp. L.
Bg (–) напръстник
De (–) Fingerhut
En (–) foxglove
Gr (1) κορακόχορτο
Ru (0) наперстянка
Sk (–) náprstník

551 *Digitalis grandiflora* Mill.
Al (–) luletogzi i dyshimtë; luletogzi i verdhë
Cs (1) náprstník velkokvětý
De (–) Großblütiger Fingerhut
En (–) yellow foxglove
Hu (0) sárga gyűszűvirág
Pl (1) naparstnica zwyczajna
Ro (0) degetar galben
Ru (0) наперстянка крупноцветковая
Sk (–) náprstník veľkokvetý

552 *Digitalis lanata* Ehrh.
De (–) Wolliger Fingerhut
En (–) woolly foxglove
Ru (0) наперстянка шерстистая
Sh (1) pustenasti naprstak
Sk (–) náprstník vlnatý

553 *Digitalis purpurea* L.
Al (–) luletogzi i purpurtë
Bg (0) червен напръстник
Cs (1) náprstník červený

De (–) Roter Fingerhut
En (–) common foxglove
Hu (0) piros gyűszűvirág
Pl (1) naparstnica purpurowa
Ro (0) degetar roşu
Ru (0) наперстянка пурпурная
Sh (1) crveni naprstak
Sk (–) náprstník červený

* **Digitaria aegyptica** (Retz.) Willd. → *Digitaria sanguinalis*

554 **Digitaria ciliaris** (Retz.) Koeler
Cs (0) rosička brvitá
De (–) Wimper-Fingerhirse
En (–) hairy finger-grass
Ru (1) росичка гребенчатая
Sh (1) trepljasta svračica
Sk (–) prstovka brvitá

* **Digitaria filiformis** auct. → *Digitaria ischaemum*

* **Digitaria glabra** (Schrad.) Beauv. → *Digitaria ischaemum*

555 **Digitaria ischaemum** (Schreb.) Muhl.
Al (–) meli formë fili
Cs (1) rosička lysá
De (–) Faden-Fingerhirse
En (–) smooth finger-grass
Hu (2) lecsepült ujjasmuhar
Pl (1) palusznik nitkowaty
Ru (2) росичка обыкновенная
Sk (–) prstovka holá

556 **Digitaria sanguinalis** (L.) Scop.
Al (–) meli i përgjakur
Bg (2) кръвно просо
Cs (2) rosička krvavá
De (–) Blut-Fingerhirse
En (–) red finger-grass
Gr (3) αιματόχορτο
Hu (3) pirók ujjasmuhar
Pl (2) palusznik krwawy
Ro (3) meişor
Ru (3) росичка кровеостанавливающая
Sh (1) crvena svračica
Sk (–) prstovka krvavá

* **Digraphis arundinacea** (L.) Trin. → *Phalaris arundinacea*

557 **Diplotaxis spp.** DC.
Bg (1) диплотаксис
De (–) Doppelsame
En (–) wall-rocket
Gr (1) διπλόταξη
Ru (1) двурядка
Sk (–) dvojsemä

558 **Diplotaxis muralis** (L.) DC.
Al (–) diplotaksa muresh
Bg (1) стенен диплотаксис
Cs (1) křez zedni
De (–) Mauer-Doppelsame
En (–) annual wall-rocket
Hu (3) fali kányazsázsa
Pl (2) dwurząd murowy
Ro (2) puturoasă
Ru (1) двурядка стенная
Sh (1) batvičasti dvoredac; obična mirunka
Sk (–) dvojsemä múrové

559 **Diplotaxis tenuifolia** (L.) DC.
Al (–) diplotaksa gjethehollë
Bg (0) теснолистен диплотаксис
Cs (–) křez tenkolistý
De (–) Schmalblättriger Doppelsame
En (–) perennial wall-rocket
Hu (1) évelő kányazsázsa
Pl (0) dwurząd wąskolistny
Ro (1) puturoasă
Ru (0) двурядка тонколистная
Sk (–) dvojsemä tenkolisté

560 *Diplotaxis viminea* (L.) DC.
Bg (1) диплотаксис
De (–) Ruten-Doppelsame
Ru (0) двурядка прутяная
Sk (–) dvojsemä šachorovité

561 *Dipsacus* spp. L.
Bg (0) лугачка
De (–) Karde
En (–) teasel
Gr (1) δίψακος
Ru (1) ворсянка
Sk (–) štetka

562 *Dipsacus fullonum* L.
Al (–) berunza e pyllit; ferrca pushëbutë
Bg (0) горска лугачка
Cs (1) štětka soukenická
De (–) Wilde Karde
En (–) wild teasel
Hu (1) erdei mácsonya
Pl (2) szczeć pospolita
Ru (1) ворсянка лесная
Sh (1) divlja češljugovina; šumska češljuga
Sk (–) štetka obyčajná

* ***Dipsacus fullonum*** Mill. → *Dipsacus sativus*

563 *Dipsacus laciniatus* L.
Bg (0) нарязанолистна лугачка
Cs (0) štětka dřípatá
De (–) Gelappte Karde
En (–) cutleaf teasel
Hu (1) héjakútmácsonya
Pl (1) szczeć wykrawana
Ro (1) scaiete
Ru (1) ворсянка разрезная
Sh (1) češljasta češljugovina; obična češljuga
Sk (–) štetka vykrojená

564 *Dipsacus pilosus* L.
Cs (0) hlavatka chlupatá
De (–) Behaarte Karde
En (–) small teasel
Hu (1) erdei fejvirág
Pl (0) szczeć owłosiona
Ro (0) scăiuş
Ru (–) ворсянка волосистая
Sk (–) štetočka chlpatá

565 *Dipsacus sativus* (L.) Honck.
Bg (0) обикновена лугачка
Cs (1) štětka soukenická
De (–) Weber-Karde
En (–) fuller's teasel
Ru (0) ворсянка посевная
Sk (–) štetka siata

* ***Dipsacus sylvestris*** Huds. → *Dipsacus sativus*

566 *Diptychocarpus strictus* (Fisch.) Trautv.
Ru (1) двоякоплодник прижатый

567 *Dittrichia graveolens* (L.) Greut.
Al (–) omani erërëndë
De (–) Stinkender Alant
En (–) stink aster
Gr (2) μικρή ακονυζιά
Sk (–) omam smradľavý

568 *Dittrichia viscosa* (L.) Greut.
Al (–) omani veshtullak
De (–) Klebriger Alant
En (–) sticky aster
Gr (2) μεγάλη ακονυζιά
Sk (–) omam lepkavý

569 *Dodartia orientalis* L.
Ru (1) додарция восточная

570 *Doronicum* spp. L.
De (–) Gemswurz
En (–) leopard's-bane
Gr (1) δωρόνικο
Ro (–) cujdă

Ru (0) дороникум
Sk (–) kamzičník

* ***Dorycnium germanicum*** (Grem.) Rik.→ *Dorycnium pentaphyllum*

* ***Dorycnium herbaceum*** Vill. → *Dorycnium pentaphyllum*

571 ***Dorycnium hirsutum*** (L.) Ser.
Al (–) dorikna leshtore; droshti
Cs (1) bílojetel chlupatý
De (–) Zottiger Backenklee
Sk (–) ďatelina biela chlpatá

572 ***Dorycnium pentaphyllum*** Scop.
Al (–) dorikna me pesë gjethe
Bg (1) тревисто звездиче
Cs (0) bílojetel pětilistý
De (–) Backenklee
Pl (0) szyplin jedwabisty
Sh (1) zeljasta beloglavica; zeljasta bjeloglavica
Sk (–) ďatelina biela päťlistá

* ***Draba krockeri*** (Andrz.) Hayek → *Erophila verna*

* ***Draba majuscula*** (Jord.) Hayek & Wib. → *Erophila verna*

573 ***Draba muralis*** L.
Al (–) draba muresh
Cs (2) chudina zední
De (–) Mauer-Felsenblümchen
En (–) wall whitlow-grass
Gr (1) δράβα
Hu (0) kövi daravirág
Pl (0) głodek murowy
Ru (1) крупка стенная
Sh (1) zidni dragušac
Sk (–) chudôbka múrová

* ***Draba obconica*** (De Bary) Hayek → *Erophila verna*

* ***Draba verna*** L. → *Erophila verna*

574 ***Dracocephalum moldavica*** L.
Cs (0) včelník multánský
De (–) Türkischer Drachenkopf
En (–) dragonhead
Pl (0) pszczelnik mołdawski
Ro (0) mătăciune
Ru (1) змееголовник молдавский
Sk (–) včelník moldavský

575 ***Dracocephalum thymiflorum*** L.
Cs (0) včelník douškokvětý
De (–) Thymianblütiger Drachenkopf
Pl (0) pszeczelnik macierzankowy
Ru (1) змееголовник тимьяноцветный
Sk (–) včelník dúškokvetý

576 ***Dracunculus vulgaris*** Schott
En (–) dragon arum
Gr (1) φιδόχορτο

577 ***Dryopteris cristata*** (L.) Gray
Cs (0) kapraď hřebenitá
De (–) Kamm-Wurmfarn
En (–) crested buckler-fern
Hu (0) tarajos pajzsika
Pl (1) narecznica grzebieniasta
Ru (0) щитовник гребенчатый
Sk (–) papraď hrebenitá

578 ***Dryopteris filix-mas*** (L.) Schott
Al (–) fieri mashkull
Bg (0) мъжка папрат
Cs (1) kapraď samec
De (–) Gemeiner Wurmfarn
En (–) male-fern
Gr (1) αρρενόφτερη
Hu (0) erdei pajzsika
Pl (1) narecznica samcza
Ro (0) ferigă
Ru (1) щитовник мужской
Sh (1) crvasta paprat; obična navala
Sk (–) papraď samčia

E

579 ***Ecballium elaterium*** (L.) Richard
Al (–) kungullushka
Bg (0) луда краставица
Cs (0) tykvice srdčitá
De (–) Gewöhnliche Spritzgurke
En (–) squirting cucumber
Gr (2) πικραγγουριά
Hu (1) magrúgó
Ro (0) plesnitoare
Ru (0) бешенный огурец обыкновенный
Sh (1) obična štrcalica
Sk (–) tekvičník srdcolistý

580 ***Echinochloa crus-galli*** (L.) Beauv.
Al (–) kostreva; muhari
Bg (3) нисък дараджан
Cs (3) ježatka kuří noha
De (–) Hühnerhirse
En (–) cockspur
Gr (3) μουχρίτσα
Hu (3) közönséges kakaslábfű
Pl (3) chwastnica jednostronna
Ro (3) mohor lat
Ru (3) ежовник обыкновенный
Sh (3) korovsko proso; obačni koštan
Sk (–) ježatka kuria noha

* ***Echinochloa crus-galli*** (L.) Beauv. **var. oryzicola Ohwi** → *Echinochloa oryzoides*

* ***Echinochloa macrocarpa*** Vasin. → *Echinochloa oryzoides*

* ***Echinochloa oryzicola*** Vasin. → *Echinochloa oryzoides*

581 ***Echinochloa oryzoides*** (Ard.) Frit.
Bg (2) воден дараджан
En (–) rice cockspur
Hu (2) nagyszemű kakaslábfű
Ro (2) mohor alb
Ru (2) ежовник рисовидный
Sk (–) ježatka veľkoplodá

* ***Echinochloa phyllopogon*** (Stapf) Vasc. → *Echinochloa oryzoides*

582 ***Echinocystis lobata*** (Michx) Torr. & Gray
Cs (0) štětinec laločnatý
De (–) Gelappte Stachelgurke
En (–) wild cucumber
Hu (1) süntök
Ru (1) ехиноцистис шиповатый
Sk (–) ježatec laločnatý

583 ***Echinops* spp.** L.
Bg (1) челядник
De (–) Kugeldistel
Gr (1) κεφαλάγκαθο
Ru (0) мордовник
Sk (–) ježibaba

584 ***Echinops bannaticus*** Roch.
Bg (1) банатски челядник
Sk (–) ježibaba banátska

585 ***Echinops microcephalus*** Sibth. & Sm.
Bg (1) малък челядник
Sk (–) ježibaba drobnohlavá

586 ***Echinops ritro*** L.
Bg (1) бодливолистен челядник
Cs (0) bělotrn ritro
En (–) globe-thistle
Ru (0) мордовник обыкновенный
Sh (1) trepavičavi glavoč
Sk (–) ježibaba modrá

* ***Echinops ruthenicus*** Bieb. → *Echinochloa crus-galli*

587 *Echinops sphaerocephalus* L.
Cs (1) bělotrn kulatohlavý
De (–) Blaue Kugeldistel
En (–) pale globe-thistle
Hu (1) fehér szamárkenyér
Pl (1) pogorzan kulisty
Ro (0) tătarnică
Ru (0) мордовник круглоголовый
Sk (–) ježibaba guľatohlavá

* ***Echinospermum lappula*** (L.) Lehm. → *Lappula squarrosa*

588 *Echium* spp. L.
Bg (1) кучи език
De (–) Natternkopf
Gr (1) βοϊδόγλωσσα
Ru (1) синяк
Sk (–) hadinec

589 *Echium italicum* L.
Al (–) ushqerëza e Italisë
Bg (1) италиански кучи език
Cs (0) hadinec italský
De (–) Hochwüchsiger Natternkopf
En (–) pale bugloss
Hu (1) magas kígyószisz
Ru (1) синяк итальянский
Sh (1) talijanska lisičina
Sk (–) hadinec taliansky

590 *Echium plantagineum* L.
Al (–) ushqerëza si dejç
De (–) Wegerich-Natternkopf
En (–) purple viper's-bugloss
Ru (0) синяк кривоцветный
Sk (–) hadinec skorocelovitý

591 *Echium russicum* Gmel.
Hu (0) piros kígyószisz
Ru (–) синяк русский
Sh (1) crvena lisičine
Sk (–) hadinec ruský

* ***Echium violaceum*** L. → *Echium plantagineum*

592 *Echium vulgare* L.
Al (–) ushqerëza e rëndomtë
Bg (1) кучи език обикновен
Cs (2) hadinec obecný
De (–) Gemeiner Natternkopf
En (–) viper's-bugloss
Hu (1) közönséges kígyószisz
Pl (1) żmijowiec zwyczajny
Ro (1) iarba şarpelui
Ru (1) синяк обыкновенный
Sh (1) obična lisičina
Sk (–) hadinec obyčajný

593 *Elatine alsinastrum* L.
Bg (1) прешленест водник
Cs (0) úpor kuřičkovitý
De (–) Quirl-Tannel
Gr (1) ελατίνη
Hu (1) pocsolya látonya
Pl (1) nadwodnik okółkowy
Ru (1) повойничек мокричный
Sk (–) úpor kuričkovitý

594 *Elatine hexandra* (Lapierre) DC.
Cs (0) úpor šestimužný
De (–) Sechsmänniger Tännel
Pl (1) nadwodnik sześciopręcikowy
Ru (0) повойничек шестилепестный
Sk (–) úpor šesťmužný

595 *Elatine triandra* Schk.
Bg (1) тритичинков водник
Cs (0) úpor trojmužný
De (–) Dreimänniger Tännel
Hu (1) kétéltű látonya
Pl (1) nadwodnik trójpręcikowy
Ru (1) повойничек трёхтычинковый
Sh (1) polegla pobarica
Sk (–) úpor trojmožný

* ***Elatinoides elatine*** (L.) Wett. → *Kickxia elatine*

* ***Elatinoides spuria*** (L.) Wett. → *Kickxia spuria*

596 ***Eleocharis* spp.** R Br.
De (–) Sumpfried
En (–) spike-rush
Gr (1) σύφα
Ru (0) болотница
Sk (–) bahnička

597 ***Eleocharis acicularis*** (L.) Roem. & Schultes
Cs (0) bahnička jehlovitá
De (–) Nadel-Sumpfried
En (–) needle spike-rush
Hu (1) apró csetkáka
Ru (0) болотница игольчатая
Sh (1) igličasta zukva
Sk (–) bahnička ihlovitá

598 ***Eleocharis palustris*** (L.) Roem. & Schultes

Al (–) heleoharzë kënetash
Bg (1) блатница
Cs (0) bahnička bahenní
De (–) Gemeines Sumpfried
En (–) common spike-rush
Hu (1) mocsári csetkáka
Pl (1) ponikło błotne
Ro (1) pipiriguț
Ru (1) болотница болотная
Sh (1) močvarna jezernica; močvarna zukva
Sk (–) bahnička močiarna

599 ***Elodea canadensis*** Michx
Bg (1) водна чума
Cs (2) vodní mor kanadský
De (–) Kanadische Wasserpest
En (–) Canadian pondweed
Hu (0) átokhínár
Pl (2) moczarka kanadyjska
Ro (2) ciuma apei
Ru (2) водяная зараза
Sh (1) kanadska vodena kuga
Sk (–) vodomor kanadský

600 ***Elsholtzia ciliata*** (Thunb.) Hyl.
Ru (1) эльсгольция реснитчатая
Sk (–) elsholtzia brvitá

601 ***Elymus* spp.** L.
Bg (1) пирей
De (–) Haargerste
Gr (1) έλυμος
Ru (1) колосняк
Sk (–) pýr

* ***Elymus arenarius*** L. → *Leymus arenarius*

* ***Elymus europaeus*** L. → *Hordelymus europaeus*

602 ***Elymus hispidus*** (Opiz) Meld.
Al (–) grami bregdetar; grami ndërmjetës
Bg (1) среден пирей
Cs (0) pýr prostřední
De (–) Graugrüne Quecke
En (–) sea couch-grass
Hu (0) deres tarackbúza
Pl (1) perz silny
Sh (1) osrednja pirika; srednja pirevina
Sk (–) pýr prostredný chlpatý

603 ***Elymus repens*** (L.) Gould
Al (–) grami zvarritës
Bg (2) пълзящ пирей
Cs (3) pýr plazivý
De (–) Gemeine Quecke
En (–) common couch-grass
Gr (2) έλυμος
Hu (3) tarackbúza

Pl (3) perz właściwy
Ro (3) pir
Ru (3) пырей ползучий
Sh (3) obična pirevina; obična pirika
Sk (–) pýr plazivý

* ***Elytrigia repens*** (L.) Nev. → *Elymus repens*

604 ***Epilobium*** **spp.** L.
Bg (–) върбовка
De (–) Weidenröschen
En (–) willowherb
Gr (1) επιλόβιο
Ru (1) кипрей
Sk (–) vŕbovka

605 ***Epilobium adenocaulon*** Haussk.
Ru (1) кипрей железисто-стебельный
Sk (–) vŕbovka cudzia

* ***Epilobium adnatum*** Gris. → *Epilobium tetragonum*

606 ***Epilobium angustifolium*** L.
Al (–) epiloba gjethengushtë
Bg (1) теснолистна върбовка
Cs (3) vrbka úzkolistá
De (–) Schmalblättriges Weidenröschen
En (–) rosebay willowherb
Hu (0) erdei deréce
Pl (1) wierzbówka kiprzyca
Ro (1) zburătoare
Ru (1) кипрей узколистный
Sh (1) uskolisna vrbolika
Sk (–) vŕbovka úzkolistá

607 ***Epilobium hirsutum*** L.
Al (–) epiloba qimeashpër
Bg (1) влакнеста върбовка
Cs (0) vrbovka chlupatá
De (–) Zottiges Weidenröschen
En (–) great willowherb
Hu (1) borzas füzike
Pl (1) wierzbownica kosmata
Ru (1) кипрей мохнатый
Sh (1) runjava vrbovica
Sk (–) vŕbovka chlpatá

* ***Epilobium lamyi*** Schultz → *Epilobium tetragonum*

608 ***Epilobium lanceolatum*** Sebast. & Mauri
Cs (0) vrbovka kopinatá
De (–) Lanzettblättriges Weidenröschen
En (–) spear-leaved willowherb
Hu (0) lándzsás füzike
Ro (1) pufuliță
Sk (–) vŕbovka kopijovitá
Ru (0) кипрей ланцетный

609 ***Epilobium montanum*** L.
Al (–) epiloba e maleve
Cs (0) vrbovka horská
De (–) Berg-Weidenröschen
En (–) broad-leaved willowherb
Hu (1) erdei füzike
Pl (1) wierzbownica górska
Ru (1) кипрей горный
Sh (1) svilasta vrbovica
Sk (–) vŕbovka horská

610 ***Epilobium palustre*** L.
Cs (0) vrbovka bahenni
De (–) Sumpf-Weidenröschen
En (–) marsh willowherb
Hu (0) mocsári füzike
Pl (1) wierzbownica błotna
Ru (1) кипрей болотный
Sk (–) vŕbovka močiarna

611 ***Epilobium parviflorum*** Schreb.
Al (–) epiloba lulevogël
Bg (1) дребноцветна върбовка
Cs (0) vrbovka málokvětá
De (–) Kleinblütiges Weidenröschen
En (–) hoary willowherb

Hu (1) kisvirágú füzike
Pl (–) wierzbownica drobnokwiatowa
Ru (1) кипрей бедноцветковый
Sk (–) vŕbovka malokvetá

612 ***Epilobium roseum*** Schreb.
Cs (0) vrbovka růžová
De (–) Rosarotes Weidenröschen
En (–) pale willowherb
Hu (1) rózsa füzike
Pl (1) wierzbownica bladoróżowa
Ru (1) кипрей розовый
Sk (–) vŕbovka ružová

613 ***Epilobium tetragonum*** L.
Bg (1) стъблосрастнала върбовка
Cs (0) vrbovka čtyřhranná
De (–) Vierkantiges Weidenröschen
En (–) square-stalked willowherb
Hu (1) gyakori füzike
Pl (1) wierzbownica czworoboczna
Ru (1) кипрей четырёхгранный
Sh (1) četvrtasta vrbovica
Sk (–) vŕbovka štvorhranná pravá

614 ***Equisetum arvense*** L.
Al (–) këputja e arave
Bg (2) полски хвощ
Cs (3) přeslička rolní
De (–) Acker-Schachtelhalm
En (–) field horsetail
Gr (1) αλογοουρά
Hu (3) mezei zsurló
Pl (3) skrzyp polny
Ro (2) barba ursului
Ru (3) хвощ полевой
Sh (2) poljska preslica; poljski rastavić
Sk (–) praslička roľná

* ***Equisetum elongatum*** Willd. → *Equisetum ramosissimum*

615 ***Equisetum hyemale*** L.
Al (–) këputja e dimrit
Cs (0) přeslička zimní
De (–) Winter-Schachtelhalm
En (–) rough horsetail
Hu (0) téli zsurló
Pl (1) skrzyp zimowy
Ro (0) pipirig de munte
Ru (1) хвощ зимующий
Sk (–) praslička jesenná

616 ***Equisetum palustre*** L.
Al (–) këputja e kënetave
Bg (1) блатен хвощ
Cs (1) přeslička bahenní
De (–) Sumpf-Schachtelhalm
En (–) marsh horsetail
Hu (–) mocsári zsurló
Pl (2) skrzyp błotny
Ro (1) barba ursului de bahne
Ru (1) хвощ болотный
Sh (1) močvarna preslica; žablji rastavić
Sk (–) preslička močiarna

617 ***Equisetum pratense*** Ehrh.
Al (–) këputja e livadheve
Cs (0) přeslička luční
De (–) Wiesen-Schachtelhalm
En (–) meadow horsetail
Pl (1) skrzyp łąkowy
Ro (0) luioane
Ru (1) хвощ луговой
Sh (1) livadski rastavić
Sk (–) praslička lúčna

618 ***Equisetum ramosissimum*** Desf.
Al (–) këputja me shumë degë
Bg (1) клонест хвощ
Cs (0) přeslička větevnatá
De (–) Ästiger Schachtelhalm
Hu (3) hosszú zsurló
Pl (1) skrzyp gałęzisty
Ru (2) хвощ ветвистый
Sk (–) praslička vetevnatá

619 ***Equisetum sylvaticum*** L.
Cs (0) přeslička lesní

De (–) Wald-Schachtelhalm
En (–) wood horsetail
Pl (2) skrzyp leśny
Ro (0) ruşinea ursului
Ru (1) хвощ лесной
Sk (–) praslička lesná

620 ***Equisetum telmateia*** Ehrh.
Al (–) këputja e madhe
Cs (0) přeslička největší
De (–) Riesen-Schachtelhalm
En (–) great horsetail
Hu (1) óriás zsurló
Ro (0) coada calului
Ru (0) хвощ большой
Sk (–) praslička obrovská

* ***Equisetum umbrosum*** Meyer → *Equisetum pratense*

621 ***Eragrostis cilianensis*** (All.) Hub.
Al (–) eragrosta kallimadhe
Bg (1) едрокласа полевичка
Cs (0) milička větší
De (–) Großähriges Liebesgras
En (–) stinking lovegrass
Hu (2) nagy tőtippan
Ru (1) полевичка крупноколосковая
Sh (1) mnogocvjetna kosmatka; obična kosmatka
Sk (–) milota veľkoklásová

* ***Eragrostis major*** Host → *Eragrostis cilianensis*

* ***Eragrostis megastachya*** (Koeler) Link → *Eragrostis cilianensis*

622 ***Eragrostis minor*** Host
Al (–) eragrosti i vogël
Bg (1) дребна полевичка
Cs (–) milička lipnicovitá
De (–) Kleines Liebesgras
En (–) small lovegrass
Hu (3) kis tőtippan
Pl (2) miłka drobna
Ro (2) iarbă sură
Ru (2) полевичка малая
Sh (1) mala kosmatka
Sk (–) milota menšia

623 ***Eragrostis pilosa*** (L.) Beauv.
Al (–) eragrosti me qime
Bg (1) овлакнена полевичка
Cs (0) milička chlupatá
De (–) Behaartes Liebesgras
En (–) hairy lovegrass
Hu (2) szőrös tőtippan
Ru (1) полевичка волосистая
Sh (1) pršljenasta kosmatka
Sk (–) milota chlpatá

* ***Eragrostis poaeformis*** Link → *Eragrostis minor*

* ***Eragrostis poaeoides*** Beauv. → *Eragrostis minor*

* ***Eragrostis verticillata*** Beauv. → *Eragrostis pilosa*

* ***Erianthus ravennae*** (L.) Beauv. → *Saccharum ravennae*

624 ***Erica* spp.** L.
De (–) Heide
En (–) heath
Gr (2) ρείκι
Ru (1) эрика
Sk (–) vresovec

625 ***Erica herbacea*** L.
Al (–) grathata
De (–) Schnee-Heide
En (–) spring heath
Sh (1) crvenkasta crnjuša; obična erika
Sk (–) vresovec spletaný

626 *Erica tetralix* L.
Al (–) shqopa kënetore
Cs (0) vřesovec čtyřřadý
De (–) Glocken-Heide
En (–) cross-leaved heath
Pl (1) wrzosiec bagienny
Ru (0) эрика крестолистная
Sk (–) vresovec štvorradý

627 *Erigeron* spp. L.
Cs (–) turan
De (–) Berufkraut
En (–) fleabane
Gr (1) ερίγερο
Ru (1) многолепестник
Sk (–) turan

628 *Erigeron acer* L.
Cs (1) turan ostrý
De (–) Scharfes Berufkraut
En (–) blue fleabane
Hu (1) bóbitás küllőrojt
Pl (1) przymiotno ostre
Ro (1) bunghişor
Ru (1) многолепестник едкий
Sh (1) gruba hudoletnica; gruba hudoljetnica
Sk (–) turan ostrý

629 *Erigeron alpinus* L.
De (–) Alpen-Berufkraut
En (–) alpine fleabane
Pl (1) przymiotno alpejskie
Ru (0) многолепестник альпийский
Sk (–) turan alpínsky

630 *Erigeron annuus* (L.) Pers.
Cs (0) hvězdník roční
De (–) Zweijähriges Berufkraut
En (–) annual fleabane
Hu (1) seprence
Pl (1) przymiotno białe
Ru (1) многолепестник однолетний
Sh (1) jednogodišnja hudoletnica; jednogodišnja hudoljetnica
Sk (–) turan ročný

* ***Erigeron canadensis*** L. → *Conyza canadensis*

* ***Erigeron crispus*** Pour. → *Conyza bonariensis*

* ***Erigeron strigosus*** Muhl. → *Erigeron annuus*

631 *Eriophorum angustifolium* Honck.
Cs (1) suchopýr úzkolistý
De (–) Schmalblättriges Wollgras
En (–) common cottongrass
Hu (0) keskenylevelű gyapjúsás
Pl (2) wełnianka wąskolistna
Ro (0) bumbăcariţă
Ru (0) пушица узколистная
Sk (–) páperník úzkolistý

632 *Eriophorum gracile* Koch
Cs (0) suchopýr štíhlý
De (–) Schlankes Wollgras
En (–) slender cottongrass
Hu (0) vékony gyapjúsás
Pl (1) wełnianka delikatna
Ru (0) пушица стройная
Sk (–) páperník útly

633 *Eriophorum latifolium* Hoppe
Al (–) eriofora gjethegjërë
Cs (0) suchopýr širokolistý
De (–) Breitblättriges Wollgras
En (–) broad-leaved cottongrass
Hu (0) széleslevelű gyapjúsás
Pl (1) wełnianka szerokolistna
Ru (0) пушица широколистная
Sh (1) širokolisna suhoperka
Sk (–) páperník širokolistý

634 *Erodium* spp. L'Hér.
Bg (0) човка
De (–) Reiherschnabel
En (–) stork's-bill
Gr (1) βελονίδα
Ru (0) аистник
Sk (–) bocianik

635 *Erodium ciconium* (L.) L'Hér.
Bg (0) щъркелова човка
Cs (0) pumpava čapí
De (–) Großer Reiherschnabel
Hu (1) gerelyes gémorr
Ru (1) аистник аистовый
Sk (–) bocianik obyčajný

636 *Erodium cicutarium* (L.) L'Hér.
Al (–) gjilpërqyqja; sqep-çafke
Bg (1) часовниче
Cs (2) pumpava rozpuková
De (–) Schierlings-Reiherschnabel
En (–) common stork's-bill
Gr (2) βελονίδα
Hu (1) bürökgémorr
Pl (3) iglica pospolita
Ro (1) pliscul cocorului
Ru (1) аистник цикутовый
Sh (1) obični čapljan
Sk (–) bocianik rozpukovitý

* ***Erodium cicutifolium*** Salisb. → *Erodium cicutarium*

* ***Erophila praecox*** (Stev.) DC. → *Erophila verna*

637 *Erophila verna* (L.) Chevall.
Al (–) draba pranverore
Bg (1) пролетна гладница
Cs (2) osívka jarni
De (–) Frühlings-Hungerblümchen
En (–) common whitlow-grass
Gr (1) ερωφίλη
Hu (1) ködvirág; tavaszi daravirág
Pl (3) wiosnówka pospolita
Ro (1) flămînzică
Ru (1) веснянка весенняя
Sh (1) prolećna nemoćnica; proljetni gladuš
Sk (–) jarmila jarná

* ***Erophila vulgaris*** DC. → *Erophila verna*

* ***Eruca sativa*** Mill. → *Eruca vesicaria*

638 *Eruca vesicaria* (L.) Cav.
Bg (1) посевна ерука
Cs (0) roketa setá
De (–) Garten-Senfrauke
En (–) common rocket
Gr (2) ρόκα
Hu (1) borsmustár
Pl (1) rokietta siewna
Ru (1) индау посевная
Sh (1) kulturna riga; usjevna riga
Sk (–) eruka siata

639 *Erucastrum gallicum* (Willd.) Schulz
Cs (0) ředkevník galský
De (–) Französische Hundsrauke
En (–) hairy rocket
Hu (1) francia nyurgaszál
Pl (–) rukwiślad francuski
Ro (0) muştar alb sălbatic
Ru (0) капустник галльский
Sk (–) reďkovník galský

* ***Ervum hirsutum*** L. → *Vicia hirsuta*

* ***Ervum tetraspermum*** L. → *Vicia tetrasperma*

640 *Eryngium amethystinum* L.
De (–) Stahlblaue Mannstreu
En (–) blue eryngo
Gr (1) παπαδίτσα
Hu (1) kék iringó; kék ördögszekér

Ro (1) spin vînăt
Ru (–) синеголовник Биберштейна
Sk (–) kotúč riečny

641 *Eryngium campestre* L.
Al (–) gjembardhi i fushës
Bg (2) ветрогон
Cs (2) máčka ladní
De (–) Feld-Mannstreu
En (–) field eryngo
Gr (1) φειδάγκαθο
Hu (1) mezei iringó; mezei ördögszekér
Pl (1) mikołajek polny
Ro (1) scaiul dracului
Ru (1) синеголовник полевой
Sh (1) obični kotrljan; perastolisni kotrljan
Sk (–) kotúč poľný

642 *Eryngium creticum* Lam.
Al (–) gjembardhi i Kretës
En (–) small-headed eryngo
Gr (1) σφαλάγκαθο
Ru (0) синеголовник критский
Sk (–) kotúč krétsky

643 *Eryngium maritimum* L.
De (–) Stranddistel
En (–) sea-holly
Gr (1) γαλανόχορτο
Ru (0) синеголовник приморский
Sk (–) kotúč prímorský

644 *Eryngium planum* L.
Cs (0) máčka plocholistá
De (–) Flachblättrige Mannstreu
Hu (1) kék iringó; kék ördögszekér
Pl (1) mikołajek płaskolistny
Ro (1) spin vînăt
Ru (1) синеголовник плосколистный
Sk (–) kotúč plocholistý

* ***Eryngium trifidum*** L. → *Eryngium campestre*

645 *Erysimum* spp. L.
Bg (1) боянка
De (–) Schöterich
Gr (1) σκυλοβρούβα
Ru (1) желтушник
Sk (–) horčičník

646 *Erysimum cheiranthoides* L.
Cs (1) trýzel chejrovitý
De (–) Acker-Schöterich
En (–) treacle mustrad
Hu (1) violás repcsény
Pl (2) pszonak drobnokwiatowy
Ru (1) желтушник левкойный
Sh (1) batvičasti šeboj; batvasti trižalj
Sk (–) horčičník cheirantovitý

* ***Erysimum cheiri*** (L.) Crantz → *Cheiranthus cheiri*

647 *Erysimum crepidifolium* Reichenb.
Cs (1) trýzel škardolistý
De (–) Bleicher Schöterich
Hu (0) sziklai repcsény
Sh (1) obični trižalj
Sk (–) horčičník škardolistý

648 *Erysimum diffusum* Ehrh.
Cs (1) trýzel rozvětvený
De (–) Grauer Schöterich
Hu (0) szürke repcsény
Ru (1) желтушник раскидистый
Sk (–) horčičník šedavý

649 *Erysimum hieracifolium* L.
Ru (1) желтушник ястребинколистный
Sk (–) horčičník jastrabníkolistý

* ***Erysimum orientale*** auct. → *Sisymbrium orientale*

* ***Erysimum ramosissimum*** Cran. → *Erysimum repandum*

650 *Erysimum repandum* L.
Bg (1) разперена боянка
Cs (1) trýzel vykrajovaný
De (–) Sperriger Schöterich
Hu (2) füstös repcsény
Pl (1) pszonak obłączysty
Ro (1) bărbuşoară
Ru (1) желтушник выемчато-зубчатый
Sh (1) jednogodišnji trižalj
Sk (–) horčičník rozložitý

* ***Erythraea pulchella*** (Sweet) Fries → *Centaurium pulchellum*

651 *Erythronium dens-canis* L.
Al (–) dhëmb qeni; kuqëla dhëmbqeni
Bg (0) конски зъб
Cs (0) kandík psí zub
De (–) Hunds-Zahnlilie
En (–) dog's-tooth violet
Hu (0) kakasmandikó
Pl (0) psi ząb liliowy
Ro (0) măseaua ciutei
Ru (0) кандык европейский
Sh (1) obični košutac; pasji zub
Sk (–) kandík psí zub

652 *Euclidium syriacum* (L.) R Br.
Bg (1) пикливче
Cs (0) rukevníček syrský
De (–) Syrisches Schnabelschötchen
Hu (1) madárköröm
Pl (0) porczak syryjski
Ru (0) крепкоплодник сирийский
Sk (–) blahobyľ sýrska

653 *Euonymus europaeus* L.
Al (–) herdhaqeni i Evropës
Bg (0) ръбест чашкодрян
Cs (0) brslen evropský
De (–) Pfaffenhütchen
En (–) spindle-tree
Hu (0) csíkos kecskerágó
Pl (2) trzmielina zwyczajna
Ro (0) salbă moale
Ru (1) бересклет европейский
Sk (–) bršlen obyčajný

654 *Eupatorium cannabinum* L.
Al (–) kërpi ujës
Bg (1) див коноп
Cs (0) sadec konopáč
De (–) Wasserdost
En (–) hemp-agrimony
Hu (2) sédkender
Pl (1) sadziec konopiasty
Ro (0) cînepa codrului
Ru (1) посконник коноплевидный
Sh (1) obična konopljuša
Sk (–) konopáč obyčajný

655 *Euphorbia* spp. L.
Bg (1) млечка
De (–) Wolfsmilch
En (–) spurge
Gr (1) φλώμος
Ru (1) молочай
Sk (–) mliečnik

* ***Euphorbia acuminata*** Lam. → *Euphorbia falcata*

656 *Euphorbia agraria* Bieb.
Al (–) qumështore fusharake; rryelli fusharak
Ro (1) laptele cucului
Ru (0) молочай пашенный

657 *Euphorbia aleppica* L.
Bg (1) алепска млечка
Ru (1) молочай алеппский
Sk (–) mliečnik sitinovitý

658 *Euphorbia amygdaloides* L.
Al (–) rryelli si bajame
Bg (1) бадемовидна млечка
Cs (0) pryšec mandloňovitý
De (–) Mandelblättrige Wolfsmilch

En (–) wood spurge
Hu (0) erdei kutyatej
Ro (0) alior
Ru (0) молочай миндалевидный
Sh (1) šumska mlečika; šumska mlječika
Sk (–) mliečnik mandľovitý

659 *Euphorbia chamaesyce* L.
Al (–) rryelli fik i ulët
Bg (1) пълзяща млечка
Cs (0) pryšec fíkolistý
De (–) Zwerg-Wolfsmilch
Ru (0) молочай мелкосмоковник
Sk (–) mliečnik figovníkolistý

660 *Euphorbia cyparissias* L.
Al (–) rryelli selvi; tamëlçak i idhët
Bg (1) кипарисова млечка
Cs (2) pryšec chvojka
De (–) Zypressen-Wolfsmilch
En (–) cypress spurge
Hu (1) farkas kutyatej
Pl (2) wilczomlecz sosnka
Ru (1) молочай кипарисовый
Sh (1) obična mlečika; smrekasta mlječika
Sk (–) mliečnik chvojkový

661 *Euphorbia epithymoides* L.
Al (–) rryelli shumëngjyrësh
Bg (1) многобагрена млечка
Cs (0) pryšec mnohobarvý
De (–) Vielfarbige Wolfsmilch
Hu (0) színeváltó kutyatej
Ru (–) молочай многоцветный
Sk (–) mliečnik mnohofarebný

662 *Euphorbia esula* L.
Al (–) qumështore e rëndomtë
Bg (1) остра млечка
Cs (2) pryšec obecný
De (–) Esels-Wolfsmilch
En (–) leafy spurge
Hu (2) útszéli kutyatej
Pl (2) wilczomlecz lancetowaty
Ru (2) молочай острый
Sh (1) ljuta mlečika; ljuta mlječika
Sk (–) mliečnik obyčajný

663 *Euphorbia exigua* L.
Al (–) rryelli gjethengushtë
Bg (1) нищожна млечка
Cs (2) pryšec drobný
De (–) Kleine Wolfsmilch
En (–) dwarf spurge
Hu (2) vetési kutyatej
Pl (2) wilczomlecz drobny
Ru (1) молочай маленький
Sk (–) mliečnik drobný

664 *Euphorbia falcata* L.
Bg (1) сърповидна млечка
Cs (0) pryšec srpovitý
De (–) Sichel-Wolfsmilch
En (–) sickle spurge
Hu (3) tarlókutyatej
Pl (1) wilczomlecz sierpowy
Ru (1) молочай серповидный
Sh (1) srpasta mlečika; srpasta mlječika
Sk (–) mliečnik kosáčikovitý

* ***Euphorbia gerardiana*** Jacq. → *Euphorbia seguierana*

* ***Euphorbia graeca*** Boiss. & Sprun. → *Euphorbia taurinensis*

665 *Euphorbia helioscopia* L.
Al (–) rryelli diellpamës
Bg (1) слънчева млечка
Cs (3) pryšec kolovratec
De (–) Sonnen-Wolfsmilch
En (–) sun spurge
Gr (2) μεγάλη γαλατσίδα
Hu (2) napraforgó kutyatej
Pl (3) wilczomlecz obrotny
Ro (2) laptele cucului
Ru (2) молочай солнцегляд

Sh (1) sunceljubiva mlečika;
sunceljubiva mlječika
Sk (–) mliečnik kolovratec

666 *Euphorbia humifusa* Willd.
Cs (1) pryšec položený
De (–) Niederliegende Wolfsmilch
Hu (0) kúszó kutyatej
Pl (1) wilczomlecz rozesłany
Ru (1) молочай приземистый
Sk (–) mliečnik rozprestretý

667 *Euphorbia lathyris* L.
Bg (0) кръстолистна млечка
Cs (0) pryšec skočcový
De (–) Kreuzblättrige Wolfsmilch
En (–) caper spurge
Ru (0) молочай чины
Sh (1) grahorasta mlečika; grahorasta
mlječika
Sk (–) mliečnik križmolistý

668 *Euphorbia lucida* Waldst. & Kit.
Cs (0) pryšec lesklý
De (–) Glänzende Wolfsmilch
En (–) shiny-leaved spurge
Hu (1) fényes kutyatej
Pl (1) wilczomlecz błyszczący
Ro (0) laptele cîinelui
Ru (1) молочай глянцевитый
Sh (1) visoka mlečika; visoka mlječika
Sk (–) mliečnik lesklý

669 *Euphorbia maculata* L.
Bg (1) петниста млечка
Cs (0) pryšec skvrnitý
De (–) Gefleckte Wolfsmilch
En (–) prostrate spurge
Hu (1) bókoló kutyatej
Ru (1) молочай пятнистый
Sh (1) pegava mlečika; tockasta mlječika
Sk (–) mliečnik škvrnitý

670 *Euphorbia myrsinites* L.
Al (–) rryelli mërsinë
Bg (1) миртолистна млечка
Cs (–) pryšec myrtový
Ru (1) молочай миртолистный
Sh (1) rudinska mlečika; rudinska
mlječika
Sk (–) mliečnik myrtovitý

671 *Euphorbia nutans* Lag.
Cs (–) pryšec nící
De (–) Nickende Wolfsmilch
Hu (1) bókoló kutyatej
Ru (0) молочай поникающий
Sk (–) mliečnik kývajúci

672 *Euphorbia palustris* L.
Bg (0) блатна млечка
Cs (0) pryšec bahenní
De (–) Sumpf-Wolfsmilch
En (–) marsh spurge
Hu (1) mocsári kutyatej
Pl (1) wilczomlecz błotny
Ru (1) молочай болотный
Sh (1) močvarna mlečika; močvarna
mlječika
Sk (–) mliečnik močiarny

673 *Euphorbia peplus* L.
Bg (1) градинска млечка
Cs (2) pryšec okrouhlý
De (–) Garten-Wolfsmilch
En (–) petty spurge
Hu (1) vézna kutyatej
Pl (3) wilczomlecz ogrodowy
Ru (2) молочай огородный
Sh (1) loptasta mlečika; loptasta mlječika
Sk (–) mliečnik okrúhlolistý

674 *Euphorbia platyphyllos* L.
Al (–) rryelli gjethegjërë
Bg (1) широколистна млечка
Cs (0) pryšec plocholistý

De (–) Breitblättrige Wolfsmilch
En (–) broad-leaved spurge
Hu (1) nagylevelű kutyatej
Pl (1) wilczomlecz szerokolistny
Ru (1) молочай расширеннолистный
Sh (1) širokolisna mlečika; širokolisna mlječika
Sk (–) mliečnik plocholistý

* ***Euphorbia polychroma*** Kerner → *Euphorbia epithymoides*

* ***Euphorbia pseudagraria*** Smirnov → *Euphorbia esula*

675 ***Euphorbia salicifolia*** Host
Bg (1) върболистна млечка
Cs (0) pryšec vrbolistý
De (–) Weidenblättrige Wolfsmilch
Hu (1) fűzlevelű kutyatej
Pl (0) wilczomlecz wierzbolistny
Ru (0) молочай иволистный
Sh (1) vrbolisna mlečika; vrbolisna mlječika
Sk (–) mliečnik vŕbolistý

676 ***Euphorbia seguierana*** Neck.
Bg (1) Сегиерова млечка
Cs (0) pryšec Seguierův
De (–) Steppen-Wolfsmilch
Hu (1) pusztai kutyatej
Pl (0) wilczomlecz Seguiera
Ro (1) laptele cucului
Ru (1) молочай Сегиерова
Sh (1) stepska mlečika; stepska mlječika
Sk (–) mliečnik sivý menši

677 ***Euphorbia serrulata*** Thuill.
Bg (1) теснолистна млечка
Cs (0) pryšec stažený
De (–) Steife Wolfsmilch
En (–) upright spurge
Hu (0) merev kutyatej
Ru (–) молочай прямой
Sk (–) mliečnik stiahnutý

* ***Euphorbia stricta*** L. → *Euphorbia serrulata*

678 ***Euphorbia taurinensis*** All.
Bg (1) гръцка млечка
Hu (1) görög kutyatej
Ru (1) молочай таврический
Sk (–) mliečnik turínsky

679 ***Euphorbia villosa*** Waldst. & Kit.
Bg (1) вълнеста млечка
Cs (0) pryšec huňatý
De (–) Zottige Wolfsmilch
En (–) hairy spurge
Hu (1) bozontos kutyatej
Pl (1) wilczomlecz włosisty
Ru (1) молочай мохнатый
Sh (1) dlakava mlečika; dlakava mlječika
Sk (–) mliečnik huňatý

* ***Euphorbia virgata*** Waldst. & Kit. → *Euphorbia esula*

680 ***Euphrasia arctica*** Lange
Cs (1) světlík krátkochlupý
Pl (1) świetlik krótkogruczołkowy
Ru (1) очанка поздняя
Sk (–) očianka polávna

* ***Euphrasia brevipila*** auct. → *Euphrasia arctica*

681 ***Euphrasia minima*** Jacq.
Bg (1) малка очанка
Cs (1) světlík nejmenší
De (–) Zwerg-Augentrost
Pl (0) świetlik maleńki
Ru (1) очанка татринская
Sk (–) očianka najmenšia

* ***Euphrasia odontites*** L. → *Odontites verna*

682 ***Euphrasia officinalis*** L.
En (–) common eyebright
Gr (1) ευφρασία
Ru (1) очанка лекарственная

683 ***Euphrasia rostkoviana*** Hayne
Bg (0) Ростковианова очанка
Cs (0) světlík lékařský
De (–) Gemeiner Augentrost
Hu (0) orvosi szemvidító
Pl (3) świetlik łąkowy
Ru (0) очанка Ростковиуса
Sk (–) očianka lekárska

684 ***Euphrasia stricta*** Wolff
Al (–) eufrazia e drejtë
Bg (0) твърда очанка
Cs (0) světlík tuhý
De (–) Steifer Augentrost
Hu (0) közönséges szemvidító
Pl (2) świetlik wyprężony
Ru (1) очанка ожатая
Sh (1) kruta očanica; obični vidac
Sk (–) očianka tuhá

* ***Evonymus europaeus*** L. → *Euonymus europaeus*

F

* ***Faba narbonensis*** Schur → *Vicia narbonensis*

* ***Fagopyrum convolvulus*** (L.) Gross → *Bilderdykia convolvulus*

685 ***Fagopyrum esculentum*** Moen.
Al (–) fagopira; hejda e rëndomtë
Bg (0) гречиха
Cs (0) pohanka střelovitá
De (–) Echter Buchweizen
En (–) buckwheat
Hu (0) hajdina
Pl (1) gryka zwyczajna
Ro (0) hrişcă
Ru (0) гречиха посевная
Sk (–) pohánka siata

686 ***Fagopyrum tataricum*** (L.) Gaertn.
Al (–) hejda e Siberisë
Cs (0) pohanka tatarská
De (–) Tatarischer Buchweizen
En (–) Siberian buckwheat
Pl (1) gryka tatarka; gryka zielona
Ro (0) hrişcă tătărească
Ru (1) гречиха татарская
Sk (–) pohánka tatárska

* ***Falcaria rivini*** Host → *Falcaria vulgaris*

* ***Falcaria sioides*** Asch. → *Falcaria vulgaris*

687 ***Falcaria vulgaris*** Bernh.
Bg (1) дорник
Cs (2) srpek obecný
De (–) Gemeine Sichelmöhre
En (–) longleaf
Gr (1) φαλκάρια
Hu (2) sarlófű
Pl (1) sierpica pospolita
Ro (1) dornic
Ru (1) резак поручейниковый
Sh (1) obična srpica
Sk (–) kosáčik obyčajný

* ***Fallopia convolvulus*** (L.) Löve → *Bilderdykia convolvulus*

* ***Fallopia dumetorum*** (L.) Holub → *Bilderdykia dumetorum*

688 ***Ferula communis*** L.
Al (–) ferula e rëndomtë
De (–) Riesen-Fenchel
En (–) giant fennel
Gr (1) μεγάλος άρθηκας
Ru (0) ферульник обыкновенный
Sk (–) ferul'a obyčajná

* ***Ferula ferulago*** L. → *Ferulago campestris*

* ***Ferula glauca*** L. → *Ferula communis*

689 ***Ferulago campestris*** (Bess.) Grec.
Al (–) ferulaga e fushës
De (–) Knotenblütige Birkwurz
Gr (1) μικρός άρθηκας
Pl (0) zapalnicznik polny
Ru (–) ферульник равнинный
Sk (–) metlovka ladná

* ***Ferulago galbanifera*** Koch → *Ferulago campestris*

690 ***Festuca spp.*** L.
Bg (1) власатка
De (–) Schwingel
En (–) fescue
Gr (1) φέστουκα
Ru (1) овсяница
Sk (–) kostrava

691 ***Festuca arundinacea*** Schreb.
Al (–) bimla si kallam; bishtpelëza si kallam
Bg (1) тръстиковидна власатка
Cs (0) kostřava rákosovitá
De (–) Rohr-Schwingel
En (–) tall fescue
Ru (1) овсяница тростниковая
Sk (–) kostrava trsťovitá

* ***Festuca capillata*** Lam. → *Festuca tenuifolia*

* ***Festuca duriuscula*** auct. → *Festuca ovina*

692 ***Festuca gigantea*** (L.) Vill.
Al (–) bimla vigane; bishtpelëza vigane
Bg (1) гигантска власатка
Cs (0) kostřava obrovská
De (–) Riesen-Schwingel
En (–) giant fescue
Hu (0) óriás csenkesz
Ru (0) овсяница высокая
Sk (–) kostrava obrovská

693 ***Festuca heterophylla*** Lam.
Al (–) bishtpelëza me gjethe të ndryshme
Bg (1) разнолистна власатка
Cs (0) kostřava různolistá
De (–) Verschiedenblättriger Schwingel
En (–) various-leaved fescue
Hu (0) felemáslevelű csenkesz
Ru (0) овсяница разнолистная
Sk (–) kostrava rôznolistá

* ***Festuca myuros*** L. → *Vulpia myuros*

694 ***Festuca ovina*** L.
Al (–) bishtpelëza leshtake; bishtpelëza pak e fortë
Bg (1) овча власатка
Cs (2) kostřava ovčí
De (–) Schaf-Schwingel
En (–) sheep's fescue
Hu (0) juhcsenkesz
Pl (1) kostrzewa owcza
Ro (0) păiuşul oilor
Ru (1) овсяница овечья
Sh (1) jančarica vlasulja; obični vijuk
Sk (–) kostrava ovčia

* ***Festuca phleoides*** Vill. → *Lophochloa cristata*

695 ***Festuca pratensis*** Huds.
Al (–) bishtpelëza e lartë; bishtpelëza e livadheve
Bg (1) ливадна власатка
Cs (1) kostřava luční
De (–) Wiesen-Schwingel
En (–) meadow fescue
Hu (0) réti csenkesz
Ro (0) păiuş de livezi
Ru (1) овсяница луговая
Sk (–) kostrava lúčna

696 ***Festuca rubra*** L.
Al (–) bishtpelëza e kuqe
Bg (1) червена власатка
Cs (2) kostřava červená
De (–) Roter Schwingel
En (–) red fescue
Hu (0) veres csenkesz
Pl (3) kostrzewa czerwona
Ro (0) păiuş roşu
Ru (1) овсяница красная
Sk (–) kostrava červená

697 ***Festuca rupicola*** Heuff.
Al (–) bishtpelëza me vija
Cs (1) kostřava žlábkovitá
De (–) Furchen-Schwingel
Hu (0) pusztai csenkesz
Pl (2) kostrzewa bruzdkowana
Ru (1) овсяница бороздчатая
Sk (–) kostrava žľabkovitá

* ***Festuca sulcata*** (Hack.) Nyman → *Festuca rupicola*

698 ***Festuca tenuifolia*** Sibth.
Al (–) bishtpelëza gjethehollë
Cs (0) kostřava vláskovitá
De (–) Haar-Schwingel
En (–) fine-leaved sheep's-fescue
Hu (0) fonalas csenkesz
Pl (1) kostrzewa nitkowata
Ru (0) овсяница тонколистная
Sk (–) kostrava vláskovitá

699 ***Festuca valesiaca*** Schleich.
Cs (0) kostřava walliská
De (–) Walliser Schwingel
Hu (0) vékony csenkesz
Pl (1) kostrzewa walezyjska
Ru (0) овсяница валлисская
Sk (–) kostrava walliská

* ***Ficaria verna*** Huds. → *Ranunculus ficaria*

700 ***Filaginella uliginosa*** (L.) Opiz
Al (–) gnafala e ligatinës
Cs (2) protěž močálová
De (–) Sumpf-Ruhrkraut
En (–) marsh cudweed
Hu (2) iszapgyopár
Pl (2) szarota błotna
Ru (1) сушеница болотная
Sh (1) grančikasti runolistak; mrki srcopuc
Sk (–) plesnivček barinný

701 ***Filago* spp.** L.
Bg (–) бяла свещица
De (–) Fadenkraut
En (–) cudweed
Gr (1) φιλάγκο
Ru (1) жабник
Sk (–) belolist

* ***Filago arvensis*** L. → *Logfia arvensis*

* ***Filago germanica*** L. → *Filago vulgaris*

* ***Filago minima*** (Sm.) Pers. → *Logfia minima*

* ***Filago paniculata*** Moen. → *Logfia arvensis*

702 ***Filago pyramidata*** L.
Bg (1) бяла свещица пирамидална
De (–) Spatelblättriges Fadenkraut
En (–) broad-leaved cudweed
Ru (0) жабник пирамидальный
Sk (–) belolist ihlancovitý

* ***Filago spathulata*** Presl → *Filago pyramidata*

703 ***Filago vulgaris*** Lam.
Al (–) filago e zakonshme
Bg (1) германска бяла свещица
Cs (0) bělolist německý
De (–) Deutsches Fadenkraut
En (–) common cudweed
Hu (1) német penészvirág
Pl (1) nicennica niemiecka
Ru (1) жабник германский
Sh (1) germanska maljavica; piramidalni bjelolist
Sk (–) belolist nemecký

* ***Filipendula hexapetala*** Gilib. → *Filipendula vulgaris*

704 **Filipendula ulmaria** (L.) Max.
Al (–) shtalbësa e egër
Bg (0) брястолистно орехче
Cs (2) tužebník jilmový
De (–) Echtes Mädesüß
En (–) meadowsweet
Hu (1) réti legyezőfű
Pl (1) wiązówka błotna

Ro (0) creţuşcă
Ru (1) лабазник вязолистный
Sh (1) močvarna končara
Sk (–) túžobník brestový

705 ***Filipendula vulgaris*** Moen.
Al (–) filipendula e rëndomtë
Bg (0) шестовенечно орехче
Cs (1) tužebník obecný
De (–) Kleines Mädesüß
En (–) dropwort
Hu (0) koloncos legyezőfű
Pl (1) wiązówka bulwkowa
Ro (0) aglică
Ru (1) лабазник шестилепестный
Sh (1) obična končara
Sk (–) túžobník obyčajný

706 ***Fimbristylis annua*** (All.) Roem. & Schultes
Al (–) zuba e ndarë dysh
De (–) Einjährige Fransenbinse
Gr (1) φιμπριστνλίδα
Ru (0) фимбристилис однолетний
Sk (–) šachorinka ročná

* ***Foeniculum piperitum*** (Uc.) Sweet → *Foeniculum vulgare*

707 ***Foeniculum vulgare*** Mill.
Al (–) maraja
Bg (0) морач
Cs (0) fenykl obecný
De (–) Echter Fenchel
En (–) common fennel
Gr (1) μάραθο
Hu (0) édesköménу
Ro (0) molură
Ru (1) фенхель обыкновенный
Sh (1) pravi komorač
Sk (–) fenikel obyčajný

708 ***Fragaria vesca*** L.
Al (–) dredhëza; luleshtrydhja
Bg (1) обикновена ягода
Cs (1) jahodník obecný
De (–) Wald-Erdbeere
En (–) wild strawberry
Gr (1) αγριοφράουλα
Hu (0) erdei szamóca
Pl (1) poziomka pospolita
Ro (0) fragi de păduri
Ru (1) земляника лесная
Sh (1) šumska jagoda
Sk (–) jahoda lesná

709 ***Fragaria viridis*** Duchesne
Cs (–) jahodník trávnice
De (–) Knackelbeere
En (–) small wild-strawberry
Ru (0) земляника зелёная
Sh (1) pucavica jagoda
Sk (–) jahoda trávnica

710 ***Fritillaria meleagris*** L.
Bg (0) шахматовидна ведрица
Cs (0) řebčík kostkovaný
De (–) Schachblume
En (–) snake's-head fritillary
Hu (0) kockásliliom
Pl (0) szachownica kostkowata
Ro (0) bibilică
Ru (0) рябчик шахматный
Sh (1) obična kockavica
Sk (–) rebčík strakatý

711 ***Fumaria* spp.** L.
Bg (1) росопас
De (–) Erdrauch
En (–) fumitory
Gr (1) καπνόχορτο
Ro (2) fumăriţă
Ru (2) дымянка
Sk (–) zemedym

712 ***Fumaria capreolata*** L.
Cs (0) zemědým úponkatý
De (–) Rankender Erdrauch

En (–) white ramping-fumitory
Pl (0) dymnica wielkokwiatowa
Ru (1) дымянка козья
Sk (–) zemedym úpončitý

* ***Fumaria media*** Mer. → *Fumaria muralis*

713 *Fumaria muralis* Sond.
Bg (1) стенен росопас
De (–) Mauer-Erdrauch
En (–) common ramping-fumitory
Sk (–) zemedym múrový

714 *Fumaria officinalis* L.
Al (–) fomi mjekësor
Bg (1) лечебен росопас
Cs (2) zemědým lékařský
De (–) Gemeiner Erdrauch
En (–) common fumitory
Gr (3) καπνόχορτο
Hu (1) orvosi füstike
Pl (3) dymnica pospolita
Ru (2) дымянка аптечная
Sh (2) obična dimnjača
Sk (–) zemedym lekársky

715 *Fumaria parviflora* Lam.
Cs (0) zemědým malokvětý
De (–) Kleinblütiger Erdrauch
En (–) fine-leaved fumitory
Hu (0) kisvirágú füstike
Pl (1) dymnica drobnokwiatowa
Ru (1) дымянка мелкоцветковая
Sk (–) zemedym málokvetý

716 *Fumaria rostellata* Knaf
Bg (1) остроплоден росопас
Cs (0) zemědým zobánkatý
De (–) Geschnäbelter Erdrauch
Hu (–) csőrös füstike
Pl (0) dymnica szerokodziałkowa
Sk (–) zemedym zobákovitý

717 *Fumaria schleicheri* Soy.
Bg (1) Шлайхеров росопас
Cs (0) zemědým Schleicherův
De (–) Dunkler Erdrauch
Hu (3) parlagi füstike
Pl (1) dymnica różowa
Ru (1) дымянка Шлейхера
Sk (–) zemedym Schleicherov

718 *Fumaria vaillantii* Loisel.
Al (–) fomi i Vailantit
Bg (1) Ваилантов росопас
Cs (1) zemědým Vaillantův
De (–) Buschiger Erdrauch
En (–) alkafoun fumitory
Hu (–) szürke füstike
Pl (1) dymnica drobnokwiatowa
Ru (1) дымянка Вайланта
Sh (1) ružičasta dimnjača
Sk (–) zemedym Vaillantov

G

719 *Gagea* spp. Salisb.
Bg (–) гарвански лук
De (–) Gelbstern
En (–) star-of-Bethlehem
Gr (1) γαγέα
Ru (1) гусиный лук
Sk (–) krivatec

720 *Gagea arvensis* (Pers.) Dumort.
Bg (1) полски пачи крак
Cs (1) křivatec rolní
De (–) Acker-Gelbstern
En (–) hairy star-of-Bethlehem
Hu (1) ugari tyúktaréj
Pl (1) złoć polna
Ro (1) scînteiuţă
Ru (1) гусиный лук полевой
Sh (1) njivsko baloče; vunasto baloče
Sk (–) krivatec poľný

721 *Gagea lutea* (L.) Ker-G.
Al (–) gagea e verdhë
Bg (1) жълт пачи крак
Cs (1) křivatec žlutý
De (–) Gemeiner Gelbstern
En (–) yellow star-of-Bethlehem
Hu (0) sárga tyúktaréj
Pl (1) złoć żółta
Ro (0) laptele păsării
Ru (1) гусиный лук жёлтый
Sh (1) žuto baloče
Sk (–) krivatec žltý

722 *Gagea minima* (L.) Ker-G.
Al (–) gagea e vogël
Bg (1) дребен гарвански лук
Cs (0) křivatec nejmenší
De (–) Kleiner Gelbstern
En (–) little star-of-Bethlehem
Hu (0) apró tyúktaréj
Pl (1) złoć mała
Ro (0) viorele galbene
Ru (0) гусиный лук малый
Sh (1) malo baloče
Sk (–) krivatec najmenší

723 *Gagea pratensis* (Pers.) Dumort.
Bg (1) ливаден гарвански лук
Cs (0) křivatec luční
De (–) Wiesen-Gelbstern
En (–) meadow star-of-Bethlehem
Hu (1) mezei tyúktaréj
Pl (1) złoć łąkowa
Ro (1) ceapa ciorii
Ru (1) гусиный лук луговой
Sk (1) krivatec lúčny

724 *Gagea pusilla* (Schm.) Schultes
Cs (0) křivatec maličký
De (–) Zwerg-Gelbstern
En (–) small star-of-Bethlehem
Hu (1) kis tyúktaréj
Ru (0) гусиный лук низкий
Sk (–) krivatec maličký

* ***Gagea stenopetala*** Reichenb. → *Gagea pratensis*

* ***Gagea villosa*** (Bieb.) Duby → *Gagea arvensis*

725 *Galanthus nivalis* L.
Al (–) boçëbora; lulebora
Bg (0) кокиче
Cs (1) sněženka podsněžník
De (–) Schneeglöckchen
En (–) snowdrop
Gr (1) γάλανθος
Pl (1) śnieżyczka przebiśnieg
Ro (0) ghiocel
Ru (1) подснежник белоснежный
Sk (–) snežienka jarná pravá

726 *Galega officinalis* L.
Al (–) qerbashi mjekësor

Bg (1) жаблек
Cs (0) jestřabina lékařská
De (–) Echte Geißraute
En (–) goat's-rue
Gr (1) ἡμεροτρίφυλλο
Hu (1) kecskeruta
Pl (1) rutwica lekarska
Ro (1) ciumărea
Ru (1) галега лекарственная
Sh (1) ljekoviti orlovac; obični orlovac
Sk (–) jastrabina lekárska

* ***Galeobdolon amplexicaule*** Moen. →
Lamium amplexicaule

727 ***Galeopsis angustifolia*** Ehrh.
Cs (2) konopice úzkolistá
De (–) Schmalblättriger Hohlzahn
En (–) narrow-leaved hemp-nettle
Hu (1) keskenylevelű kenderkefű
Pl (1) poziewnik wąskolistny
Ro (1) faţa mîţei
Ru (1) пикульник узколистный
Sk (–) konopnica úzkolistá

728 ***Galeopsis bifida*** Boenn.
Cs (0) konopice polní dvouklaná
De (–) Zweispaltiger Hohlzahn
Hu (1) kicsípett kenderkefű
Pl (2) poziewnik dwudzielny
Ru (1) пикульник двунадрезный
Sk (–) konopnica napuchnutá dvojklaná

* ***Galeopsis grandiflora*** Roth →
Galeopsis segetum

729 ***Galeopsis ladanum*** L.
Bg (1) бъзовлек
Cs (0) konopice širokolistá
De (–) Breitblättriger Hohlzahn
En (–) red hemp-nettle
Hu (3) piros kenderkefű
Pl (3) poziewnik polny
Ro (2) tapoşnic
Ru (2) пикульник ладонниковый
Sh (1) širokolisna srba
Sk (–) konopnica širokolistá

* ***Galeopsis neglecta*** Schultes →
Galeopsis tetrahit

* ***Galeopsis ochroleuca*** Lam. →
Galeopsis segetum

730 ***Galeopsis pubescens*** Bess.
Cs (2) konopice pýřitá
De (–) Weichhaariger Hohlzahn
En (–) downy hemp-nettle
Hu (2) pelyhes kenderkefű
Pl (2) poziewnik miękkowłosy
Ru (2) пикульник пушистый
Sh (1) pahuljava srba
Sk (–) konopnica pýritá

731 ***Galeopsis segetum*** Neck.
Cs (0) konopice bledožlutá
De (–) Gelber Hohlzahn
En (–) yellow hemp-nettle
Hu (1) szőrös kenderkefű
Sk (–) konopnica bledožltá

732 ***Galeopsis speciosa*** Mill.
Bg (1) красива бударица
Cs (0) konopice zdobná
De (–) Bunter Hohlzahn
En (–) large-flowered hemp-nettle
Hu (1) szőrös kenderkefű
Pl (2) poziewnik pstry
Ro (1) zebră
Ru (1) пикульник заметный
Sh (1) šarenkasta srba
Sk (–) konopnica úhľadná

733 ***Galeopsis tetrahit*** L.
Bg (2) петниста бударица
Cs (3) konopice polní
De (–) Gemeiner Hohlzahn
En (–) common hemp-nettle

Hu (2) tarka kenderkefű
Pl (3) poziewnik szorstki
Ro (3) lungurică
Ru (2) пикульник обыкновенный
Sh (2) čvorasti smrdelj; smrdljiva kopriva
Sk (–) konopnica napuchnutá pravá

* **Galeopsis versicolor** Curt. → *Galeopsis speciosa*

* **Galinsoga caracasana** auct. → *Galinsoga ciliata*

734 Galinsoga ciliata (Rafin.) Blake
Bg (2) влакнеста галинзога
Cs (3) pěťour srstnatý
De (–) Behaartes Franzosenkraut
En (–) shaggy soldier
Hu (1) bozontos gombvirág
Pl (2) żółtlica owłosiona
Ru (2) галинсога четырёхлучевая
Sk (–) pätúr srstnatý

* **Galinsoga hispida** Benth. → *Galinsoga ciliata*

735 Galinsoga parviflora Cav.
Bg (2) дребноцветна галинзога
Cs (3) pěťour maloúborný
De (–) Kleinblütiges Franzosenkraut
En (–) gallant soldier
Hu (3) gyakori gombvirág
Pl (3) żółtlica drobnokwiatowa
Ro (3) busuioc de cîmp
Ru (3) галинсога мелкоцветковая
Sh (2) obična konica
Sk (–) pätúr drobnokvetý

* **Galinsoga quadriradiata** auct. → *Galinsoga ciliata*

736 Galium spp. L.
Bg (1) лепка
De (–) Labkraut
Gr (1) κολλητσίδα
Ru (2) подмаренник
Sk (–) lipkavec

737 Galium aparine L.
Al (–) jevga; ngjitësja rodhe
Bg (1) обикновена лепка
Cs (3) svízel přítula
De (–) Kletten-Labkraut
En (–) cleavers
Gr (2) μεγαλόκαρπη κολλητσίδα
Hu (3) ragadós galaj
Pl (3) przytulia czepna
Ro (3) turiță
Ru (3) подмаренник цепкий
Sh (3) lepuša broćika; priljepača broćika
Sk (–) lipkavec obyčajný

* **Galium hercynicum** auct. → *Galium saxatile*

738 Galium humifusum Bieb.
Ru (1) подмаренник распростёртый
Sk (–) lipkavec ležatý

739 Galium mollugo L.
Al (–) ngjitësja mollugë
Bg (1) меко еньовче
Cs (3) svízel povázka
De (–) Wiesen-Labkraut
En (–) hedge bedstraw
Hu (1) közönséges galaj
Pl (1) przytulia pospolita
Ro (0) sînziene albe
Ru (1) подмаренник мягкий
Sh (2) obična broćika
Sk (–) lipkavec poviazka

740 Galium palustre L.
Al (–) ngjitësja e kënetave
Bg (1) блатна лепка
Cs (0) svízel bahenní
De (–) Sumpf-Labkraut
En (–) common marsh-bedstraw

Hu (0) mocsári galaj
Pl (1) przytulia błotna
Ru (1) подмаренник болотный
Sh (1) močvarna broćika
Sk (–) lipkavec močiarny

741 ***Galium parisiense*** L.
Al (–) ngjitësja e Parizit
Bg (1) парижко еньовче
Cs (0) svízel pařížský
De (–) Pariser Labkraut
En (–) wall bedstraw
Hu (1) párizsi galaj
Ru (1) подмаренник парижский
Sk (–) lipkavec parížsky

* ***Galium pedemontanum*** (Bella.) All. → *Cruciata pedemontana*

742 ***Galium pumilum*** Murray
Cs (0) svízel nízký
De (–) Niederes Labkraut
En (–) slender bedstraw
Hu (0) törpe galaj
Pl (1) przytulia szorstkoowockowa
Ru (0) подмаренник малорослый
Sk (–) lipkavec nízky

743 ***Galium rotundifolium*** L.
Bg (1) кръглолистно еньовче
Cs (0) svízel drsný
De (–) Rundblättriges Labkraut
En (–) round-leaved bedstraw
Hu (0) kereklevelű galaj
Pl (–) przytulia okragła
Ru (0) подмаренник шероховатый
Sk (–) lipkavec okrúhlolistý

744 ***Galium rubioides*** L.
Cs (0) svízel mořenovitý
De (–) Krappartiges Labkraut
Hu (0) réti galaj
Pl (0) przytulia szerokolistna
Ro (1) drăgaică mare
Ru (0) подмаренник мареновидный
Sk (–) lipkavec morenovitý

745 ***Galium saxatile*** L.
Cs (0) svízel nízký hercynský
De (–) Felsen-Labkraut
En (–) heath bedstraw
Pl (1) przytulia hercyńska
Ru (0) подмаренник герцинский
Sk (–) lipkavec skalný

746 ***Galium spurium*** L.
Cs (1) svízel nepravý
De (–) Acker-Labkraut
En (–) false cleavers
Gr (3) μικρόκαρπη κολλητσίδα
Hu (1) gyanús galaj
Pl (2) przytulia fałszywa
Ru (2) подмаренник ложный
Sk (–) lipkavec prietulý nepravý

747 ***Galium sylvaticum*** L.
Cs (–) svízel lesní
De (–) Wald-Labkraut
En (–) wood bedstraw
Sh (1) šumska broćika
Sk (–) lipkavec lesný

* ***Galium tricorne*** Stok. → *Galium tricornutum*

748 ***Galium tricornutum*** Dandy
Bg (2) трирога лепка
Cs (0) svízel trojrohý
De (–) Dreihörniges Labkraut
En (–) corn cleavers
Gr (2) κυρτόκαρπη κολλητσίδα
Hu (2) vetési galaj
Pl (2) przytulia trójrożna
Ru (1) подмаренник трёхрогий
Sh (1) divlja broćika
Sk (–) lipkavec trojrohý

749 ***Galium uliginosum*** L.
Bg (1) мочурно еньовче
Cs (0) svízel močálový
De (–) Moor-Labkraut
En (–) fen bedstraw
Hu (0) lápi galaj
Ru (0) подмаренник топяной
Sk (–) lipkavec slatinný

* ***Galium vernum*** Scop. → *Cruciata glabra*

750 ***Galium verum*** L.
Al (–) ngjitësja e vërtetë
Bg (1) обикновено еньовче
Cs (2) svízel syřišťový
De (–) Echtes Labkraut
En (–) lady's bedstraw
Hu (1) tejoltó galaj
Pl (1) przytulia właściwa
Ro (0) sînziene galbene
Ru (1) подмаренник настоящий
Sh (1) ivanjska broćika
Sk (–) lipkavec syrišťový pravý

751 ***Genista* spp.** L.
Bg (0) жълтуга
De (–) Ginster
Gr (1) γενίστα
Ru (0) дрок
Sk (–) kručinka

752 ***Genista acanthoclada*** DC.
Gr (2) αφάνα
Sk (–) kručinka tŕnistá

753 ***Genista germanica*** L.
Al (–) gjineshtra e Gjermanisë
Bg (0) немска жълтуга
Cs (1) kručinka německá
De (–) Deutscher Ginster
En (–) German greenweed
Pl (1) janowiec ciernisty
Ru (0) дрок германский
Sk (–) kručinka nemecká

754 ***Genista pilosa*** L.
Cs (–) kručinka chlupatá
De (–) Behaarter Ginster
En (–) hairy greenweed
Ru (0) дрок волосистый
Sh (1) dlakava žutica; dlakava žutilica
Sk (–) kručinka chlpatá

* ***Genista procumbens*** Waldst. & Kit. → *Cytisus procumbens*

* ***Genista sagittalis*** L. → *Chamaespartium sagittale*

755 ***Genista tinctoria*** L.
Al (–) gjineshtra ngjyruese
Bg (0) висока жълтуга
Cs (0) kručinka barvířská
De (–) Färber-Ginster
En (–) dyer's greenweed
Pl (1) janowiec barwierski
Ro (1) drobiţă
Ru (1) дрок красильный
Sh (1) obična žutica; obična žutilica
Sk (–) kručinka farbiarska

756 ***Gentiana* spp.** L.
De (–) Enzian
En (–) gentian
Gr (1) γεντιανή
Ru (0) горечавка
Sk (–) horec

* ***Gentiana ciliata*** L. → *Gentianella ciliata*

757 ***Gentiana lutea*** L.
Al (–) sanza
Cs (0) hořec žlutý
De (–) Gelber Enzian
En (–) yellow gentian
Pl (0) goryczka żółta
Ro (0) ghinţură galbenă
Ru (0) горечавка жёлтая

Sh (1) obična lincura; žuta širištara
Sk (–) horiepnik žltý

758 *Gentianella ciliata* (L.) Borkh.
Bg (0) ресничеста тинтявка
Cs (0) hořeček brvitý
De (–) Gefranster Enzian
En (–) fringed gentian
Hu (0) prémes tárnics
Pl (1) goryczka orzęsiona
Ru (0) горечавник реснитчатый
Sh (1) trepavičasta širištara
Sk (–) horec brvitý

759 *Geranium* spp. L.
Bg (1) здравец
De (–) Storchschnabel
En (–) cranesbill
Gr (1) γεράνι
Ru (1) герань
Sk (–) pakost

* ***Geranium carolinianum*** L. → *Geranium dissectum*

* ***Geranium cicutarium*** L. → *Erodium cicutarium*

760 *Geranium columbinum* L.
Al (–) kamaroshe pëllumbi; rruka pëllumbi
Bg (1) гълъбов здравед
Cs (1) kakost holubiči
De (–) Stein-Storchschnabel
En (–) long-stalked cranesbill
Hu (1) galambláb gólyaorr
Pl (1) bodziszek gołębi
Sh (1) gola iglica; golubiji zdravac
Ru (1) герань голубиная
Sk (–) pakost holubičí

761 *Geranium dissectum* L.
Al (–) kamaroshe e prerë
Bg (1) насеченолистен здравец
Cs (0) kakost dvousečný
De (–) Schlitzblättriger Storchschnabel
En (–) cut-leaved cranesbill
Hu (1) sallangos gólyaorr
Pl (2) bodziszek porozcinany
Ru (1) герань рассечённая
Sh (1) rasečena iglica; rasječena iglica
Sk (–) pakost prstnatosečný

762 *Geranium divaricatum* Ehrh.
Bg (1) разклонен здравец
Cs (0) kakost rozsochatý
De (–) Spreizender Storchschnabel
Hu (1) berzedt gólyaorr
Pl (1) bodziszek rozłożysty
Ru (1) герань раскидистая
Sk (–) pakost rozložitý

763 *Geranium macrorrhizum* L.
Al (–) kamaroshe rrënjëgjatë
Bg (1) обикновен здравец
Cs (0) kakost velkokořenný
De (–) Felsen-Storchschnabel
En (–) rock cranesbill
Ro (0) priboi
Ru (0) герань крупнокорневищная
Sk (–) pakost pakoreňatý

764 *Geranium molle* L.
Al (–) kamaroshe butëloshe; rruka e butë
Bg (1) мек здравец
Cs (0) kakost měkký
De (–) Weicher Storchschnabel
En (–) dove's-foot cranesbill
Hu (1) puha gólyaorr
Pl (2) bodziszek kosmaty
Ru (1) герань нежная
Sh (1) mekana iglica
Sk (–) pakost mäkký

765 *Geranium palustre* L.
Bg (0) блатен здравец
Cs (0) kakost bahenní
De (–) Sumpf-Storchschnabel

En (–) marsh cranesbill
Hu (0) mocsári gólyaorr
Pl (1) bodziszek błotny
Ru (0) герань болотная
Sk (–) pakost močiarny

766 ***Geranium phaeum*** L.
Bg (0) кафяв здравец
Cs (0) kakost hnědočervený
De (–) Brauner Storchschnabel
En (–) dusky cranesbill
Hu (0) fodros gólyaorr
Pl (1) bodziszek żałobny
Ro (0) pălăria cucului
Ru (0) герань красно-бурая
Sk (–) pakost hnedočervený

767 ***Geranium pratense*** L.
Al (–) rruka e livadheve
Bg (0) ливаден здравец
Cs (2) kakost luční
De (–) Wiesen-Storchschnabel
En (–) meadow cranesbill
Hu (0) mezei gólyaorr
Pl (2) bodziszek łąkowy
Ro (0) ciocul berzei
Ru (1) герань луговая
Sh (1) livadski zdravac; zavinuta iglica
Sk (–) pakost lúčny

768 ***Geranium pusillum*** L.
Bg (1) малък здравец
Cs (1) kakost maličký
De (–) Kleiner Storchschnabel
En (–) small-flowered cranesbill
Hu (1) apró gólyaorr
Pl (2) bodziszek drobny
Ro (1) buchet
Ru (1) герань маленькая
Sh (1) majušna iglica
Sk (–) pakost maličký

769 ***Geranium pyrenaicum*** Burm.
Al (–) rruka e Pirinejeve
Bg (1) пиренейски здравец
Cs (0) kakost pyrenejský
De (–) Berg-Storchschnabel
En (–) hedgerow cranesbill
Hu (1) déli gólyaorr
Pl (1) bodziszek pirenejski
Ru (0) герань пиренейская
Sk (–) pakost pyrenejský

770 ***Geranium robertianum*** L.
Al (–) kamaroshe e Robertit
Bg (0) зловонен здравец
Cs (1) kakost smrdutý
De (–) Ruprechtskraut
En (–) herb Robert
Hu (0) nehézszagú gólyaorr
Pl (1) bodziszek cuchnący
Ro (0) năprasnică
Ru (0) герань Роберта
Sk (–) pakost smrdľavý

771 ***Geranium rotundifolium*** L.
Al (–) kamaroshe me gjethe të rrumbullakëta
Bg (1) кръглолистен здравец
Cs (0) kakost okrouhlolisty
De (–) Rundblättriger Storchschnabel
En (–) round-leaved cranesbill
Hu (1) kereklevelű gólyaorr
Pl (2) bodziszek okrągłolistny
Ru (1) герань круглолистная
Sh (1) mirišljavi zdravac
Sk (–) pakost okrúhlolistý

772 ***Geranium sanguineum*** L.
Al (–) kamaroshe me ngjyrë gjaku
Bg (0) кървавочервен здравец
Cs (0) kakost krvavý
De (–) Blutroter Storchschnabel
En (–) bloody cranesbill
Hu (0) piros gólyaorr
Pl (1) bodziszek czerwony
Ru (0) герань кровяно-красная
Sk (–) pakost krvavý

773 *Geranium sibiricum* L.
Ru (1) герань сибирская
Sk (–) pakost sibírsky

774 *Geranium sylvaticum* L.
Al (–) kamaroshe e pyllit
Bg (0) горски здравец
Cs (0) kakost lesní
De (–) Wald-Storchschnabel
En (–) wood cranesbill
Pl (1) bodziszek leśny
Ro (0) fratele priboiului
Ru (0) герань лесная
Sk (–) pakost lesný

775 *Geranium tuberosum* L.
Bg (1) клубеноподобен здравец
En (–) tuberous cranesbill
Ru (0) герань клубневая
Sk (–) pakost hľazovitý

776 *Geum aleppicum* Jacq.
Ru (1) гравилат прямой
Sk (–) kuklík veľkokvetý

777 *Geum rivale* L.
Bg (0) поточно омайниче
Cs (1) kuklík potoční
De (–) Bach-Nelkenwurz
En (–) water avens
Pl (2) kuklik zwisły
Ro (0) călţunul doamnei
Ru (1) гравилат речной
Sh (1) potočni blaženak
Sk (–) kuklík vodný

778 *Geum urbanum* L.
Al (–) mlaka qytetesh
Bg (0) градско омайниче
Cs (2) kuklík městský
De (–) Echte Nelkenwurz
En (–) wood avens
Hu (1) erdei gyömbérgyökér
Pl (2) kuklik pospolity
Ro (1) cerenţel
Ru (1) гравилат городской
Sh (1) obični blaženak
Sk (–) kuklík mestský

* ***Githago segetum*** auct. → *Agrostemma githago*

779 *Gladiolus illyricus* Koch
Al (–) gladiola e Ilirisë; thiksa e egër
De (–) Illyrische Siegwurz
En (–) wild gladiolus
Gr (1) μικρή μαχαιρίδα
Sk (–) mečík žilnatý

780 *Gladiolus imbricatus* L.
Bg (0) покритоцветно петльово перо
Cs (0) mečík střechovitý
De (–) Wiesen-Siegwurz
Hu (0) réti kardvirág
Pl (1) mieczyk dachówkowaty
Ro (0) săbiuţă
Ru (0) шпажник черепичный
Sh (1) crepovita sabljičica; crijepovita sabljičica
Sk (–) mečík strechovitý

781 *Gladiolus italicus* Mill.
Al (–) gladiola e grunjëve
Bg (1) посевно петльово перо
Gr (1) μεγάλη μαχαιρίδα
Ru (0) шпажник итальянский
Sk (–) mečík dvojradý

782 *Gladiolus palustris* Gaudin
Al (–) gladiola e kënetave
Cs (0) mečík bahenní
De (–) Sumpf-Siegwurz
En (–) marsh gladiolus
Hu (0) mocsári kardvirág
Pl (1) mieczyk błotny
Ru (0) шпажник болотный
Sk (–) mečík močiarny

* **Gladiolus segetum** Ker-G. → *Gladiolus italicus*

783 Glaucium corniculatum (L.) Rud.
Bg (1) рогчест глауциум
Cs (0) rohatec růžkatý
De (–) Roter Hornmohn
En (–) common horned-poppy
Gr (1) γλαύσιο
Hu (1) veres szarumák
Pl (0) siwiec pomaranczowy
Ro (1) mac cornut
Ru (1) глауциум рогатый
Sh (1) obična makovica
Sk (–) rohatec rožkatý

784 Glaucium flavum Cran.
Bg (1) жълт глауциум
Cs (0) rohatec žluty
De (–) Gelber Hornmohn
En (–) yellow horned-poppy
Gr (1) γλαύσιο
Hu (1) sárga szarumák
Pl (0) siwiec żółty
Ru (0) глауциум жёлтый
Sh (1) žuta makovica
Sk (–) rohatec žltý

785 Glechoma hederacea L.
Al (–) glekoma në tratjë urthi
Bg (1) бръшлянова самовойка
Cs (2) popenec břečťanovitý
De (–) Gundermann
En (–) ground-ivy
Gr (1) χαμοκισσός
Hu (2) kerek repkény
Pl (2) bluszczyk kurdybanek
Ro (1) rotungioară
Ru (2) будра плющевидная
Sh (1) obična dobričica
Sk (–) zádušník brečtanovity

786 Glechoma hirsuta Waldst. & Kit.
Al (–) glekoma qimeashpër
Bg (1) космата самовойка
Cs (0) popenec břečťanovitý chlupatý
Hu (2) kerek repkény
Pl (1) bluszczyk kosmaty
Ru (1) будра волосистая
Sk (–) zádušník chlpatý

787 Glyceria fluitans (L.) R Br.
Al (–) gliqeri fluronjës
Bg (0) речна росица
Cs (1) zblochan vzplývavý
De (–) Manna-Schwaden
En (–) floating sweet-grass
Hu (1) réti harmatkása
Pl (1) manna jadalna
Ro (1) rourică
Ru (1) манник наплывающий
Sh (1) potočna pirevina
Sk (–) steblovka plávajúca

788 Glyceria maxima (Hartm.) Holm.
Cs (1) zblochan vodní
De (–) Großer Schwaden
En (–) reed sweet-grass
Hu (1) vízi harmatkása
Ro (1) mană de apă
Ru (1) манник крупный
Sh (1) bijeli šaš; velika sladika
Sk (–) steblovka trsťovitá

789 Glyceria plicata (Fries) Fries
Bg (0) сдиплена росица
Cs (0) zblochan řasnatý
De (–) Falten-Schwaden
Hu (0) fodros harmatkása
Pl (1) manna fałdowana
Ru (0) манник складчатый
Sk (–) steblovka riasnatá

790 Glycyrrhiza spp. L.
Bg (0) сладник
De (–) Süßholz
Gr (1) γλυκόρριζα
Ru (0) солодка
Sk (–) sladké drievko

791 *Glycyrrhiza echinata* L.
Bg (1) четинест сладник
De (–) Russisches Süßholz
En (–) spiny-fruited liquorice
Hu (1) ednámkóró
Ro (1) lemn dulce
Ru (0) солодка щетинистая
Sh (1) bodljasti sladić; čekinjasti sladić
Sk (–) sladké drievko bodľavé

792 *Glycyrrhiza glabra* L.
Al (–) jamballi
Bg (0) обикновен сладник
Cs (0) lékořice lysá
De (–) Lakritze
En (–) root liquorice
Gr (1) γλυκόρριζα
Hu (0) igazi édesgyökér
Ru (1) солодка голая
Sh (1) glatki sladić
Sk (–) sladké drievko hladkoplode

* ***Gnaphalium arvense*** L. → *Logfia arvensis*

* ***Gnaphalium germanicum*** L. → *Filago vulgaris*

793 *Gnaphalium luteo-album* L.
Bg (1) жълтобял смил
Cs (0) protěž žlutobílá
De (–) Gelbes Ruhrkraut
En (–) Jersey cudweed
Hu (2) halvány gyopár
Pl (1) szarota żółtobiała
Ru (1) сушеница желтовато-белая
Sk (–) plesnivec piesočný

* ***Gnaphalium sylvaticum*** L. → *Omalotheca sylvatica*

* ***Gnaphalium uliginosum*** L. → *Filaginella uliginosa*

794 *Goldbachia laevigata* (Bieb.) DC.
Ru (1) гольдбахия гладкая

795 *Gratiola officinalis* L.
Al (–) dernoku mjekësor
Bg (1) сиротица
Cs (0) konitrud lékařský
De (–) Gemeines Gnadenkraut
En (–) hedge hyssop
Gr (1) γρατιόλα
Hu (1) csikorgófű
Pl (1) konitrut błotny
Ro (1) veninariță
Ru (1) аврон лекарственный
Sh (1) trorebrasta milica
Sk (–) konitrúd lekársky

796 *Grindelia squarrosa* (Pursh) Dunal
Cs (0) grindelie kostrbatá
En (–) gumweed
Ru (1) гринделия растопыренная
Sk (–) grindelia kostrbatá

797 *Groenlandia densa* (L.) Fourr.
De (–) Dichtblättriges Laichkraut
En (–) opposite-leaved pondweed
Hu (0) sűrűlevelű békaszőlő
Ru (0) рдест густолистный
Sh (1) zubati mresnjak; zubati mrijesnjak
Sk (–) červeňovec hustolistý

798 *Gymnadenia conopsea* (L.) R Br.
Al (–) gjimnadena si rrjetë
Bg (0) комаровидна гимнадения
Cs (0) petiprstka žežulník
De (–) Mücken-Händelwurz
En (–) fragrant orchid
Hu (0) szúnyoglábú bibircsvirág
Pl (1) gółka długoostrogowa
Ro (0) ură
Ru (0) кокушник комариковый
Sk (–) päťprstnica stredná

* ***Gypsophila arvensis*** Burkh. → *Gypsophila muralis*

799 *Gypsophila muralis* L.
Bg (1) стенна мишорка
Cs (1) šater zední
De (–) Mauer-Gipskraut
En (–) wall gypsophila
Gr (1) γυψόφιλο
Hu (3) mezei fátyolvirág
Pl (1) łyszczec polny
Ru (2) качим постенный
Sh (1) obična sadarka
Sk (–) šater múrový

800 *Gypsophila paniculata* L.
Ru (1) качим метельчатый
Sk (–) šater latnatý

* ***Gypsophila purpurea*** Gilib. → *Gypsophila muralis*

H

801 ***Hainardia cylindrica*** (Willd.) Greut.
Ru (0) монерма цилиндрическая
Sh (1) cilindrična tankorepka

* ***Haplophyllum biebersteinii*** Spach. →
Haplophyllum suaveolens

802 ***Haplophyllum suaveolens*** (DC.) Don
Bg (1) ароматен цялолист
Ru (0) цельнолистник цветоножковый
Sk (–) ruta voňavá

* ***Haynaldia villosa*** (L.) Schur →
Dasypyrum villosum

803 ***Hedera helix*** L.
Al (–) urthi
Bg (1) бршлян
Cs (1) břečťan popínavý
De (–) Gemeiner Efeu
En (–) ivy
Gr (1) κισσός
Hu (0) borostyán
Pl (1) bluszcz pospolity
Ro (0) iederă
Ru (1) плющ обыкновенный
Sk (–) brečtan popínavý

804 ***Hedypnois cretica*** (L.) Dum.-C.
Al (–) hedipnoi i Kretes
De (–) Röhrchenkraut
Gr (1) γλυκοπνοή
Ru (0) гединопноис критский

* ***Hedypnois polymorpha*** DC. →
Hedypnois cretica

* ***Heleocharis palustris*** R Br. →
Eleocharis palustris

* ***Heleochloa alopecuroides*** (Pill. & Mitt.) Host → *Crypsis alopecuroides*

805 ***Helianthemum*** **spp.** Mill.
Bg (0) желтак
De (–) Sonnenröschen
En (–) rock-rose
Gr (1) ηλιάνθεμο
Ro (0) mălăoaie
Ru (1) солнцецвет
Sk (–) devätorník

* ***Helianthemum nitidum*** Clem. →
Helianthemum nummularium

806 ***Helianthemum nummularium*** (L.) Mill.
Al (–) heleantema e zakonshme
Bg (0) обикновен желтак
Cs (0) devaterník penízkovitý
De (–) Gemeines Sonnenröschen
En (–) common rock-rose
Hu (0) közönséges napvirág
Pl (1) posłanek kutnerowaty
Ro (0) iarba osului
Ru (0) солнцецвет монетолистный
Sk (–) devätorník obyčajný

* ***Helianthemum ovatum*** (Viv.) Dunal →
Helianthemum nummularium

807 ***Helianthus annuus*** L.
Al (–) luledielli
Bg (0) слънчоглед
Cs (0) slunečnice roční
De (–) Gemeine Sonnenblume
En (–) common sunflower
Gr (1) ήλιος
Hu (0) közönséges napraforgó
Pl (0) słonecznik zwyczajny
Ro (0) floarea soarelui
Ru (1) подсолнечник однолетний
Sh (1) divlji suncokret
Sk (–) slnečnica ročná

808 ***Helianthus tuberosus*** L.
Al (–) molla e dheut; repë
Bg (0) земна ябълка
Cs (1) slunečnice topinambur
De (–) Erdbirne; Topinambur
En (–) Jerusalem artichoke
Gr (1) κολοκάσι
Hu (0) csicsóka
Pl (1) topinambur
Ro (1) nap porcesc
Ru (1) подсолнечник клубненосный
Sk (–) topinambur hľuzovitý; slečnica ambulovitá

809 ***Helichrysum* spp.** Mill.
Bg (0) жълт смил
De (–) Strohblume
Gr (1) ελίχρυσο
Ru (0) цмин
Sk (–) slamiha

810 ***Helichrysum arenarium*** (L.) Moen.
Al (–) akësi; sumbulli
Bg (0) пясъчен жълт смил
Cs (0) smil písečný
De (–) Sand-Strohblume
Hu (0) homoki szalmagyopár
Pl (2) kocanka piaskowa
Ro (0) siminoc
Ru (0) цмин песчаный
Sk (–) slamiha piesočná

811 ***Heliotropium* spp.** L.
Bg (1) знойка
De (–) Sonnenwende
En (1) heliotrope
Gr (1) ηλιοτρόπιο
Ru (1) гелиотроп
Sk (–) skrutec

812 ***Heliotropium ellipticum*** Ledeb.
Ru (1) гелиотроп эллиптический
Sk (–) skrutec elipsovitý

813 ***Heliotropium europaeum*** L.
Al (–) dielldridhësi i Evropës
Bg (1) обикновена знойка
Cs (0) otočník evropský
De (–) Sonnenwende
En (–) common heliotrope
Gr (2) κοινό ηλιοτρόπιο
Hu (3) parlagi kunkor
Ro (1) vanilie sălbatică
Ru (1) гелиотроп европейский
Sh (1) obični posunac
Sk (–) skrutec európsky

814 ***Heliotropium suaveolens*** Bieb.
Bg (1) миризлива знойка
Ru (0) гелиотроп душистый
Sk (–) skrutec voňavý

815 ***Helleborus cyclophyllus*** Boiss.
Gr (1) σκάρφη

816 ***Helleborus dumetorum*** Waldst. & Kit.
Cs (0) čemeřice křovištní
De (–) Hecken-Nieswurz
En (–) shrubby hellebore
Hu (0) kisvirágú hunyor
Pl (1) ciemiernik gajowy
Ro (0) cutcurig
Sk (–) čemerica krovinatá

817 ***Helleborus niger*** L.
Al (–) shpendra e zezë
Cs (0) čemeřice černá
De (–) Schwarze Nieswurz
En (–) Christmas rose
Ro (0) spînz
Ru (0) морозник чёрный
Sh (1) crni kukurek; crni kukurijek
Sk (–) čemerica čierna

818 ***Helleborus odorus*** Waldst. & Kit.
Al (–) kukereta; shpendra
Bg (0) кукуряк

Cs (0) čemeřice vonná
De (–) Wohlriechende Nieswurz
Hu (0) illatos hunyor
Ro (0) cutcurig
Ru (0) морозник душистый
Sh (1) mirisavi kukurek; mirisavi kukurijek
Sk (–) čemerica voňavá

* **Helleborus pallidus** Host → *Helleborus dumetorum*

* **Helminthotheca echioides** (L.) → *Picris echioides*

* **Helmintia echioides** (L.) Gaertn. → *Picris echioides*

* **Heracleum pollinianum** Bertol. → *Heracleum sphondylium*

819 *Heracleum sphondylium* L.
Al (–) heraklea rrotullë
Bg (1) девесил
Cs (3) bolševník obecný
De (–) Wiesen-Bärenklau
En (–) hogweed
Gr (1) σφοντύλι
Hu (1) medvetalp
Pl (2) barszcz zwyczajny
Ro (1) brînca ursului
Ru (1) борщевик обыкновенный
Sh (1) medvedji dlan; medvjedji dlan
Sk (–) boľševník obyčajný

820 *Herniaria* spp. L.
De (–) Bruchkraut
En (–) rupturewort
Gr (1) ερνιάρια
Ru (0) грыжник
Sk (–) prietržník

821 *Herniaria glabra* L.
Al (–) herniaria e shogët
Bg (1) голо белило
Cs (0) průtržník lysý
De (–) Kahles Bruchkraut
En (–) smooth rupturewort
Hu (1) kopasz porcika
Pl (1) połonicznik nagi
Ro (0) feciorică
Ru (1) грыжник гладкий
Sh (1) gola priputnica
Sk (–) prietržník holý

822 *Herniaria hirsuta* L.
Al (–) herniaria qimeashpër
Bg (1) космато белило
Cs (0) průtržník chlupatý
De (–) Bahaartes Bruchkraut
En (–) hairy rupturewort
Hu (2) borzas porcika
Pl (1) połonicznik kosmaty
Ru (1) грыжник волосистый
Sk (–) prietržník chlpatý

823 *Herniaria incana* Lam.
Bg (0) сиво изсипливче
Cs (0) průtržník šedivý
De (–) Graues Bruchkraut
Hu (1) szürke porcika
Pl (1) połonicznik siwy
Ro (0) săpunas
Ru (0) грыжник седоватый
Sk (–) prietržník šedivý

824 *Hesperis* spp. L.
Bg (0) вечерпик
De (–) Nachtviole
Gr (2) εσπερίδα
Ru (0) вечерница
Sk (–) večernica

825 *Hesperis matronalis* L.
Al (–) hesperis matronale
Bg (0) обикновен вечерник
Cs (0) večernice vonná
De (–) Gemeine Nachtviole

En (–) dame's-violet
Hu (0) hölgyestike
Pl (1) wieczornik damski
Ro (0) nopticoasă
Ru (0) ночная фиалка
Sk (–) večernica voňavá

826 ***Hesperis tristis*** L.
De (–) Trübe Nachtviole
Ru (0) вечерница мрачная
Sh (1) sitni jorgovančić
Sk (–) večernica smútočná

827 ***Heteracia szovitsii*** Fisch. & Meyer
Ru (1) гетерация Шовица

* ***Hibiscus ternatus*** Cav. → *Hibiscus trionum*

828 ***Hibiscus trionum*** L.
Al (–) bamja e qeve
Bg (1) либеничник
Cs (1) ibišek trojdílný
De (–) Gelbe Stundenblume
En (–) flower-of-an-hour
Gr (1) αγριοϊβίσκος
Hu (3) varjúmák
Pl (0) ketmia południowa
Ro (3) zămoşiţă
Ru (2) гибискус тройчатый
Sh (2) njivska lubeničarka; njivska ljubeničarka
Sk (–) ibištek trojdielny

* ***Hibiscus vesicarius*** Bess. → *Hibiscus trionum*

829 ***Hieracium* spp.** L.
Bg (0) рунянка
De (–) Habichtskraut
En (–) hawkweed
Gr (1) γεράκι
Hu (0) hölgymál
Pl (–) jastrzębiec
Ro (–) vulturică
Ru (1) ястребинка
Sk (–) jastrabník

* ***Hieracium bauhinii*** Bess. → *Hieracium praealtum*

830 ***Hieracium caespitosum*** Dumort.
Cs (0) jestřabník luční
De (–) Wiesen-Habichtskraut
En (–) yellow hawkweed
Hu (0) réti hölgymál
Pl (2) jastrzębiec łąkowy
Ru (1) ястребинка луговая
Sk (–) chlpánik trstnatý

831 ***Hieracium echioides*** Lumn.
Cs (0) jestřábník hadincovitý
De (–) Natternkopfblättriges Habichtskraut
Hu (0) szúrós hölgymál
Pl (1) jastrzębiec żmijowcowy
Ru (0) ястребинка румянковая
Sk (–) jastrabník hadincovitý

832 ***Hieracium pilosella*** L.
Al (–) këmashna me pak lesh
Bg (1) миши уши
Cs (1) jestřábník chlupáček
De (–) Mausohr-Habichtskraut
En (–) mouse-ear hawkweed
Hu (1) ezüstös hölgymál
Pl (2) jastrzębiec kosmaczek
Ro (0) vulturică
Ru (1) ястребинка волосистая
Sh (1) obična runjika
Sk (–) chlpánik lekársky

833 ***Hieracium praealtum*** Vill.
Al (–) këmashna me shumë flatra
Cs (0) jestřábník Bauhinův
Ru (0) ястребинка превысокая
Sh (1) Bauhinijeva runjika; vrežasta runjika
Sk (–) chlpánik vysoký

* ***Hieracium pratense*** Taus. → *Hieracium caespitosum*

834 *Hieracium racemosum* Waldst. & Kit.
Al (–) këmashna me vile
Cs (0) jestřábník hroznatý
De (–) Traubiges Habichtskraut
Hu (0) fürtös hölgymál
Pl (1) jastrzębiec gałęzisty
Ru (1) ястребника крупноватая
Sk (–) jastrabník hroznovitý

835 *Hieracium sabaudum* L.
Cs (0) jestřábník savojský
De (–) Savoyer Habichtskraut
Hu (0) olasz hölgymál
Pl (1) jastrzębiec sabaudzki
Ru (1) ястребинка блуждающая
Sk (–) jastrabník savojský

* ***Hieracium stoloniferum*** Waldst. & Kit. → *Hieracium pilosella*

836 *Hieracium umbellatum* L.
Cs (0) jestřábník okoličnatý
De (–) Dolden-Habichtskraut
En (–) leafy hawkweed
Hu (0) ernyős hölgymál
Pl (1) jastrzębiec baldaszkowy
Ro (1) iarba vulturului
Ru (1) ястребинка зонтичная
Sh (1) štitkasta runjika
Sk (–) jastrabník okolíkatý

837 *Hippocrepis* spp. L.
Al (–) hipokrepëza
Bg (0) подковниче
Cs (–) podkovka
De (–) Hufeisenklee
En (–) horseshoe-vetch
Gr (1) ιπποκρεπίδα
Hu (0) patkócím
Pl (–) konikleca
Ru (0) гиппокрепис
Sk (–) podchovka

838 *Hippuris vulgaris* L.
Al (–) bishtkali ujës
Bg (–) обикновена водна борика
Cs (0) prustka obecná
De (–) Gemeiner Tannenwedel
En (–) mare's-tail
Gr (1) ιππουρίδα
Hu (0) vízilófark
Pl (1) przęstka pospolita
Ro (0) coada calului
Ru (0) хвостник обыкновенный
Sh (1) obični borak
Sk (–) trúskavec obyčajný

839 *Hirschfeldia incana* (L.) Lag.
Cs (0) ředkevník huseniček
De (–) Weißhaariger Bastardsenf
En (–) hoary mustard
Gr (1) βρούβα
Ru (0) гиршфельдия серая
Sk (–) hirschfeldka šedá

840 *Holcus lanatus* L.
Al (–) belisha leshatake
Bg (1) медовница
Cs (3) medyněk vlnatý
De (–) Wolliges Honiggras
En (–) Yorkshire-fog
Gr (1) τριχωτός ολκός
Hu (1) pelyhes selyemperje
Pl (2) kłosówka wełnista
Ro (1) flocoşică
Ru (1) бухарник шерстистый
Sh (2) vunasta medunika; vunasta pahulja
Sk (–) medúnok vlnatý

841 *Holcus mollis* L.
Al (–) belisha
Bg (1) мека медовница
Cs (3) medyněk měkký
De (–) Weiches Honiggras
En (–) creeping soft-grass
Gr (1) μαλακός ολκός

Hu (0) lágy selyemperje
Pl (1) kłośowka miękka
Ru (1) бухарник мягкий
Sh (1) meka medunika; mekana medunika
Sk (–) medúnok mäkký

* **Holoschoenus romanus** (L.) Frit. → *Scirpus holoschoenus*

* **Holosteum glutinosum** (Bieb.) Fisch. & Meyer → *Holosteum umbellatum*

842 **Holosteum umbellatum** L.
Bg (1) плевул
Cs (1) plevel okoličnatý
De (–) Doldige Spurre
En (–) jagged chickweed
Gr (1) ολόστεο
Hu (1) olocsán
Pl (2) mokrzycznik baldaszkowy
Ro (2) cuişoriţă
Ru (1) костенец зонтичный
Sh (1) štitkasti plevulj; štitkasti plijevulj
Sk (–) pleveľ okolíkatý

843 **Hordelymus europaeus** (L.) Harz
Cs (0) ječmenka evropská
De (–) Wald-Haargerste
En (–) wood barley
Hu (0) hajperje
Pl (1) wydmuchrzyca zwyczajna
Ru (0) кювьера европейская
Sk (–) jačmienka európska

844 **Hordeum bulbosum** L.
Al (–) elbi me qepujkë
Bg (1) луковичен ечемик
En (–) bulbous barley
Gr (2) βολβοκρίθαρο
Ru (1) ячмень луковичный
Sk (–) jačmeň guľatý

* **Hordeum caput-medusae** (L.) Cosson & Durieu → *Taeniatherum caputmedusae*

* **Hordeum crinitum** (Schreb.) Desf. → *Taeniatherum caput-medusae*

* **Hordeum glaucum** Steud. → *Hordeum murinum* ssp. *murinum*

845 **Hordeum jubatum** L.
Ru (1) ячмень гривастый
Sk (–) jačmeň hrivnatý

* **Hordeum leporinum** Link → *Hordeum murinum* ssp. *leporinum*

846 **Hordeum marinum** Huds.
Al (–) elbi bregdetar
Bg (1) морски ечемик
Cs (0) ječmen přímořský
De (–) Strand-Gerste
En (–) sea barley
Ru (1) ячмень приморский
Sh (1) primorski ječam
Sk (–) jačmeň prímorský

* **Hordeum maritimum** Stok. → *Hordeum marinum*

847 **Hordeum murinum** L. ssp. **leporinum** (Link) Arc.
En (–) wild barley-grass
Hu (0) egérárpa
Ru (1) ячмень дикий
Sk (–) jačmeň zajačí

848 **Hordeum murinum** L. ssp. **murinum**
Al (–) elbi i egër
Bg (1) миши ечемик
Cs (2) ječmen myší
De (–) Mäuse-Gerste
En (–) wall barley

Gr (2) αγριοκρίθαρο
Hu (0) egérárpa
Pl (1) jęczmień płonny
Ro (1) orzul şoarecilor
Ru (1) ячмень мышиный
Sh (1) divlji ječam
Sk (–) jačmeň myší

849 *Hordeum secalinum* Schreb.
Al (–) elbi thekror
Cs (0) ječmen uzlatý
De (–) Roggen-Gerste
En (–) meadow barley
Pl (1) jęczmień żytni
Ru (0) ячмень ржаной
Sk (–) jačmeň žitný

* ***Horminium verticillatum*** Mill. → *Salvia verticillata*

850 *Hottonia palustris* L.
Bg (0) хотоня
Cs (0) žebratka bahenní
De (–) Sumpf-Wasserfeder
En (–) water-violet
Hu (0) békaliliom
Pl (1) okrężnica bagienna
Ru (0) турча болотная
Sk (–) perutník močiarny

851 *Humulus lupulus* L.
Al (–) sumbullbari
Bg (0) хмел
Cs (1) chmel otáčivý
De (–) Gemeiner Hopfen
En (–) wild hop
Gr (1) λυκίσκος
Hu (0) komló
Pl (1) chmiel zwyczajny
Ro (0) hamei
Ru (1) хмель обыкновенный
Sk (–) chmeľ obyčajný

852 *Hydrocharis morsus-ranae* L.
Al (–) lapagrethi
Bg (0) водянка
Cs (0) voďanka žabí
De (–) Froschbiß
En (–) frogbit
Hu (0) békatutaj
Pl (3) zabiściek pływający
Ro (1) iarba broaştelor
Ru (1) водокрас обыкновенный
Sh (1) beli žabogriz; bijeli žabogriz
Sk (–) vodnianka žabia

853 *Hyoscyamus* spp. L.
Bg (1) попадийка
De (–) Bilsenkraut
En (–) henbane
Gr (1) γέροντας
Ru (1) белена
Sk (–) blen

854 *Hyoscyamus albus* L.
Bg (1) бяла попадийка
Cs (0) blín bílý
De (–) Weißes Bilsenkraut
En (–) white henbane
Ro (1) măselariţă albă
Ru (1) белена белая
Sh (1) bela bunika; bijela bunika
Sk (–) blen biely

* ***Hyoscyamus bohemicus*** Schmidt → *Hyoscyamus niger*

855 *Hyoscyamus niger* L.
Al (–) matërgona e zezë
Bg (1) черна попадийка
Cs (2) blín černý
De (–) Schwarzes Bilsenkraut
En (0) common henbane
Hu (1) beléndek
Pl (1) lulek czarny
Ro (1) măselariţă
Ru (1) белена чёрная
Sh (1) obična bunika
Sk (–) blen čierny

856 ***Hyparrhenia hirta*** (L.) Stapf
Gr (2) υπαρένια
Sk (–) fúzatka srstnatá

* ***Hypecoum grandiflorum*** Benth. → *Hypecoum imberbe*

857 ***Hypecoum imberbe*** Sibth. & Sm.
Bg (1) едроцветен хипекоум
Gr (1) υπήκοο
Sk (–) čunica veľkokvetá

858 ***Hypecoum pendulum*** L.
De (–) Hängendes Gelbäugelchen
Ru (1) гипекоум вислоплодный
Sk (–) čunica visiaca

* ***Hypericum alpinum*** auct. → *Hypericum richeri*

* ***Hypericum crispum*** L. → *Hypericum triquetrifolium*

859 ***Hypericum hirsutum*** L.
Al (–) lulebasani qimeashpër
Bg (1) космат кантарион
Cs (0) třezalka chlupatá
De (–) Behaartes Johanniskraut
En (–) hairy Saint John's-wort
Hu (0) borzas orbáncfű
Pl (1) dziurawiec kosmaty
Ru (0) зверобой жестковолосый
Sk (–) ľubovník chlpatý

860 ***Hypericum humifusum*** L.
Cs (0) třezalka položená
De (–) Niederliegendes Johanniskraut
En (–) trailing Saint John's-wort
Hu (1) heverő orbáncfű
Pl (1) dziurawiec rozesłany
Ru (1) зверобой распростёртый
Sk (–) ľubovník rozprestretý

861 ***Hypericum maculatum*** Cran.
Al (–) lulebasani me njolla
Bg (0) четириръбест кантарион
Cs (0) třezalka skvrnitá
De (–) Geflecktes Johanniskraut
En (–) imperforate Saint John's-wort
Hu (0) pettyes orbáncfű
Pl (1) dziurawiec czteroboczny
Ru (0) зверобой четырёхгранный
Sk (–) ľubovník škvrnitý

862 ***Hypericum perforatum*** L.
Al (–) lulebasani
Bg (1) жълт кантарион
Cs (3) třezalka tečkovaná
De (–) Tüpfel-Johanniskraut
En (–) common Saint John's-wort
Gr (1) βάλσαμο
Hu (1) közönséges orbáncfű
Pl (2) dziurawiec zwyczajny
Ro (1) pojarniţă
Ru (1) зверобой продырявленный
Sh (1) obični kantarion; rupičasta pljuskavica
Sk (–) ľubovník bodkovaný

* ***Hypericum quadrangulum*** auct. → *Hypericum maculatum*

863 ***Hypericum richeri*** Vill.
Al (–) lulebasani i Rikerit
Bg (0) алпийски кантарион
De (–) Alpen-Johanniskraut
En (–) alpine Saint John's-wort
Pl (1) dziurawiec alpejski
Sk (–) ľubovník Richerov

864 ***Hypericum tetrapterum*** Fries
Al (–) lulebasani katërflatrash
Bg (0) крилат кантарион
Cs (0) třezalka čtyřkřídlá
De (–) Flügel-Johanniskraut
En (–) square-stalked Saint John's-wort
Hu (0) mocsári orbáncfű
Pl (1) dziurawiec skrzydełkowaty

Ru (0) зверобой острый
Sk (–) ľubovník štvorkrídly

865 *Hypericum triquetrifolium* Turra
Gr (2) αγούδουρας
Sk (–) ľubovník trojhranokvetý

* ***Hypericum veronense*** Schrank → *Hypericum perforatum*

866 *Hypochoeris* spp. L.
Bg (1) свиняк
De (–) Ferkelkraut
En (–) cat's-ear
Gr (1) υποχορίδα
Ru (1) пазник
Sk (–) prasatník

867 *Hypochoeris glabra* L.
Al (–) hipohera e shogët
Bg (1) гол свиняк
Cs (0) prasetník lysý
De (–) Kahles Ferkelkraut
En (–) smooth cat's-ear
Pl (1) prosienicznik gładki
Ru (1) пазник голый
Sh (1) glatki jastrebljak
Sk (–) prasatník holý

868 *Hypochoeris maculata* L.
Al (–) hipohera me njolla
Bg (1) петнист свиняк
Cs (0) náholník plamatý
De (–) Geflecktes Ferkelkraut
En (–) spotted cat's-ear
Hu (0) foltos véreslapu
Pl (1) prosienicznik plamisty
Ro (0) iarbă împuşcată
Ru (1) пазник крапчатый
Sh (1) pegavi jastrebljak; pjegavi jastrebljak
Sk (–) prasatník škvrnitý

869 *Hypochoeris radicata* L.
Al (–) hipohera me rrënjë
Bg (1) коренест свиняк
Cs (0) prasetník kořenatý
De (–) Gemeines Ferkelkraut
En (–) common cat's-ear
Hu (0) kacúros véreslapu
Pl (1) prosienicznik szorstki
Ro (1) buruiană porcească
Ru (1) пазник стержнекорневой
Sh (1) zrakasti jastrebljak
Sk (–) prasatník koreňatý

I

870 *Ilex aquifolium* L.
Al (–) ashja
Bg (0) самодивски чимшир
Cs (0) cesmina ostrolistá
De (–) Gewöhnliche Stechpalme
En (–) holly
Gr (1) λιόπρινο
Ro (0) laur
Ru (0) падуб остролистный
Sk (–) cesmína ostrolistá

871 *Illecebrum verticillatum* L.
Cs (0) nehtovec přeslenitý
De (–) Quirlige Knorpelblume
Pl (1) goździeniec okółkowy
Sk (–) nechtovec praslenatý

872 *Impatiens noli-tangere* L.
Bg (0) горска слабонога
Cs (1) netýkavka nedůtklivá
De (–) Großes Springkraut
En (–) touch-me-not balsam
Hu (0) erdei nenyúljhozzám
Pl (1) niecierpek pospolity
Ro (0) slăbănog
Ru (1) недотрога обыкновенная
Sk (–) netýkavka nedotklivá

873 *Impatiens parviflora* DC.
Cs (3) netýkavka malokvětá
De (–) Kleinblütiges Springkraut
En (–) small balsam
Hu (0) kisvirágú nenyúljhozzám
Pl (2) niecierpek drobnokwiatowy
Ru (1) недотрога мелкоцветковая
Sk (–) netýkavka drobnokvetá

874 *Imperata cylindrica* (L.) Rae.
Al (–) imperata cilindrike
De (–) Silberhaargras
En (–) cogon-grass
Gr (2) δεματόχορτο
Ru (1) аланг-аланг
Sk (–) lalang valcovitý

875 *Inula* spp. L.
Bg (1) оман
De (–) Alant
Gr (1) ινούλα
Ru (1) девясил
Sk (–) omam

876 *Inula britannica* L.
Al (–) omani i Britanisë
Bg (1) ливаден оман
Cs (0) oman britský
De (–) Wiesen-Alant
Hu (1) réti peremizs
Pl (1) oman łąkowy
Ro (1) sovîrvariţă
Ru (1) девясил британский
Sh (1) obični oman
Sk (–) omam britský

877 *Inula caspica* Blume
Ru (1) девясил каспийский
Sk (–) omam kaspický

* ***Inula dysenterica*** L. → *Pulicaria dysenterica*

878 *Inula ensifolia* L.
Bg (1) мечолистен оман
Cs (0) oman mečolistý
De (–) Schwert-Alant
En (–) narrow-leaved fleabane
Hu (0) kardos peremizs
Pl (1) omam wąskolistny
Ru (0) девясил мечелистный
Sk (–) omam mečolistý

879 *Inula germanica* L.
Al (–) omani i Gjermanisë
Bg (0) германски оман
Cs (0) oman německý

De (–) Deutscher Alant
Hu (0) hengeres peremizs
Pl (1) oman niemiecki
Ru (1) девясил германский
Sk (–) omam nemecký

* **Inula graveolens** (L.) Desf. → *Dittrichia graveolens*

880 ***Inula helenium*** L.
Al (–) omani
Bg (0) бял оман
Cs (0) oman pravý
De (–) Echter Alant
En (–) elecampane
Hu (0) örménygyökér
Pl (1) orman wielki
Ro (1) iarbă mare
Sh (1) pravi oman
Sk (–) omam lekársky

881 ***Inula hirta*** L.
Bg (0) космат оман
Cs (0) oman srstnatý
De (–) Behaarter Alant
En (–) hairy fleabane
Hu (–) borzas peremizs
Pl (1) oman szorstki
Ru (1) девясил жестковолосый
Sh (1) hrapavi oman; kosmati oman
Sk (–) omam srstnatý

882 ***Inula oculus-christi*** L.
Al (–) oman syri
Bg (0) очен оман
Cs (0) oman oko kristovo
De (–) Christus-Auge
Hu (1) selymes peremizs
Ru (0) девясил глазковый
Sk (–) omam oko Kristovo

883 ***Inula salicina*** L.
Cs (0) oman vrbolistý
De (–) Weidenblättriger Alant
En (–) Irish fleabane
Hu (0) fűzlevelű peremizs
Pl (1) oman wierzbolistny
Ru (1) девясил иволистный
Sh (1) vrbolisni oman
Sk (–) omam vŕbolistý

* **Inula semiamplexicaulis** Reut. → *Inula salicina*

* **Inula viscosa** (L.) Aiton → *Dittrichia viscosa*

884 ***Iris pseudacorus*** L.
Al (–) badra e ujit
Bg (0) батна перуника
Cs (0) kosatec žlutý
De (–) Sumpf-Schwertlilie
En (–) yellow iris
Hu (0) sárga nőszirom
Pl (2) kosaciec zolty
Ro (1) stînjenei de baltă
Ru (0) касатик жёлтый
Sh (1) močvarna perunika
Sk (–) kosatec žltý

885 ***Iris sibirica*** L.
Bg (0) сибирска перуника
Cs (1) kosatec sibiřský
De (–) Sibirische Schwertlilie
Hu (0) szibériai nőszirom
Pl (0) kosaciec syberyjski
Ru (0) касатик сибирский
Sk (–) kosatec sibírsky

886 ***Iris sintenisii*** Janka
Al (–) shpatorja e Sintenisit
Bg (1) синтенизова перуника

887 ***Isatis tinctoria*** L.
Al (–) isatisi ngjyrues
Bg (0) багрилна сърпица
Cs (0) boryt barvířský
De (–) Färber-Waid
En (–) woad
Hu (1) festő csülleng

Pl (0) urzet barwierski
Ro (0) drobuşor
Ru (0) вайда красильная
Sh (1) obični sač; sinji vrbovnik
Sk (–) boryt farbiarsky

* ***Isolepis setacea*** (L.) R Br. → *Scirpus setaceus*

888 *Iva xanthifolia* Nutt.
Cs (2) pouva řepňolistá
De (–) Schlagkraut
En (–) marsh-elder
Hu (1) íva
Ru (1) циклахена дурнишниколистная
Sh (1) obična iva
Sk (–) iva voškovníkovitá

J

889 *Jasione montana* L.
Bg (0) планинско вятърче
Cs (0) pavinec horský
De (–) Berg-Sandglöckchen
En (–) sheep's-bit scabious
Hu (0) kékcsillag
Pl (1) jasieniec piaskowy
Ru (0) букашник горный
Sh (1) brdski pavinac; obični prisadnik
Sk (–) pavinec horský

* ***Juncellus serotinus*** (Rottb.) Clarke → *Cyperus serotinus*

890 *Juncus* spp. L.
Bg (1) дзука
De (–) Binse
En (–) rush
Gr (2) βούρλο
Ru (2) ситник
Sk (–) sitina

891 *Juncus acutiflorus* Ehrh.
Al (–) kulmaku lulemprehtë
Bg (1) остроцветна дзука
Cs (0) sítina ostrokvětá
De (–) Spitzblütige Binse
En (–) sharp-flowered rush
Hu (1) fekete szittyó
Pl (0) sit ostrokwiatowy
Ro (0) pipirig de pădure
Ru (0) ситник остроцветковый
Sk (–) sitina ostrokvetá

* ***Juncus alpinoarticulatus*** Chaix → *Juncus articulatus*

892 *Juncus articulatus* L.
Al (–) kulmaku me nyje; zhuka me nyje
Bg (1) членеста дзука
Cs (0) sítina článkovaná
De (–) Glanzfrüchtige Binse
En (–) jointed rush
Hu (1) fülemüleszittyó
Pl (1) sit członowaty
Ru (–) ситник членистый
Sk (–) sitina článkovaná

893 *Juncus bufonius* L.
Al (–) kulmaku i kënetave; kulmaku thithlopë
Bg (1) клонеста дзука
Cs (3) sítina žabí
De (–) Kröten-Binse
En (–) toad rush
Hu (2) varangyszittyó
Pl (3) sit dwudzielny
Ro (1) iarba broaştei
Ru (1) ситник жабий
Sh (1) lisnata sita; samocvjetni sit
Sk (–) sitina žabia

894 *Juncus bulbosus* L.
Cs (0) sítina cibulkatá
De (–) Rasen-Binse
En (–) bulbous rush
Hu (1) sziki szittyó
Pl (2) sit drobny
Ru (1) ситник луковичный
Sk (–) sitina cibuľkatá pravá

895 *Juncus capitatus* Weigel
Bg (1) главеста дзука
Cs (0) sítina hlavatá
De (–) Kopf-Binse
En (–) dwarf rush
Hu (1) fejecskés szittyó
Pl (2) sit główkowaty
Ru (1) ситник головчатый
Sk (–) sitina hlavatá

896 *Juncus compressus* Jacq.
Al (–) kulmaku i rrasur
Bg (1) сплескана дзука
Cs (0) sítina smáčknutá

De (–) Zusammengedrückte Binse
En (–) round-headed rush
Hu (1) réti szittyó
Pl (1) sit ściśniony
Ru (1) ситник сплюснутый
Sk (–) sitina stisknutá

897 ***Juncus conglomeratus*** L.
Bg (1) сбита дзука
Cs (2) sítina klubkatá
De (–) Knäuel-Binse
En (–) compact rush
Hu (1) csomós szittyó
Pl (1) sit skupiony
Ru (0) ситник скученный
Sh (1) gronjasta sita; klupčasti sit
Sk (–) sitina klbková

898 ***Juncus effusus*** L.
Al (–) kulmaku i përhapur; zhuka e butë
Bg (1) разперена дзука
Cs (0) sítina rozkladitá
De (–) Flatter-Binse
En (–) soft rush
Hu (1) békaszittyó
Pl (1) sit rozpierzchły
Ro (1) spetează
Ru (1) ситник расходящийся
Sh (2) gola sita; rahlocvjetni sit
Sk (–) sitina rozprestretá

899 ***Juncus filiformis*** L.
Bg (1) нишковидна дзука
Cs (0) sítina niťovitá
De (–) Faden-Binse
En (–) thread rush
Pl (0) sit cienki
Ru (0) ситник нитевидный
Sk (–) sitina vláknitá

900 ***Juncus inflexus*** L.
Cs (0) sítina sivá
De (–) Graugrüne Binse
En (–) hard rush
Hu (1) deres szittyó
Pl (1) sit siny
Ro (1) pipirig
Ru (1) ситник склоняющийся
Sk (–) sitina sivá

* ***Juncus lampocarpus*** Ehrh. → *Juncus articulatus*

901 ***Juncus maritimus*** Lam.
Al (–) kulmaku bregdetar
Bg (1) морска дзука
De (–) Strand-Binse
En (–) sea rush
Hu (1) tengeri szittyó
Pl (0) sit morski
Ro (0) pipirig de mare
Ru (0) ситник морской
Sk (–) sitina pobrežná

902 ***Juncus sphaerocarpus*** Nees
Ru (1) ситник круглоплодный
Sk (–) sitina guľatoplodá

903 ***Juncus squarrosus*** L.
Cs (0) sítina kostrbatá
De (–) Sparrige Binse
En (–) heath rush
Pl (1) sit sztywny
Ru (0) ситник растопыренный
Sk (–) sitina kostrbatá

904 ***Juncus subnodulosus*** Schrank
Bg (0) възлеста дзука
Cs (0) sítina uzlíkatá
De (–) Stumpfblütige Binse
En (–) blunt-flowered rush
Hu (1) nagy szittyó
Pl (0) sit tępokwiatowy
Sk (–) sitina tupokvetá

905 ***Juncus tenageia*** L.
Bg (0) плитководна дзука
Cs (0) sítina rybničná

De (–) Sand-Binse
Hu (1) gömbös szittyó
Pl (1) sit błotny
Ru (1) ситник мелководный
Sk (–) sitina rybničná

906 *Juncus tenuis* Willd.
Ru (1) ситник тонкий
Sk (–) sitina tenká

907 *Juniperus communis* L.
Bg (1) обикновена хвойна
Cs (0) jalovec obecný
De (–) Gemeiner Wacholder
En (–) common juniper
Gr (1) κέδρος
Hu (0) közönséges boróka
Pl (1) jałowiec pospolity
Ro (1) ienupăr
Ru (1) можжевельник обыкновенный
Sk (–) borievka obyčajná

908 *Jurinea mollis* (L.) Reichenb.
Bg (0) обикновен миск
Cs (0) sinokvět měkký
Gr (1) γιουρινέα
Pl (0) jurinea miękka
Sk (–) sinokvet mäkký pravý

K

909 *Karelinia caspia* (Pallas) Less.
Ru (1) карелиния каспийская

910 *Kickxia elatine* (L.) Dumort.
Al (–) kicksia me gjethe bredhi
Bg (1) лош вятър
Cs (1) úporek hrálovitý
De (–) Echtes Tännelkraut
En (–) sharp-leaved fluellen
Hu (3) cseplesz tátika
Pl (1) lnica oszczepowata
Ro (1) avrămeasă
Ru (1) киксия настоящая
Sh (1) žuta kiksija; žuti lanilist
Sk (–) kiksia hralovitá

911 *Kickxia spuria* (L.) Dumort.
Al (–) kicksia e rreme
Bg (1) трескавиче
Cs (0) úporek nepravý
De (–) Unechtes Tännelkraut
En (–) round-leaved fluellen
Hu (3) kétszínű tátika
Pl (1) lnica zgiętoostrogowa
Ro (1) avrămeasă
Ru (1) киксия ложная
Sh (1) šarena kiksija
Sk (–) kiksia pochybná

912 *Knautia* spp. L.
Bg (1) кнауция
De (–) Witwenblume
Gr (1) κουφολάχανο
Ru (1) короставник
Sk (–) chrastavec

913 *Knautia arvensis* (L.) Coult.
Al (–) knauta e arës
Bg (1) полско краставче
Cs (2) chrastavec rolní
De (–) Acker-Witwenblume
En (–) field scabious
Hu (1) mezei varfű
Pl (1) świerzbnica polna
Ro (1) muşcatul dracului
Ru (1) короставник полевой
Sh (1) obična prženica; plava udovica
Sk (–) chrastavec roľný

914 *Knautia integrifolia* (L.) Bertol.
Al (–) knauta gjetheplotë
Bg (0) целолистна кнауция
Sh (1) celolisna prženica; cijelolisna prženica
Sk (–) chrastevec celolistý

915 *Knautia macedonica* Gris.
Al (–) knauta e Maqedonisë
Bg (1) македонска кнауция
Ru (0) короставник македонский
Sk (–) chrastevec macedónsky

916 *Kochia laniflora* (Gmel.) Borb.
Cs (0) bytel vlnokvětý
De (–) Sand-Radmelde
Hu (1) homoki seprőfű
Ru (0) кохия шерстистоцветная
Sk (–) kochia vlnatá

917 *Kochia scoparia* (L.) Schrad.
Al (–) micukthi si fshesë
Bg (1) дворна метла
Cs (1) bytel metlovitý
De (–) Besen-Radmelde
En (–) mock cypress
Gr (1) νεροκυπάρισσο
Hu (1) kerti seprőfű
Pl (0) mietelnik żakula
Ro (0) mături
Ru (1) прутняк веничный
Sh (1) obična metla
Sk (–) kochia metlovitá

918 *Koeleria* spp. Pers.
Bg (1) тънкокрак

De (–) Schillergras
Gr (1) κοελέρια
Ru (1) келерия
Sk (–) ometlina

* ***Koeleria gracilis*** Pers. → *Koeleria macrantha*

919 *Koeleria macrantha* (Ledeb.) Schultes
Al (–) këleria e hajthme
Bg (1) нежен тънкокрак
Cs (0) smělek štíhlý
En (–) crested hair-grass
Pl (1) strzęplica nadobna
Ru (1) келерия тонкая
Sk (–) ometlina štíhla pravá

920 *Koeleria pyramidata* (Lam.) Beauv.
Bg (0) пирамидален тънкокрак
Cs (1) smělek jehlancovitý
De (–) Pyramiden-Schillergras
En (–) pyramidal hair-grass
Pl (1) strzęplica piramidalna
Ru (0) келерия пирамидная
Sk (–) ometlina ihlancovitá

921 *Koelpinia linearis* Pallas
Ru (–) кельпиния линейнолистная

L

922 ***Lactuca altaica*** Fisch. & Meyer
Ru (–) латук алтайский

* ***Lactuca cretica*** Desf. → *Steptorhamphus tuberosus*

923 ***Lactuca saligna*** L.
Al (–) marule me gjethe shelgu
Bg (1) теснолистна салата
Cs (0) locika vrbovitá
De (–) Weiden-Lattich
En (–) Chinese lettuce
Gr (1) μακρόφυλλο αγριομάρουλο
Hu (2) szálas saláta
Pl (0) sałata solna
Ro (1) lăptuci
Ru (1) латук солончаковый
Sh (1) uskolisna salata; vrbolisna ločika
Sk (–) locika vŕbolistá

* ***Lactuca scariola*** L. → *Lactuca serriola*

924 ***Lactuca serriola*** L.
Al (–) marulë skariole; ogrishta
Bg (1) компасна салата
Cs (3) locika kompasová
De (–) Stachel-Lattich
En (–) prickly lettuce
Gr (1) αγριομάρουλο
Hu (2) keszeg saláta
Pl (1) sałata kompasowa
Ro (1) salată sălbatică
Ru (1) латук компасный
Sh (1) divlja salata
Sk (–) locika kompasová

925 ***Lactuca sibirica*** (L.) Maxim.
Ru (1) молокан сибирский
Sk (–) locika sibírska

926 ***Lactuca tatarica*** (L.) Meyer
De (–) Tataren-Lattich
Ru (1) латук татарский
Sk (–) locika tatárska

927 ***Lactuca viminea*** (L.) Presl
Bg (1) скална салата
Cs (0) locika prutnatá
De (–) Ruten-Lattich
Gr (1) αγνόφυλλο αγριομάρουλο
Sk (–) locika prútnatá

928 ***Lactuca virosa*** L.
Cs (0) locika jedovatá
De (–) Gift-Lattich
En (–) opium lettuce
Gr (1) τοξικό αγριομάρουλο
Pl (–) sałata jadowita
Ro (0) lăptucă veninoasă
Ru (0) молокан ядовитый
Sk (–) locika jedovatá

* ***Laelia orientalis*** Desv. → *Bunias orientalis*

929 ***Lagurus ovatus*** L.
Al (–) bishtlepuri vezak
Bg (0) овален лагурус
Cs (0) zaječí ocásek vejčitý
De (–) Sammetgras
En (–) hare's-tail
Gr (1) γατάκι
Ru (0) зайцехвостник яйцевидный
Sh (1) jajasta baršunka
Sk (–) zajačí chvost vajcovitý

930 ***Lamiastrum galeobdolon*** (L.) Ehrend. & Polat.
Al (–) hithëbuta Galeobdelon
Cs (1) hluchavka žlutá
De (–) Goldnessel
En (–) yellow dead-nettle
Hu (0) sárga árvacsalán
Pl (1) jasnota gajowiec
Ro (0) gălbeniţă
Sk (–) pitulník žltý

931 *Lamium album* L.
Bg (1) бяла мъртва коприва
Cs (3) hluchavka bílá
De (–) Weiße Taubnessel
En (–) white dead-nettle
Gr (1) λευκό λάμιο
Hu (1) fehér árvacsalán
Pl (2) jasnota biała
Ro (1) urzică moartă
Ru (1) яснотка белая
Sk (–) hluchavka biela

932 *Lamium amplexicaule* L.
Al (–) hithëbuta kërcellpushtuese
Bg (1) стъблообхватна мъртва коприва
Cs (3) hluchavka objímavá
De (–) Stengelumfassende Taubnessel
En (–) henbit dead-nettle
Gr (1) δωδεκάνθι
Hu (3) bársonyos árvacsalán
Pl (3) jasnota różowa
Ro (3) sugel
Ru (3) яснотка стеблеобъемлющая
Sh (1) njivska mrtva kopriva
Sk (–) hluchavka objímová

933 *Lamium bifidum* Cyr.
Gr (1) δίλοβο λάμιο
Sk (–) hluchavka dvojdomá

* ***Lamium galeobdolon*** (L.) L. → *Lamiastrum galeobdolon*

934 *Lamium hybridum* Vill.
De (–) Bastard-Taubnessel
En (–) cut-leaved dead-nettle
Ru (1) яснотка гибридная
Sk (–) hluchavka hybridná

935 *Lamium maculatum* L.
Al (–) hithëbuta me njolla
Cs (1) hluchavka skvrnitá
De (–) Gefleckte Taubnessel
En (–) spotted dead-nettle
Gr (1) στικτό λάμιο
Hu (0) foltos árvacsalán
Pl (1) jasnota plamista
Ro (0) urzică moartă
Ru (1) яснотка пятнистая
Sh (1) pegava mrtva kopriva; pjegava mrtva kopriva
Sk (–) hluchavka škvrnitá

936 *Lamium molluccellifolium* Fries
En (–) northern dead-nettle
Ru (1) яснотка молюцеллолистная

* ***Lamium paczoskianum*** Vorosch. → *Lamium amplexicaule*

937 *Lamium purpureum* L.
Al (–) hithëbuta e purpurtë
Bg (1) червена мъртва коприва
Cs (3) hluchavka nachová
De (–) Rote Taubnessel
En (–) red dead-nettle
Gr (1) πορφυρό λάμιο
Hu (2) piros árvacsalán
Pl (3) jasnota purpurowa
Ro (3) sugel puturos
Ru (3) яснотка пурпурная
Sh (3) crvena mrtva kopriva
Sk (–) hluchavka purpurová

* ***Lamium tetrahit*** Cran. → *Galeopsis tetrahit*

* ***Lappa communis*** Cosson & Germ. → *Arctium tomentosum*

* ***Lappa minor*** Hill → *Arctium minus*

938 *Lappula barbata* (Bieb.) Gürke
Ru (1) липучка бородчатая
Sk (–) lopuštík bradatý

* ***Lappula consanguinea*** (Fisch. & Meyer) Gürke → *Lappula squarrosa*

* **Lappula echinata** Frit. → *Lappula squarrosa*

939 **Lappula marginata** (Bieb.) Gürke
Hu (0) berzedt koldustetű
Ru (1) липучка пониклая
Sk (–) lopuštík rozložitý

* **Lappula myosotis** Moen. → *Lappula squarrosa*

* **Lappula patula** (Lehm.) Menyh. → *Lappula marginata*

940 **Lappula squarrosa** (Retz.) Dumort.
Bg (1) таралежовидни ушици
Cs (0) strošek pomněnkovitý
De (–) Kletten-Igelsame
En (–) bur forget-me-not
Hu (1) parlagi mizsót
Pl (1) lepnik zwyczajny
Ro (1) lipici
Ru (1) липучка ежевидная
Sh (1) obična čičkovica; obični čičkovac
Sk (–) lopuštík nezábudkový

941 **Lapsana communis** L.
Al (–) lapsana e rëndomtë
Bg (1) обикновен сърбун
Cs (3) kapustka obecna
De (–) Rainkohl
En (–) nipplewort
Gr (2) λαψάνα
Hu (1) bojtorjánsaláta
Pl (2) łoczyga pospolita
Ro (1) zgrăbunţică
Ru (1) бородавник обыкновенный
Sh (1) obična ognjičina; sitnoglavka ognjičica
Sk (–) kapustička obyčajná

942 **Laser trilobum** (L.) Borkh.
Cs (–) timoj trojlaločný
De (–) Dreilappiger Roßkummel
Ru (0) лазурник трёхлопастный
Sh (1) planinski gladaš
Sk (–) timoj trojlaločny

943 **Laserpitum latifolium** L.
Cs (0) hladýs širolistý
De (–) Breitblättriges Laserkraut
En (–) sermountain
Hu (0) széleslevelű bordamag
Pl (1) okrzyn szerokolistny
Ro (0) smeoaică
Ru (0) гладыш широколистный
Sh (1) širokolisni gladac
Sk (–) hladýš širokolistý

944 **Lathraea squamaria** L.
Bg (1) горска майка
Cs (1) podbílek šupinatý
De (–) Rötliche Schuppenwurz
En (–) toothwort
Gr (1) λαθραία
Hu (–) vicsorgó
Pl (1) łuskiewnik różowy
Ro (0) muma pădurii
Ru (1) Петров крест чешуйчатый
Sh (1) ljuskasta potajnica; obična potajnica
Sk (–) zubovník šupinatý

945 **Lathyrus spp.** L.
Bg (1) секирче
De (–) Platterbse
Gr (1) αγϱιολαθούϱι
Ru (1) чина
Sk (–) hrachor

946 **Lathyrus angulatus** L.
Bg (1) сферично секирче
De (–) Ziegelrote Kicher
Ru (0) чина угловатая
Sk (–) hrachor hranatý

947 **Lathyrus annuus** L.
Bg (1) едногодишно секирче

Cs (0) hrachor roční
En (–) annual yellow-vetchling
Ru (0) чина однолетняя
Sk (–) hrachor ročny

948 ***Lathyrus aphaca*** L.
Al (–) sygjarpëri
Bg (2) безлистно секирче
Cs (0) hrachor pačočkový
De (–) Ranken-Platterbse
En (–) yellow vetchling
Hu (1) levéltelen lednek
Ro (1) linte galbenă
Ru (1) чина безлисточковая
Sh (1) bezlisni grahor; obični grahor
Sk (–) hrachor bezlistý

* ***Lathyrus cicer*** Gaudin → *Lathyrus cicera*

949 ***Lathyrus cicera*** L.
Al (–) vingjra qiqër
Bg (1) нахутно секирче
Cs (0) hrachor cizrnovitý
De (–) Rote Platterbse
En (–) red vetchling
Hu (0) csicseri borsó
Ru (0) чина нутовая
Sk (–) hrachor cicerovitý

* ***Lathyrus coccineus*** All. → *Lathyrus sphaericus*

* ***Lathyrus erectus*** Lag. → *Lathyrus inconspicuus*

950 ***Lathyrus hirsutus*** L.
Al (–) vingjra qimeashpër
Bg (1) косматo секирче
Cs (0) hrachor chlupatý
De (–) Haarige Platterbse
En (–) hairy vetchling
Hu (1) borzas lednek
Pl (2) groszek kosmatostrąkowy
Ru (1) чина волосистая
Sh (1) maljavi grahor
Sk (–) hrachor chlpatý

* ***Lathyrus hirtus*** L. → *Lathyrus hirsutus*

* ***Lathyrus hispidus*** Boiss. → *Lathyrus inconspicuus*

951 ***Lathyrus inconspicuus*** L.
Bg (1) изправено секирче
Ru (0) чина незаметная
Sk (–) hrachor drobnokvetý

952 ***Lathyrus latifolius*** L.
Bg (1) широколистно секирче
Cs (1) hrachor lesní širokolistý
De (–) Breitblättrige Platterbse
En (–) everlasting pea
Hu (1) nagyvirágú lednek
Pl (0) groszek szerokolistny
Ro (1) mănerei de pădure
Ru (1) чина широколистная
Sh (1) veliki grahor
Sk (–) hrachor širokolistý

953 ***Lathyrus niger*** (L.) Bernh.
Cs (–) hrachor černý
De (–) Schwarzwerdende Platterbse
En (–) black bitter-vetch
Ru (0) чина чёрная
Sh (1) crni grahor
Sk (–) lecha čierna

954 ***Lathyrus nissolia*** L.
Al (–) vingjra e Nisolit
Bg (1) Нисолово секирче
Cs (0) hrachor Nissoliův
De (–) Gras-Platterbse
En (–) grass vetchling
Hu (1) kacstalan lednek
Pl (0) groszek liściakowy
Ru (0) чина Ниссолия

Sh (1) uskolisni grahor
Sk (–) hrachor Nissoliov

* **Lathyrus paluster** auct. → *Lathyrus palustris*

955 Lathyrus palustris L.
Al (–) vingjra e moçaleve
Bg (1) блатно секирче
Cs (0) hrachor bahenní
De (–) Sumpf-Platterbse
En (–) marsh pea
Hu (0) mocsári lednek
Pl (1) groszek błotny
Ru (0) чина болотная
Sk (–) hrachor močiarny

* **Lathyrus phylloideus** St.-L. → *Lathyrus nissolia*

956 Lathyrus pratensis L.
Al (–) vingjra e livadheve
Bg (1) ливадно секирче
Cs (3) hrachor luční
De (–) Wiesen-Platterbse
En (–) meadow vetchling
Hu (1) réti lednek
Pl (3) groszek żółty
Ro (1) lintea pratului
Ru (1) чина луговая
Sh (1) žuti grahor
Sk (–) hrachor lúčny

* **Lathyrus purpureus** Presl → *Lathyrus cicera*

957 Lathyrus sativus L.
Al (–) koçkulla
Bg (1) посевно секирче
Cs (0) hrachor setý
De (–) Saat-Platterbse
En (–) Indian pea
Hu (0) szegletes lednek
Pl (–) lędźwian siewny
Ro (0) măzăroi
Ru (0) чина посевная
Sh (1) poljski grahor
Sk (–) hrachor siaty

* **Lathyrus segetum** Lam. → *Lathyrus aphaca*

* **Lathyrus silvester** auct. → *Lathyrus sylvestris*

958 Lathyrus sphaericus Retz
Al (–) vingjra rruzullore
Bg (1) сферично секирче
De (–) Kugelige Platterbse
En (–) red hairy-vetchling
Hu (0) téglaszínű lednek
Ru (0) чина сферическая
Sh (1) submediteranski grahor
Sk (–) hrachor guľovitý

* **Lathyrus stans** Vis. → *Lathyrus cicera*

959 Lathyrus sylvestris L.
Al (–) vingjra e pyllit
Bg (0) горско секирче
Cs (1) hrachor lesní
De (–) Wald-Platterbse
En (–) flat pea
Hu (0) erdei lednek
Pl (1) groszek leśny
Ru (1) чина лесная
Sh (1) crveni grahor
Sk (–) hrachor lesný

960 Lathyrus tuberosus L.
Al (–) arashkë; vingjra me zhardhoke
Bg (2) грудкообразуващо секирче
Cs (3) hrachor hliznatý
De (–) Knollen-Platterbse
En (–) tuberous pea
Hu (3) mogyorós lednek
Pl (1) groszek bulwiasty
Ro (3) sîngele voinicului

Ru (2) чина клубневая
Sh (1) gomoljasti grahor
Sk (–) hrachor hľuznatý

* ***Lathyrus variegatus*** (Ten.) Gren. & Godr. → *Lathyrus venetus*

961 ***Lathyrus venetus*** (Mill.) Wohlf.
Al (–) vingjra e Venedikut
Bg (1) синьо секирче
Cs (0) hrachor benátský
De (0) Vielblütige Walderbse
Hu (1) tarka lednek
Ru (0) чина синеватая
Sk (–) lecha benátska

962 ***Lavatera arborea*** L.
De (–) Baumartige Strachpappel
En (0) tree mallow
Gr (1) δεντρώδης λαβατέρα
Sk (–) lavaterka stromovitá

963 ***Lavatera cretica*** L.
De (–) Strauchpappel
En (–) smaller tree-mallow
Gr (1) κρητική λαβατερα
Ru (0) хатьма критская
Sk (–) lavaterka krétska

964 ***Lavatera thuringiaca*** L.
Al (–) lavatera e Tyringjisë
Bg (1) тюрингска ружа
Cs (0) slézovec durinský
De (–) Thüringer Strauchpappel
Hu (1) madármályva
Pl (0) ślazówka turyngska
Ro (1) nalbă
Ru (1) хатьма тюрингенская
Sh (1) bledoružičasta lavatera; blijedoružičasta lavatera
Sk (–) lavaterka turynská

965 ***Ledum palustre*** L.
Al (–) rozmarina e kënetës
Cs (0) rojovník bahenní
De (–) Sumpf-Porst
En (–) Labrador-tea
Pl (1) bagno zwyczajne
Ru (0) багульник болотный
Sk (–) rojovník močiarny

966 ***Leersia oryzoides*** (L.) Swartz
Bg (1) оризов тросък
Cs (0) tajnička rýžovitá
De (–) Reisquecke
En (–) cut-grass
Hu (2) rizsfű
Pl (1) zamokrzyca ryżowa
Ro (3) orezică
Ru (2) леерсия рисовидная
Sh (1) divlja riža
Sk (–) tajnička ryžovitá

967 ***Legousia* spp.** Dura.
Bg (1) орниче
De (–) Frauenspiegel
Gr (1) λεγουζιά
Ru (1) легузия
Sk (–) zrkadlovka

968 ***Legousia falcata*** (Ten.) Frit.
De (–) Sichelförmiger Frauenspiegel
Sk (–) zrkadlovka kosákovitá

969 ***Legousia hybrida*** (L.) Delar.
Cs (0) zvonĕnka zvrhlá
De (–) Kleiner Frauenspiegel
En (–) Venus's looking-glass
Pl (1) zworitnica drobna
Ru (1) легузия гибридная
Sk (–) zrkadlovka malá

970 ***Legousia pentagonia*** (L.) Druce
Bg (1) голямо орниче
En (–) large Venus's looking-glass
Gr (1) πεντάγωνη λεγουζιά
Sk (–) zrkadlovka päťstranná

971 ***Legousia speculum-veneris*** (L.) Chaix
Al (–) leguzia pasqyrë
Bg (1) малко орниче
Cs (0) zvoněnka zrcadlovka
De (–) Gemeiner Frauenspiegel
En (–) greater Venus's looking-glass
Gr (2) αγριογιούλι
Hu (1) tükörvirág
Pl (0) zwrotnica zwierciadło Wenery
Ro (1) floarea Sfintei Vineri
Ru (0) легузия зеркало Венеры
Sh (1) obična zrcalica
Sk (–) zrkadlovka venušina

972 ***Lembotropis nigricans*** (L.) Gris.
Al (–) vjexhësa e zezëremë
Bg (0) черен зановец
Cs (1) čilimník černající
De (–) Schwärzender Geißklee
En (–) black broom
Hu (0) fürtös zanót
Pl (1) szczodrzeniec czerniejący
Ro (0) lemnul bobului
Ru (0) острокильница чернеющая
Sh (1) crna žučica
Sk (–) čilimníkovec černejúci

973 ***Lemna gibba*** L.
Bg (0) издута водна леща
Cs (0) okřehek hrbatý
De (–) Buckelige Wasserlinse
En (–) gibbous duckweed
Hu (1) púpos békalencse
Pl (0) rzęsa garbata
Ru (0) ряска горбатая
Sh (1) grbava sočivica
Sk (–) žaburinka hrbatá

974 ***Lemna minor*** L.
Al (–) manëroja më e vogël; qefil
Bg (1) малка водна леща
Cs (3) okřenek menší
De (–) Kleine Wasserlinse
En (–) common duckweed
Gr (1) νεροφακή
Hu (1) apró békalencse
Pl (3) rzęsa drobna
Ro (3) lintiţă
Ru (2) ряска маленькая
Sh (1) mala sočivica
Sk (–) žaburinka menšia

975 ***Lemna trisulca*** L.
Bg (1) тридолна водна леща
Cs (0) okřenek trojbrázdý
De (–) Dreifurchige Wasserlinse
En (–) ivy-leaved duckweed
Hu (1) keresztes békalencse
Pl (3) rzęsa trójrowkowa
Ro (1) lintiţă
Ru (1) ряска тройчатая
Sh (1) trouglasta sočivica
Sk (–) žaburinka trojbradá

976 ***Lens orientalis*** (Boiss.) Popov
Ru (1) чечевица восточная
Sk (–) šošovica východná

977 ***Leontice leontopetalum*** L.
Bg (1) земен хляб
En (–) rakaf
Gr (2) φούσκα

978 ***Leontodon* spp.** L.
Bg (1) жгълтица
De (–) Löwenzahn
En (–) hawkbit
Gr (1) παπαδούλα
Ru (0) кульбаба
Sk (–) púpavec

979 ***Leontodon autumnalis*** L.
Bg (1) есенна жълтица
Cs (–) pampeliška podzimní
De (–) Herbst-Löwenzahn
En (–) autumn hawkbit

Hu (0) őszi oroszlánfog
Pl (1) brodawnik jesienny
Ro (1) capul călugărului
Ru (1) кульбаба осенняя
Sh (1) lisnati lavlji zub
Sk (–) púpavec jesenny

980 *Leontodon hispidus* L.
Al (–) dhëmbëluani i grathët
Cs (2) pampeliška srstnatá
De (–) Rauher Löwenzahn
En (–) rough hawkbit
Hu (1) közönséges oroszlánfog
Pl (1) brodawnik zwyczajny
Ro (1) potcapul călugărului
Ru (1) кульбаба щетинистая
Sh (1) kratkodlakavi lavlji zub
Sk (–) púpavec srstnatý

981 *Leonurus cardiaca* L.
Al (–) bishtluani
Cs (2) buřina srdečník
De (–) Echtes Herzgespann
En (–) motherwort
Gr (1) πολυετές πυκνόκομο
Hu (1) szúrós gyöngyajak
Pl (1) śerdecznik pospolity
Ro (1) coada leului
Ru (1) пустырник обыкновенный
Sh (1) obična srčanica
Sk (–) srdcovník obyčajný

982 *Leonurus marrubiastrum* L.
Cs (0) buřina jablečníkovitá
De (–) Filziges Herzgespann
En (–) false motherwort
Gr (1) υδρόφιλο πυκνόκομο
Hu (1) pemete gyöngyajak
Ro (0) coada mîţei
Ru (1) гривохвост шандровый
Sk (–) srdcovník jabĺčnikovitý

* ***Leonurus quinquelobatus*** Usteri → *Leonurus cardiaca*

983 *Lepidium* spp. L.
Bg (1) горуха
De (–) Kresse
Gr (1) λεπίδιο
Ru (1) клоповник
Sk (–) žerucha

984 *Lepidium campestre* (L.) R Br.
Al (–) barbushi; djegësi i fushës
Bg (1) полска горуха
Cs (1) řeřicna ladní
De (–) Feld-Kresse
En (–) field pepperwort
Hu (2) mezei zsázsa
Pl (2) pieprzyca polna
Ro (1) hreniţă
Ru (1) клоповник полевой
Sh (1) poljska grbica; poljska renika
Sk (–) žerucha ladná

985 *Lepidium densiflorum* Schrad.
Cs (0) řeřicha hustokvětá
De (–) Dichtblütige Kresse
En (–) greenflower pepperwort
Hu (2) jövevény zsázsa
Pl (1) pieprzyca gęstokwiatowa
Ru (1) клоповник густоцветковый
Sk (–) žerucha hustokvetá

* ***Lepidium draba*** L. → *Cardaria draba*

986 *Lepidium graminifolium* L.
Al (–) djegësi gjethegrami
Bg (1) житнолистна горуха
Cs (0) řeřicha trávolistá
De (–) Grasblättrige Kresse
En (–) tall pepperwort
Hu (1) keskenylevelű zsázsa
Ru (0) клоповник злаколистный
Sk (–) žerucha trávolistá

987 *Lepidium latifolium* L.
Bg (1) широколистна горуха
Cs (0) řeřicha širolista

De (–) Breitblättrige Kresse
En (–) dittander
Pl (1) pieprzyca szerokolistna
Ro (0) hrenoasă
Ru (0) клоповник широколистный
Sk (–) žerucha širokolista

988 *Lepidium perfoliatum* L.
Al (–) djegësi gjethepushtues
Bg (1) прогаслолистна горуха
Cs (–) řeřicha prorostla
De (–) Durchwachsenblättrige Kresse
Hu (2) felemás zsázsa
Pl (1) pieprzyca przerosłolistna
Ru (1) клоповник пронзённолистный
Sh (1) ušasta renika
Sk (–) žerucha prerastaná

989 *Lepidium pinnatidifidium* Ledeb.
Ru (1) клоповник перистый

990 *Lepidium ruderale* L.
Bg (1) буренлива горуха
Cs (3) řeřicha rumni
De (–) Stink-Kresse
En (–) narrow-leaved pepperwort
Hu (1) büdös zsázsa
Pl (2) pieprzyca gruzowa
Ro (1) păduchernițã
Ru (1) клоповник мусорный
Sh (1) bezvjenčična grbica
Sk (–) žerucha rumná

991 *Lepidium sativum* L.
Bg (1) посевна горуха
Cs (0) řeřicha setá
De (–) Garten-Kresse
En (–) garden cress
Hu (0) kerti zsázsa
Pl (0) pieprzyca siewna
Ro (0) creson
Ru (0) клоповник посевной
Sk (–) žerucha siata

992 *Lepidium virginicum* L.
Cs (0) řeřicha virginská
De (–) Virginische Kresse
En (–) Virginia pepperwort
Hu (1) amerikai zsázsa
Pl (0) pieprzyca wirgińska
Ru (0) клоповник виргинский
Sk (–) žerucha virgínska

993 *Leptaleum filifolium* (Willd.) DC.
Ru (1) лепталеум нителистный

* ***Leptilon canadense*** Brit. →
Conyza canadensis

* ***Lepturus cylindricus*** (L.) Trin →
Hainardia cylindrica

994 *Leucanthemum vulgare* Lam.
Al (–) krizantema si margaritë
Bg (1) маргаритка
Cs (3) kopretina bílá
De (–) Wiesen-Wucherblume
En (–) oxeye daisy
Gr (1) λευκάνθεμο
Hu (1) réti margitvirág
Pl (2) złocień właściwy
Ro (1) mărgărită
Ru (1) нивяник обыкновенный
Sh (1) obična ivančica
Sk (–) králik biely

995 *Leucojum aestivum* L.
Al (–) bilbilbardha; pras gomari
Bg (0) блатно кокиче
Cs (0) bledule letní
De (–) Sommer-Knotenblume
En (–) summer snowflake
Gr (1) καλοκαιρινή χιονούλα
Hu (0) nyári tőzike
Ro (0) ghiocei de baltă
Ru (0) белоцветник летний
Sk (–) bleduľa letná

996 ***Leucojum autumnale*** L.
En (–) autumn snowflake
Gr (1) φθινοπωρινή χιονούλα
Sk (–) bleduľa jesenná

997 ***Leymus arenarius*** (L.) Hoch.
Al (–) elimusi ranor
Bg (1) пясъчен ечемик
Cs (0) ječmenice písečná
De (–) Strandroggen
En (–) sea lyme-grass
Pl (1) wydmuchrzyca piaskowa
Ro (1) perişor
Ru (1) колосняк песчаный
Sk (–) jačmenica piesočná

998 ***Leymus multicaulis*** (Kar. & Kir.) Tzvelev
Ru (1) колосняк многостебельный
Sk (–) jačmenica mnohobyľová

999 ***Leymus ramosus*** (Trin.) Tzvelev
Ru (2) острец ветвистый
Sk (–) jačmenica vetvistá

1000 ***Limosella aquatica*** L.
Cs (0) blatěnka vodní
De (–) Wasser-Schlammkraut
En (–) mudwort
Hu (1) iszaprojt
Pl (1) namulnik brzegowy
Ro (1) canarul bălţii
Ru (1) лужница водяная
Sh (1) obična voduška
Sk (–) blatnica vodná

1001 ***Linaria* spp.** Mill.
Bg (1) луличка
De (–) Leinkraut
En (–) toadflax
Gr (1) λινάρια
Ru (1) льнянка
Sk (–) pyštek

* ***Linaria acutiloba*** Fisch. → *Linaria vulgaris*

1002 ***Linaria angustissima*** (Loisel.) Borb.
Cs (0) lnice nejužší
De (–) Schmales Leinkraut
Hu (1) keskenylevelű gyújtoványfű
Sk (–) pyštek úzkolisty

1003 ***Linaria arvensis*** (L.) Desf.
Cs (2) lnice rolní
De (–) Acker-Leinkraut
En (–) corn toadflax
Hu (0) mezei gyújtoványfű
Pl (2) lnica polna
Ru (1) льнянка полевая
Sh (1) njivski lanilist
Sk (–) pyštek roľný

* ***Linaria dalmatica*** (L.) Mill. → *Linaria genistifolia*

* ***Linaria elatine*** (L.) Mill. → *Kickxia elatine*

1004 ***Linaria genistifolia*** (L.) Mill.
Bg (0) далматинска луличка
Cs (0) lnice kručinkolistá
De (–) Ginsterblättriges Leinkraut
Hu (0) rekettyelevelű gyújtoványfű
Pl (0) lnica janowcowa
Ru (1) льнянка дроколистная
Sh (1) žutiličasti lanilist
Sk (–) pyštek kručinkolistý

* ***Linaria italica*** Trev. → *Linaria angustissima*

* ***Linaria minor*** (L.) Desf. → *Chaenorhinum minus*

* ***Linaria spuria*** (L.) Mill → *Kickxia spuria*

1005 *Linaria vulgaris* Mill.
Al (–) peçeruku
Bg (1) обикновена луличка
Cs (2) lnice květel
De (–) Gemeines Leinkraut
En (–) common toadflax
Hu (3) közönséges gyújtoványfű
Pl (2) lnica pospolita
Ru (2) льнянка обыкновенная
Sh (2) obični lanilist
Sk (–) pyštek obyčajný

1006 *Lindernia procumbens* (Krock.) Phil.
Bg (1) делвоцветно линдерна
Cs (0) puštička pouzdernatá
De (–) Gemeines Büchsenkraut
En (–) creeping false-pimpernel
Hu (1) iszapfű
Pl (0) lindernia mułowa
Ru (1) линдерния лежачая
Sk (–) lindernia rozprestretá

* ***Lindernia pyxideria*** L. → *Lindernia procumbens*

1007 *Linum* spp.
Bg (1) лен
De (–) Lein
En (–) flax
Gr (1) αγριολίναρο
Ro (0) ineaţă
Ru (0) лён
Sk (–) ľan

1008 *Linum austriacum* L.
Al (–) liri austriak
Bg (1) австрийски лен
Cs (0) len rakouský
De (–) Österreichischer Lein
En (–) Austrian flax
Pl (0) len austriacki
Ru (0) лён австрийский
Sk (–) ľan rakúsky

1009 *Linum catharticum* L.
Al (–) liri dlirues
Bg (0) слабителен лен
Cs (0) len luční
De (–) Purgier-Lein
En (–) purging flax
Hu (1) békalen
Pl (1) len przeczyszczający
Ru (1) лён слабительный
Sh (1) čistilac lan; protulisni lan
Sk (–) ľan prečisťujúci

* ***Linum corymbosum*** Reichenb. → *Linum strictum*

* ***Linum gallicum*** L. → *Linum trigynum*

1010 *Linum hirsutum* L.
Al (–) liri qimeashpër
Bg (1) космат лен
Cs (0) len chlupatý
En (–) hairy flax
Hu (0) borzas len
Pl (1) len włochaty
Ro (0) in mare
Ru (0) лён жестковолосый
Sk (–) ľan chlpatý

1011 *Linum hologynum* Reichenb.
Sh (1) crvenkasti lan

1012 *Linum nodiflorum* L.
Al (–) liri me lule ndërnyje
Bg (1) възлоцветен лен
Ru (–) лён узловатый
Sk (–) ľan kolenkatý

1013 *Linum perenne* L.
Al (–) liri shumëvjeçar
Bg (1) многогодишен лен
Cs (0) len vytrvalý
De (–) Ausdauernder Lein
En (–) perennial flax

Hu (0) évelő len
Pl (0) len trwały
Ru (0) лён многолетний
Sk (–) ľan vytrvalý

1014 *Linum strictum* L.
Bg (1) щитовиден лен
De (–) Steifer Lein
En (–) upright yellow-flax
Ru (–) лён щиточковатый
Sk (–) ľan tuhý

1015 *Linum tenuifolium* L.
Al (–) liri gjethehollë
Bg (1) теснолистен лен
Cs (0) len tenkolistý
De (–) Schmalblättriger Lein
En (–) narrow-leaved flax
Hu (0) árvalevelű len
Ru (0) лён тонколистный
Sk (–) ľan tenkolistý

1016 *Linum trigynum* L.
Al (–) liri trigin
Bg (1) френски лен
Cs (–) len galský
De (–) Franzosen-Lein
Hu (0) francia len
Ru (0) лён трёхпестичный
Sk (–) ľan trojblizný

1017 *Lithospermum* spp. L.
Bg (0) врабчово семе
De (–) Steinsame
Gr (1) λιθόσπερμο
Ru (0) воробейник
Sk (–) kamienka

* ***Lithospermum arvense*** L. → *Buglossoides arvensis*

* ***Lithospermum incrassatum*** Guss. → *Buglossoides arvensis*

1018 *Lithospermum officinale* L.
Al (–) kokërruja mjekësore
Cs (0) kamejka lékařská
De (–) Echter Steinsame
En (–) common gromwell
Hu (0) kőmagvú gyöngyköles
Pl (1) nawrot lekarski
Ru (1) воробейник лекарственный
Sh (1) obično vrapseme
Sk (–) kamienka lekárska

1019 *Litwinowia tenuissima* (Pallas) Woronow
Ru (1) литвиновия тонкая

1020 *Logfia arvensis* (L.) Holub
Al (–) filago e arës
Bg (1) обикновена бяла свещица
Cs (1) bělolist rolní
De (–) Acker-Fadenkraut
En (–) least cudweed
Hu (1) gyapjas penészvirág
Pl (2) nicennica polna
Ro (1) flocoşele
Ru (1) жабник полевой
Sh (1) poljska maljavica; vunasti bjelolist
Sk (–) belolist roľný

1021 *Logfia minima* (Sm.) Dumort.
Bg (1) дребна свещица
Cs (0) bělolist nejmenší
De (–) Zwerg-Fadenkraut
En (–) small cudweed
Hu (1) hegyi penészvirág
Pl (2) nicennica drobna
Ru (1) жабник мелкий
Sk (–) belolist najmenší

* ***Lolium annuum*** Gilib. → *Lolium temulentum*

* **Lolium italicum** Braun → *Lolium multiflorum*

* **Lolium linicola** Braun → *Lolium remotum*

1022 **Lolium multiflorum** Lam.
Al (–) bartalioni; egjri shumëlulesh
Bg (2) италиански райграс
Cs (1) jílek mnohokvěty
De (–) Italienisches Raygras
En (–) Italian ryegrass
Gr (2) πολύανθη ήρα
Hu (1) olasz perje
Pl (2) życica wielokwiatowa
Ru (1) плевел многоцветковый
Sh (1) talijanski ljulj
Sk (–) mätonoh mnohokvetý

1023 **Lolium perenne** L.
Al (–) egjri shumëvjeçar
Bg (1) английски райграс
Cs (1) jílek vytrvalý
De (–) Deutsches Weidelgras
En (–) perennial ryegrass
Gr (1) πολυετής ήρα
Hu (2) angolperje
Pl (3) życica trwała
Ro (0) raigras englezesc
Ru (1) плевел многолетний
Sh (1) utrinac ljulj; višegodišnji ljulj
Sk (–) mätonoh trváci

1024 **Lolium remotum** Schrank
Al (–) egjri i lirit
Bg (2) ленена пиявица
Cs (2) jílek oddálený
De (–) Lein-Lolch
En (–) hardy ryegrass
Hu (1) lenvadóc
Pl (2) życica lnowa
Ro (1) sălbăţia inului
Ru (1) плевел льняной
Sk (–) mätonoh odstávajúci

1025 **Lolium rigidum** Gaudin
Bg (1) твърда глушица
Cs (0) jílek tuhy
De (–) Steifer Lolch
En (–) rigid ryegrass
Gr (2) λεπτή ήρα
Ru (1) плевел жёсткий
Sk (–) mätonoh tuhý

* **Lolium siculum** Parl. → *Lolium multiflorum*

* **Lolium strictum** Presl → *Lolium rigidum*

1026 **Lolium temulentum** L.
Al (–) egjri
Bg (1) лиявица
Cs (2) jílek mámivý
De (–) Taumel-Lolch
En (–) darnel
Gr (2) μεθυστική ήρα
Hu (1) szédítő vadóc
Pl (2) życica roczna
Ro (1) sălbăţie
Ru (1) плевел опьяняющий
Sh (1) pijani ljulj
Sk (–) mätonoh mámivý

1027 **Lophochloa cristata** (L.) Hyl.
Al (–) këleria si fleum
Bg (1) тимотейковиден тънкокрак
Cs (0) smělek bojínkovitý
Gr (1) λοφοχλόη
Ru (0) рострария гребенчатая
Sk (–) ometlina timotejkovitá

1028 **Lotus spp.** L.
Bg (1) звездан
De (–) Hornklee
En (–) birdsfoot trefoil
Gr (1) λωτός
Ru (1) лядвенец
Sk (–) ľadenec

1029 **Lotus angustissimus** L.
Al (–) thuepula shumë e ngushtë

Bg (1) дребен звездан
En (–) slender birdsfoot-trefoil
Hu (0) karcsú kerep
Ru (0) лядвенец узкий
Sk (–) ľadenec najužší

1030 ***Lotus corniculatus*** L.
Al (–) thuepula
Bg (1) звездан рогчест
Cs (1) štírovník růžkatý
De (–) Gemeiner Hornklee
En (–) common birdsfoot-trefoil
Hu (2) szarvas kerep
Pl (3) komonica rożkowa
Ro (0) ghizdei mărunt
Ru (1) лядвенец рогатый
Sh (1) obična smiljkita; žuti zvezdan
Sk (–) ľadenec rožkatý pravý

* ***Lotus pentaphyllus*** Gilib. → *Lotus corniculatus*

1031 ***Lotus tenuis*** Waldst. & Kit.
Al (–) thuepula gjetheimët
Bg (0) теснолистен звездан
De (–) Schmalblatt-Hornklee
En (–) narrow-leaved birdsfoot-trefoil
Hu (1) sziki kerep
Pl (0) komonica wąskolistna
Ru (1) лядвенец тонкий
Sk (–) ľadenec rožkatý tenkolistý

1032 ***Lotus uliginosus*** Schk.
Al (–) thuepula kënetore
Bg (0) блатен звездан
Cs (0) štírovník močálový
De (–) Sumpf-Hornklee
En (–) greater birdsfoot-trefoil
Hu (0) lápi kerep
Pl (3) komonica błotna
Ro (0) ghizdei mare
Ru (0) лядвенец топяной
Sh (1) močvarna smiljkita
Sk (–) ľadenec barinný

1033 ***Ludwigia palustris*** (L.) Ell.
Bg (1) блатна лудвигия
Cs (0) zakucelka bahenní
De (–) Sumpf-Heusenkraut
En (–) Hampshire purslane
Hu (0) mocsári tóalma
Ru (0) людвигия болотная
Sk (–) ludwigia močiarna

1034 ***Lupinus* spp.** L.
Bg (1) лупина
De (–) Lupine
En (–) lupin
Gr (1) λούπινο
Ru (0) люпин
Sk (–) vlčí bôb

1035 ***Lupinus angustifolius*** L.
Al (–) lupini i kaltër
Bg (1) теснолистна лупина
Cs (1) vlčí bob úzkolistý
De (–) Schmalblättrige Lupine
En (–) blue lupin
Hu (0) keskenylevelű csillagfürt
Pl (–) łubin wąskolistny
Ru (0) люпин узколистный
Sk (–) vlčí bôb úzkolistý

1036 ***Luzula campestris*** (L.) DC.
Al (–) luzula e fushës
Bg (1) полска светлика
Cs (2) bika ladní
De (–) Feld-Hainsimse
En (–) field wood-rush
Gr (1) λουζούλα
Hu (0) mezei perjeszittyó
Pl (1) kosmatka polna
Ro (0) mălaiul cucului
Ru (1) ожика слабоволосистая
Sh (1) obična bekica; poljska bekica
Sk (–) chlpaňa poľná

1037 ***Luzula multiflora*** (Retz.) Lej.
Al (–) luzula shumëlulshe

Bg (0) многоцветна светлика
Cs (0) bika mnohokvétá
De (–) Vielblütige Hainsimse
En (–) many-headed wood-rush
Pl (1) kosmatka licznokwiatowa
Ru (0) ожика многоцветковая
Sk (–) chlpaňa mnohokvetá

* **Lychnis dioica** L. → *Silene dioica*

1038 ***Lychnis flos-cuculi*** L.
Al (–) karafili i livadheve
Cs (1) kohoutek luční
De (–) Kuckucks-Lichtnelke
En (–) ragged-robin
Gr (1) κουκολούλουδο
Hu (1) réti kakukkszegfű
Pl (3) firletka poszarpana
Ro (0) floarea cucului
Ru (1) горицвет кукушкин
Sh (1) obična rumenika; rumena drijemina
Sk (–) kukučka lúčna

* **Lychnis githago** (L.) Scop. → *Agrostemma githago*

* **Lychnis vespertina** Sibth. → *Silene alba*

1039 ***Lychnis viscaria*** L.
De (–) Gemeine Pechnelke
Sh (1) obična lepica; obična ljepica
Sk (–) smolnička obyčajná

1040 ***Lycium barbarum*** L.
Cs (3) kustovnice cizí
De (–) Gemeiner Bocksdorn
En (–) Duke of Argyll's tea-plant
Hu (0) közönséges ördögcérna
Ro (1) cătină de garduri
Ru (1) дереза обыкновенная
Sh (1) obični licijum
Sk (–) kustovnica cudzia

1041 ***Lycium europaeum*** L.
Cs (0) kustovnice evropská
En (–) European boxthorn
Gr (1) αραμιά
Ru (1) дереза европейская
Sk (–) kustovnica európska

1042 ***Lycopodium clavatum*** L.
Al (–) likopodi
Bg (0) клиновиден плавун
Cs (0) plavuň vidlačka
De (–) Kolben-Bärlapp
En (–) clubmoss
Hu (0) kapcsos korpafű
Pl (1) widłak goździsty
Ro (0) pedicuţă
Ru (0) плаун булавовидный
Sk (–) plavúň vidličkovitý

* **Lycopsis arvensis** L. → *Anchusa arvensis*

* **Lycopsis orientalis** L. → *Anchusa arvensis*

1043 ***Lycopus europaeus*** L.
Al (–) këmbujku i Evropës
Bg (1) катушка
Cs (1) karbinec rolní
De (–) Gemeiner Wolfstrapp
En (–) gipsywort
Gr (1) λύκοπος
Hu (1) vízi peszérce
Pl (1) karbieniec pospolity
Ro (1) piciorul lupului
Ru (1) зюзник европейский
Sh (1) češljasta vučja noga; obična gagamija
Sk (–) karbinec európsky

1044 ***Lycopus exaltatus*** L.
Bg (1) висока катушка
Cs (0) karbinec statný
De (–) Hoher Wolfstrapp

Hu (1) magas peszérce
Pl (0) karbieniec wyniosły
Ro (1) ciorvană mare
Ru (1) зюзник высокий
Sh (1) cetvorna vučja noga
Sk (–) karbinec statný

* ***Lygia passerina*** (L.) Fas. →
Thymelaea passerina

1045 ***Lysimachia atropurpurea*** L.
Al (–) lisimaka purpur e errët
Bg (1) червено ленивче
Ru (1) вербейник тёмно-пурпурный
Sk (–) čerkáč nachový

1046 ***Lysimachia nummularia*** L.
Al (–) bargjergji
Bg (1) пълзящо ленивче
Cs (1) vrbina penízková
De (–) Pfennigkraut
En (–) creeping-jenny
Gr (1) πλαγιαστή λυσιμαχία
Hu (1) pénzlevelű lizinka
Pl (2) tojeść rozesłana
Ro (1) gălbăşoară
Ru (1) вербейник монетчатый
Sh (1) polegli protivak
Sk (–) čerkáč peniažtekovitý

1047 ***Lysimachia vulgaris*** L.
Al (–) lisimaka e zakonshme
Bg (1) обикновено ленивче
Cs (0) vrbina obecná
De (–) Gemeiner Gilbweiderich
En (–) yellow loosestrife
Gr (1) κοινή λυσιμαχία
Hu (1) mocsári lizinka
Pl (2) tojeść pospolita
Ru (1) вербейник обыкновенный
Sh (1) obični protivak
Sk (–) čerkáč obyčajný

1048 ***Lythrum* spp.** L.
Bg (1) блатия
De (–) Weiderich
Gr (1) λύθρο
Ru (0) дербенник
Sk (–) vrbica

1049 ***Lythrum hyssopifolia*** L.
Al (–) bargjaku me gjethe hisopi
Bg (1) хисополистна блатия
Cs (0) kyprej ysopolistý
De (–) Ysopblättriger Weiderich
En (–) grass-poly
Hu (1) alacsony füzény
Pl (2) krwawnica wąskolistna
Ru (1) дербенник иссополистный
Sk (–) vrbica ysopolistá

1050 ***Lythrum portula*** (L.) Webb
Bg (0) портуловиден пеплис
Cs (0) kalužník šruchový
De (–) Gemeiner Sumpfquendel
En (–) water purslane
Hu (1) tócsahúr
Pl (1) bebłek błotny
Ro (1) mătreaţă
Sh (1) lopatasti piličnjak
Sk (–) kalužník portulakový

1051 ***Lythrum salicaria*** L.
Al (–) bargjaku shelgjesh
Bg (1) върболистна блатия
Cs (2) kyprej vrbice
De (–) Blut-Weiderich
En (–) purple loosestrife
Hu (1) réti füzény
Pl (2) krwawnica pospolita
Ro (1) răchitan
Ru (1) дербенник иволистный
Sh (1) velika vrbica
Sk (–) vrbica vŕbolistá

* ***Lythrum tomentosum*** (L.) DC. →
Lythrum salicaria

1052 *Lythrum virgatum* L.
Bg (1) пръteстa блатия
Cs (0) kyprej prutnatý
De (–) Ruten-Weiderich
En (–) slender loosestrife
Hu (1) vesszős füzény
Pl (1) krwawnica rózgowata
Ru (1) дербенник прутовидный
Sh (1) šibasta vrbica; šibljasta vrbic
Sk (–) vrbica prutnatá

M

1053 ***Malva* spp.** L.
Bg (1) слез
De (–) Malve
En (–) mallow
Gr (1) μολόχα
Ru (1) просвирник
Sk (–) slez

1054 ***Malva alcea*** L.
Al (–) mëllagë gjethendarë
Bg (0) ружовиден слез
Cs (0) sléz léčivý
De (–) Sigmarskraut
En (–) large musk-mallow
Hu (1) érdes mályva
Pl (1) ślaz zygmarek
Ru (1) просвирник штокроза
Sk (–) slez veľkokvetý

* ***Malva ambigua*** Guss. → *Malva sylvestris*

* ***Malva crispa*** (L.) L. → *Malva verticillata*

* ***Malva decumbens*** Host → *Malva alcea*

* ***Malva italica*** Pollini → *Malva alcea*

* ***Malva mohileviensis*** Downar → *Malva verticillata*

* ***Malva montana*** Forsk. → *Malva verticillata*

1055 ***Malva neglecta*** Wallr.
Al (–) mëllaga e vogël
Bg (1) обикновен слез
Cs (2) sléz přehlížený
De (–) Weg-Malve
En (–) dwarf mallow
Hu (1) papsajtmályva
Pl (2) ślaz zaniedbany
Ro (1) nalbă măruntă
Ru (1) присвирник пренебрежный
Sh (1) mali slez; mali sljez
Sk (–) slez nebadaný

1056 ***Malva pusilla*** Sm.
Al (–) mëllaga e rëndomtë
Bg (1) дребен слез
Cs (0) sléz nizounký
De (–) Kleinblütige Malve
En (–) small mallow
Hu (1) apró mályva
Pl (2) ślaz drobnokwiatowy
Ro (1) nalbă mică
Ru (1) просвирник низкий
Sh (1) sitnolisni slez; sitnolisni sljez
Sk (–) slez nízunký

1057 ***Malva sylvestris*** L.
Al (–) mëllaga e pyllit
Bg (1) горски слез
Cs (0) sléz lesní
De (–) Wilde Malve
En (–) common mallow
Gr (2) μολόχα
Hu (1) erdei mályva
Pl (1) ślaz dziki
Ro (1) nalbă
Ru (1) просвирник лесной
Sh (1) crni slez; crni sljez
Sk (–) slez lesný

1058 ***Malva verticillata*** L.
Al (–) mëllaga qerthullake
Cs (0) sléz přeslenitý
De (–) Quirl-Malve
En (–) curled mallow
Hu (0) takarmánymályva
Pl (1) ślaz kędzierzawy
Ro (1) nalbă creaţă
Ru (1) просвирник мутовчатый
Sk (–) slez praslenitý

1059 *Marrubium peregrinum* L.
Al (–) bari i qenit
Bg (1) преселен пчелник
Cs (0) jablečník cizí
De (–) Ungarischer Andorn
En (–) branched horehound
Gr (1) σάρωμα
Hu (1) fehér pemetefű
Pl (0) szanta obca
Ro (0) cătuşnică sălbatică
Ru (1) шандра чужеземная
Sh (1) pustena macina trava; pusteni tetrljan
Sk (–) jablčník cudzí

1060 *Marrubium vulgare* L.
Al (–) maruba e rëndomtë
Bg (1) обикновен пчелник
Cs (0) jablečník obecný
De (–) Gemeiner Andorn
En (–) white horehound
Gr (1) σκουλόχορτο
Hu (1) orvosi pemetefű
Pl (1) szanta zwyczajna
Ro (1) unguraş
Ru (1) шандра обыкновенная
Sh (1) obična macina trava; obični tetrljan
Sk (–) jablčník obyčajný

1061 *Marsilea quadrifolia* L.
Al (–) marsilia katërgjethëshe
Bg (1) четирилистно разковниче
Cs (0) marsilka čtyřlistá
De (–) Vierblättriger Kleefarn
Hu (1) mételyfű
Pl (0) marsylia czterolistna
Ro (1) trifoiaş de baltă
Ru (1) марсилия четырёхлистная
Sk (–) marsilia štvorlistá

1062 *Matricaria* spp. L.
Bg (1) лайка
De (–) Kamille
En (–) mayweed
Gr (1) χαμομήλι
Ru (2) ромашка
Sk (–) rumanček

* ***Matricaria chamomilla*** L. → *Chamomilla recutita*

* ***Matricaria discoidea*** DC. → *Chamomilla suaveolens*

* ***Matricaria inodora*** L. → *Matricaria perforata*

1063 *Matricaria maritima* L.
Al (–) luleshqere bregdetare
Cs (3) heřmánkovec přímořský
De (–) Strand-Kamille
En (–) false mayweed
Hu (3) eb szikfű
Pl (1) maruna nadmorska
Ru (1) ромашка приморская
Sk (–) rumanček prímorský

* ***Matricaria matricarioides*** (Less.) Porter → *Chamomilla suaveolens*

1064 *Matricaria perforata* Mérat
Al (–) kamomili me erë të keqe
Bg (1) немиризлива лайка
Cs (3) heřmánkovec přímořský nevonný
De (–) Duftlose Kamille
En (–) scentless mayweed
Hu (3) eb szikfű
Pl (3) maruna bezwonna
Ro (3) muşeţel prost
Ru (3) ромашка непахучая
Sh (1) bezmirisna kamilica
Sk (–) parumanček prímorský nevoňavý

* ***Matricaria suaveolens*** (Pursh.) Buch → *Chamomilla suaveolens*

* ***Matricaria tenuifolia*** (Kit.) Sim. → *Matricaria trichophylla*

1065 ***Matricaria trichophylla*** (Boiss.) Boiss.
Bg (1) теснолистна лайка
De (–) Feinblättrige Kamille
Hu (1) bánsági szikfű
Sh (1) mali vratić
Sk (–) parumanček tvrdolistý

* ***Matthiola glandulosa*** Vis. → *Matthiola sinuata*

1066 ***Matthiola incana*** (L.) R Br.
Al (–) pllatka e përhimët
Bg (0) сива менекша
Cs (0) fiala šedivá
De (–) Garten-Levkoje
En (–) hairy stock
Gr (1) κοινή βιολέτα
Hu (0) kerti viola
Pl (0) lewkonia letnia
Ro (0) micsandră
Ru (0) левкой осенний
Sk (–) fiala šedivá

1067 ***Matthiola sinuata*** (L.) R Br.
Al (–) pllatka gjirake
En (–) sea stock
Gr (1) κολπωτή βιολέτα
Sk (–) fiala chobotnatá

1068 ***Medicago* spp.** L.
Bg (1) люцерна
De (–) Schneckenklee
Gr (1) μηδική
Ru (1) люцерна
Sk (–) tolica

* ***Medicago apiculata*** Willd. → *Medicago polymorpha*

1069 ***Medicago arabica*** (L.) Huds.
Al (–) jonxha me njolla
Bg (1) арабска люцерна
Cs (0) tolice arabská
De (–) Arabischer Schneckenklee
En (–) spotted medick
Hu (1) arab lucerna
Ru (1) люцерна арабская
Sh (1) pegava vija; pjegava vija
Sk (–) tolica arabská

1070 ***Medicago carstiensis*** Jacq.
Al (–) jonxha karstike
Bg (0) карстова люцерна
De (–) Karst-Schneckenklee
Sh (1) kraška dunjica

1071 ***Medicago coronata*** (L.) Bartal.
Bg (1) кореноплодна люцерна
Ru (1) люцерна увенчанная

* ***Medicago denticulata*** Willd. → *Medicago polymorpha*

1072 ***Medicago disciformis*** DC.
Bg (1) дископлодна люцерна
Sk (–) tolica terčoplodá

* ***Medicago falcata*** L. → *Medicago sativa* ssp. *falcata*

* ***Medicago gerardii*** Waldst. & Kit. → *Medicago rigidula*

* ***Medicago hispida*** Gaertn. → *Medicago polymorpha*

* ***Medicago lappacea*** Desr. → *Medicago polymopha*

1073 ***Medicago littoralis*** Roh.
Al (–) jonxha e brigjeve
Bg (1) крайбрежна люцерна
Ru (1) люцерна прибрежная

1074 ***Medicago lupulina*** L.
Al (–) jonxhë lupulini
Bg (1) хмелна люцерна
Cs (2) tolice dětelová

De (–) Hopfenklee
En (–) black medick
Hu (3) komlós lucerna
Pl (2) lucerna chmielowa
Ro (1) trifoi mărunt
Ru (1) люцерна хмелевидная
Sh (1) obična dunjica
Sk (–) tolica chmeľovitá

* **Medicago marginata** Willd. → *Medicago rigidula*

1075 Medicago minima (L.) Bartal.
Al (–) jonxha e vogël
Bg (1) дребна люцерна
Cs (1) tolice nejmenší
De (–) Zwerg-Schneckenklee
En (–) bur medick
Hu (1) apró lucerna
Pl (1) lucerna kolczastostrąkowa
Ru (1) люцерна малая
Sh (1) majušna dunjica
Sk (–) tolica najmenšia

* **Medicago mollissima** Roth → *Medicago minima*

* **Medicago nigra** (L.) Krock. → *Medicago polymorpha*

1076 Medicago orbicularis (L.) Bartal.
Al (–) jonxha e egër; jonxha rrethore
Bg (1) кръглоплодна люцерна
En (–) button medick
Hu (0) korongos lucerna
Ro (1) vîrteju pămîntului
Ru (1) люцерна округлая
Sh (1) zdjeličasta dunjica
Sk (–) tolica okrúhlolistá

* **Medicago parviflora** Gilib. → *Medicago lupulina*

1077 Medicago polymorpha L.
Al (–) jonxha e dhëmbëzuar
Bg (1) изменчива люцерна
Cs (0) tolice srstnatá
De (–) Rauher Schneckenklee
En (–) toothed medick
Ru (–) люцерна зубчатая
Sk (–) tolica čierna

* **Medicago procumbens** Bess. → *Medicago sativa* ssp. *falcata*

1078 Medicago rigidula (L.) All.
Al (–) jonxha pak e epshme
Bg (1) твърдоплодна люцерна
Cs (0) tolice přítuhá
De (–) Sammet-Schneckenklee
Hu (0) keménytövisű lucerna
Ru (0) люцерна жестковатая
Sk (–) tolica natvrdlá

1079 Medicago sativa L. **ssp. falcata** (L.) Arc.
Al (–) jonxha e verdhë
Bg (1) жълта люцерна
Cs (1) tolice srpovitá
De (–) Sichelklee
En (–) sickle medick
Hu (0) sárkerep-lucerna
Pl (2) lucerna sierpowata
Ro (0) lucernă galbenă
Ru (1) люцерна синяя
Sh (1) srpasta dunjica
Sk (–) tolica kosákovitá

1080 Medicago sativa L. **ssp. sativa**
Al (–) jonxha
Bg (1) обикновена люцерна
Cs (1) tolice vojtěška
De (–) Luzerne
En (–) lucerne
Hu (0) takarmánylucerna
Pl (1) lucerna siewna
Ro (0) lucernă

Ru (1) люцерна посевная
Sk (–) lucerna siata

1081 ***Melampyrum* spp.** L.
Bg (1) гайтанка
De (–) Wachtelweizen
En (–) cow-wheat
Gr (1) μελάμπυρο
Ru (1) марьянник
Sk (–) čermela

1082 ***Melampyrum arvense*** L.
Al (–) grurëziu i arave; lulefruthi
Bg (1) полска гайтанка
Cs (2) černýš rolní
De (–) Acker-Wachtelweizen
En (–) field cow-wheat
Hu (2) mezei csormolya
Pl (1) pszeniec polny
Ro (2) miază-noapte
Ru (1) марьянник полевой
Sh (1) njivska urodica
Sk (–) čermela roľná

1083 ***Melampyrum barbatum*** Waldst. & Kit.
Cs (–) černýš vousatý
De (–) Bart-Wachtelweizen
En (–) bearded cow-wheat
Hu (3) szakállas csormolya
Ru (1) марьянник бородчатый
Sh (1) dlakava urodica
Sk (–) čermela bradatá

1084 ***Melampyrum cristatum*** L.
Al (–) grurëlopë kreshtak; grurëziu me kreshtë
Cs (0) černýš hřebenitý
De (–) Kamm-Wachtelweizen
En (–) crested cow-wheat
Hu (0) taréjos csormolya
Pl (1) pszeniec grzebieniasty
Ro (1) ciormoiag
Ru (1) марьянник гребенчатый
Sk (–) čermela hrebenitá

1085 ***Melampyrum pratense*** L.
Al (–) grurëlopa e livadheve
Cs (0) černýš luční
De (–) Wiesen-Wachtelweizen
En (–) common cow-wheat
Hu (0) réti csormolya
Pl (1) pszeniec zwyczajny
Ru (1) марьянник голубой
Sh (1) livadska urodica
Sk (–) čermela lúčna

* ***Melandrium album*** (Mill.) Gar. → *Silene alba*

* ***Melandrium dioicum*** (L.) Cosson & Germ. → *Silene dioica*

* ***Melandrium diurnum*** (Sibth. & Sm.) Fries → *Silene dioica*

* ***Melandrium noctiflorum*** (L.) Fries → *Silene noctiflora*

* ***Melandrium pratense*** Rohl. → *Silene alba*

* ***Melandrium rubrum*** (Weigel) Gar. → *Silene dioica*

* ***Melandrium sylvestre*** (Schk.) Röhl. → *Silene dioica*

1086 ***Melica ciliata*** L.
Cs (–) strdivka brvitá
De (–) Wimper-Perlgras
En (–) hairy melick
Gr (1) βλεφαριδωτή μελίκη
Ru (1) перловник реснитчатый
Sk (–) mednička brvitá

1087 ***Melica nutans*** L.
Al (–) bjelisha lavjerëse; melika lavjërese
Bg (1) наведена бисерка

Cs (0) strdivka nící
De (–) Nickendes Perlgras
En (–) mountain melick
Hu (0) bókoló gyöngyperje
Pl (2) perłówka zwisła
Ru (0) перловник поникший
Sk (–) medníčka ovisnutá

1088 ***Melica uniflora*** Retz.
Al (–) bjelisha me një lule
Bg (1) едноцветна бисерка
Cs (0) strdivka jednokvětá
De (–) Einblütiges Perlgras
En (–) wood melick
Gr (1) μονή μελίκη
Hu (0) egyvirágú gyöngyperje
Pl (1) perłówka jednokwiatowa
Ro (0) mărgică
Ru (1) перловник одноцветковый
Sk (–) medníčka jednokvetá

1089 ***Melilotus* spp.** L.
Bg (1) комунига
De (–) Steinklee
Gr (1) μελίλωτος
Ru (1) донник
Sk (–) komonica

1090 ***Melilotus alba*** Med.
Al (–) makthi i bardhë
Bg (1) бяла комунига
Cs (1) komonice bílá
De (–) Weißer Steinklee
En (–) white melilot
Hu (2) fehér somkóró
Pl (3) nostrzyk biały
Ro (1) sulfină albă
Ru (1) донник белый
Sh (1) beli kokotac; bijeli kokotac
Sk (–) komonica biela

1091 ***Melilotus altissima*** Thuill.
Al (–) makthi shumë i lartë
Cs (0) komonice nejvyšší
De (–) Sumpf-Steinklee
En (–) tall melilot
Hu (0) réti somkóró
Pl (1) nostrzyk wyniosły
Ru (0) донник рослый
Sk (–) komonica najvyššia

* ***Melilotus arvensis*** Wallr. → *Melilotus officinalis*

* ***Melilotus caerulea*** (L.) Desr. → *Trigonella caerulea*

1092 ***Melilotus indica*** (L.) All.
Al (–) makthi indian; makthi lulëvogël
Bg (0) индийска комунига
Cs (0) komonice indická
De (–) Kleinblütiger Steinklee
En (–) small melilot
Ru (1) донник индийский
Sk (–) komonica indická

* ***Melilotus macrorrhiza*** auct. → *Melilotus altissima*

1093 ***Melilotus officinalis*** (L.) Pallas
Al (–) makthi mjekësor
Bg (1) жълта комунига
Cs (2) komonice lékařská
De (–) Echter Steinklee
En (–) ribbed melilot
Hu (2) orvosi somkóró
Pl (3) nostrzyk żółty
Ro (1) sulfină galbenă
Ru (1) донник лекарственный
Sh (1) žuti kokotac
Sk (–) komonica lekárska

* ***Melilotus rugosus*** Gilib. → *Melilotus officinalis*

1094 ***Mentha* spp.** L.
Bg (1) мента
De (–) Minze
En (–) mint

Gr (1) μέντα
Ru (–) мята
Sk (–) mäta

1095 ***Mentha aquatica*** L.
Al (–) mendra e ujit
Bg (1) водна мента
Cs (0) máta vodní
De (–) Wasser-Minze
En (–) water mint
Hu (–) vízi menta
Pl (1) mięta wodna
Ro (1) izma broaştei
Ru (1) мята водяная
Sh (1) vodena metvica; vodena nana
Sk (–) mäta vodná

1096 ***Mentha arvensis*** L.
Al (–) mendra e fushës
Bg (1) полска мента
Cs (3) máta rolní
De (–) Acker-Minze
En (–) corn mint
Hu (2) mezei menta
Pl (3) mięta polna
Ro (2) izma cerbului
Ru (2) мята полевая
Sh (3) poljska metvica; poljska nana
Sk (–) mäta roľná

* ***Mentha candicans*** Cran. → *Mentha longifolia*

* ***Mentha incana*** Willd. → *Mentha longifolia*

1097 ***Mentha longifolia*** (L.) Huds.
Al (–) mendra gjethegjatë
Bg (1) дълголистна мента
Cs (0) máta dlouholistá
De (–) Roß-Minze
En (–) horse mint
Hu (2) ló menta
Pl (2) mięta długolistna
Ro (1) izmă proastă
Ru (1) мята длиннолистная
Sh (1) dugolisna metvica; dugolisna nana
Sk (–) mäta dlholistá

1098 ***Mentha pulegium*** L.
Al (–) mendër kënete; mendra e egër
Bg (1) приятна мента
Cs (1) máta polej
De (–) Polei-Minze
En (–) pennyroyal mint
Hu (1) csombordmenta
Pl (1) mięta polej
Ro (1) busuiocul cerbilor
Ru (1) мята блошница
Sh (1) konjska nana; poljačak metvica
Sk (–) polej

1099 ***Mentha spicata*** L.
Cs (0) máta klasnatá
De (–) Grüne Minze
En (–) spearmint
Hu (2) ló menta
Pl (0) mięta zielona
Ru (0) мята колосовая
Sk (–) mäta klasnatá

* ***Mentha sylvestris*** L. → *Mentha longifolia*

1100 ***Mentha verticillata*** L.
Bg (1) прешленеста мента
Cs (0) máta vodní
Hu (0) örvös menta
Pl (1) mięta okręgowa
Ru (0) мята мутовчатая
Sk (–) mäta praslenatá

1101 ***Menyanthes trifoliata*** L.
Al (–) tërfili i kënetave
Bg (1) водна детелина
Cs (0) vachta trojlistá
De (–) Sumpf-Fieber

En (–) bogbean
Hu (0) vidrafű
Pl (1) bobrek trójlistkowy
Ro (0) trifoişte
Ru (1) вахта трёхлистная
Sh (1) dugostapna trolistica; gorska grčica
Sk (–) vachta trojlistá

1102 ***Mercurialis annua*** L.
Al (–) bar thëllënëze njëvjeçar; merkuriali njëvjeçar
Bg (1) едногодишна живолина
Cs (3) bažanka roční
De (–) Einjähriges Bingelkraut
En (–) annual mercury
Gr (2) σκαρολάχανο
Hu (3) egyéves szélfű
Pl (2) szczyr roczny
Ro (1) trepădătoare
Ru (2) пролесник однолетний
Sh (1) jednogodišnji prosinac
Sk (–) bažanka ročná

1103 ***Mercurialis perennis*** L.
Al (–) bar thëllënëze shumëvjeçar; merkuriali shumëvjeçar
Bg (1) многогодишна живолина
Cs (2) bažanka vytrvalá
De (–) Wald-Bingelkraut
En (–) dog's mercury
Hu (0) erdei szélfű
Pl (1) szczyr trwały
Ro (0) brei
Ru (1) пролесник многолетний
Sk (–) bažanka trváca

* ***Micropus erectus*** L. → *Bombycilaena erecta*

1104 ***Milium effusum*** L.
Al (–) meli i përhapur
Bg (1) разперено горско просо
Cs (0) pšeníčko rozkladité
De (–) Weiches Flattergras
En (–) wood millet
Hu (0) kásafű
Pl (1) prosownica rozpierzchła
Ro (0) meişor
Ru (0) бор развесистый
Sh (1) obična prosulja
Sk (–) pšeno rozložité

1105 ***Milium vernale*** Bieb.
Al (–) meli pranveror
Bg (1) пролетно горско просо
De (–) Rauhes Flattergras
En (–) spring millet
Gr (2) μίλιο
Ru (1) бор весенний
Sh (1) divlje proso
Sk (–) pšeno jarné

1106 ***Minuartia* spp.** L.
Bg (1) мишовка
De (–) Miere
En (–) sandwort
Gr (1) μινουάρτια
Ru (0) минуарция
Sh (–) mišjakinjica
Sk (–) kurička

1107 ***Minuartia viscosa*** (Schreb.) Schinz & Thell.
Cs (0) kuřička lepivá
De (–) Klebrige Minze
En (–) sticky sandwort
Hu (0) enyves kőhúr
Pl (1) mokrzyca lepka
Ru (0) минуарция липкая
Sk (–) kurička lepkavá

1108 ***Misopates orontium*** (L.) Rafin.
Bg (1) обикновена зейка
Cs (1) šklebivec menší
De (–) Acker-Löwenmaul
En (–) lesser snapdragon
Hu (1) vetési oroszlánszáj

Pl (2) wilżyn polny
Ru (1) мизопатес горный
Sh (1) divlja zevalica; divlja zijevalica
Sk (–) škľabinec menší

1109 ***Moehringia muscosa*** L.
Al (–) mehringia myshkore
Bg (1) мъховидна кутявка
Cs (0) mateřka mechovitá
De (–) Moos-Nabelmiere
En (–) mossy sandwort
Hu (0) mohos csitri
Sk (–) materka machovitá

1110 ***Moehringia trinervia*** (L.) Clairv.
Al (–) mehringia trinervore
Bg (1) трижилкова кутявка
Cs (0) mateřka trojžilná
De (–) Dreinervige Nabelmiere
En (–) three-nerved sandwort
Hu (0) erdei csitri
Pl (1) możylinek trójnerwowy
Ro (0) merinană
Ru (0) мерингия трёхжилковая
Sh (1) trožilna merinka
Sk (–) materka trojžilná

1111 ***Moenchia mantica*** (L.) Bartl.
Al (–) mënka si trajsë
Bg (1) поревка
Cs (0) bělička mantická
De (–) Weißmiere
Hu (2) rigószegfű
Pl (0) menchia mantejska
Sh (1) obična mašanka
Sk (–) moenchia južná

1112 ***Molinia caerulea*** (L.) Moen.
Al (–) molinea e zhveshur
Bg (1) синкава молиния
Cs (1) bezkolenec modry
De (–) Pfeifengras
En (–) purple moor-grass
Hu (1) kékperje
Pl (2) trzęślica modra
Ro (0) iarbă albastră
Ru (0) молиния голубая
Sk (–) bezkolenec belasý

* ***Molinia litoralis*** Host → *Molinia caerulea*

1113 ***Monochoria korsakowii*** Reg. & Maack
Ru (1) монохория Корсакова

1114 ***Montia fontana*** L.
Cs (0) zdrojovka prameništní
De (–) Bach-Quellkraut
En (–) blinks
Gr (1) μόντια
Hu (0) forrásfű
Ru (0) монция блестящесемянная
Sk (–) zdrojovka pramenistá

1115 ***Moricandia arvensis*** (L.) DC.
Bg (3) полски синат
De (–) Acker-Moricandie
En (–) violet cabbage

* ***Mulgedium tataricum*** (L.) DC. → *Lactuca tatarica*

1116 ***Muscari* spp.** Mill.
Bg (1) кукувиче грозде
De (–) Träubelhyazinthe, Traubenhyazinthe
En (–) grape-hyacinth
Gr (1) βορβός
Pl (–) szafirek
Ro (1) ceapa ciorii
Ru (1) гадючий лук
Sk (–) modrinec

1117 ***Muscari botryoides*** (L.) Mill.
Al (–) pështriku si vile; rrush qyqe
Bg (1) гроздовидно кукувиче грозде
Cs (0) modřenec širolistý
De (–) Kleine Träubelhyazinthe

En (–) small grape-hyacinth
Hu (0) epergyöngyike
Pl (0) szafirek drobnokwiatowy
Ro (0) porumbei
Ru (1) гадючий лук гроздиевидный
Sk (–) modrinec širokolistý

1118 *Muscari comosum* (L.) Mill.
Al (–) pështriku me baluke
Bg (1) качулато кукувиче грозде
Cs (0) modřenec chocholatý
De (–) Schopfige Träubelhyazinthe
En (–) tassel-hyacinth
Hu (–) üstökös gyöngyike
Pl (1) szafirek miękkolistny
Ro (1) ceapa ciorii
Ru (1) гадючий лук хохолковый
Sh (1) obična presličica
Sk (–) modrinec chocholatý

1119 *Muscari neglectum* Guss.
Al (–) pështriku me vile
Bg (1) кичуресто кукувиче грозде
Cs (0) modřenec hroznatý
De (–) Übersehene Träubelhyazinthe
Gr (2) βορβός
Hu (0) fürtös gyöngyike
Pl (0) szafirek groniasty
Ro (1) porumbei
Ru (1) мускари незамеченный
Sh (1) grozdasta presličica
Sk (–) modrinec zanedbaný

* ***Muscari racemosum*** (L.) Lam. & DC. → *Muscari neglectum*

1120 *Muscari tenuiflorum* Taus.
Ru (1) мускари тонкоцветковый
Sk (–) modrinec tenkokvetý

* ***Muscari tubiflorum*** Stev. → *Muscari comosum*

* ***Myagrum amplexicaule*** Moen. → *Myagrum perfoliatum*

* ***Myagrum perenne*** L. → *Rapistrum perenne*

1121 *Myagrum perfoliatum* L.
Al (–) miagrumi me gjethepushtuese
Bg (1) шупливка
Cs (0) povázka prorostlá
De (–) Pfeilblättriger Hohldotter
En (–) mitre cress
Gr (1) μύαγρο
Hu (2) légyfogó
Ro (1) golicică
Ru (1) полевка пронзённолистная
Sh (1) obična buzdika
Sk (–) myšojed prerastaný

1122 *Mycelis muralis* (L.) Dumort.
Cs (1) mléčka zední
De (–) Mauerlattich
En (–) wall lettuce
Hu (0) kakicsvirág
Pl (1) sałatnik leśny
Ro (0) susai pădureţ
Ru (0) мицелис стенной
Sk (–) mliečka múrová

1123 *Myosotis* spp. L.
Bg (1) незабравка
De (–) Vergißmeinnicht
En (–) forget-me-not
Gr (1) μυοσωτίδα
Ro (1) nu-mă-uita
Ru (1) незабудка
Sk (–) nezábudka

* ***Myosotis arenaria*** Schrad. → *Myosotis stricta*

1124 *Myosotis arvensis* (L.) Hill
Al (–) lulemiza e arës
Bg (1) полска незабравка
Cs (3) pomněnka rolní
De (–) Acker-Vergißmeinnicht
En (–) field forget-me-not

Hu (3) parlagi nefelejcs
Pl (3) niezapominajka polna
Ru (2) незабудка полевая
Sh (1) sitnocvetni spomenak; sitnocvjetni spomenak
Sk (–) nezábudka roľná

* **Myosotis caespitosus** Schultz → *Myosotis laxa*

* **Myosotis collina** Hoffm. → *Myosotis discolor*

1125 **Myosotis discolor** Pers.
Al (–) lulemiza e kodrës
Cs (0) pomněnka různobarvá
De (–) Buntes Vergißmeinnicht
En (–) changing forget-me-not
Hu (1) tarka nefelejcs
Pl (2) niezapominajka różnobarwna
Ru (1) незабудка разноцветная
Sk (–) nezábudka pestrá

* **Myosotis filiformis** Schleich. → *Myosotis discolor*

* **Myosotis hispida** Schlecht. → *Myosotis ramosissima*

* **Myosotis intermedia** Link → *Myosotis arvensis*

* **Myosotis lappula** L. → *Lappula squarrosa*

1126 **Myosotis laxa** Lehm.
Bg (1) туфеста незабравка
Cs (0) pomněnka řidká; pomněnka trsnatá
En (–) marsh forget-me-not
Hu (0) gyepes nefelejcs
Pl (1) niezapominajka darniowa
Ru (0) незабудка дернистая
Sk (–) nezábudka vlnokvetá

* **Myosotis micrantha** Pallas → *Myosotis stricta*

* **Myosotis palustris** (L.) Hill → *Myosotis scorpioides*

1127 **Myosotis ramosissima** Roch.
Bg (1) грапава незабравка
Cs (0) pomněnka drsná
De (–) Hügel-Vergißmeinnicht
En (–) early forget-me-not
Hu (1) borzas nefelejcs
Pl (1) niezapominajka pagórkowa
Ru (1) незабудка ветвистая
Sh (1) runjavi spomenak
Sk (–) nezábudka kopcová

1128 **Myosotis scorpioides** L.
Al (–) lulemiza e moçaleve
Bg (1) блатна незабравка
Cs (1) pomněnka bahenní
De (–) Sumpf-Vergißmeinnicht
En (–) water forget-me-not
Hu (0) mocsári nefelejcs
Pl (2) niezapominajka błotna
Ru (1) незабудка скорпионовидная
Sh (1) obični spomenak
Sk (–) nezábudka močiarna

1129 **Myosotis stricta** Link
Bg (1) дребноцветна незабравка
Cs (2) pomněnka drobnokvětá
De (–) Sand-Vergißmeinnicht
En (–) upright forget-me-not
Hu (1) apró nefelejcs
Pl (3) niezapominajka piaskowa
Ru (1) незабудка песчаная
Sk (–) nezábudka chudokvetá

1130 **Myosotis sylvatica** Hoffm.
Al (–) lulemiza e pyllit
Bg (1) горска незабравка
Cs (1) pomněnka lesní
De (–) Wald-Vergißmeinnicht

En (–) wood forget-me-not
Hu (0) erdei nefelejcs
Pl (1) niezapominajka leśna
Ru (1) незабудка лесная
Sk (–) nezábudka lesná

* **Myosotis verna** Opiz → *Myosotis stricta*

* **Myosotis versicolor** (Pers.) Sm. → *Myosotis discolor*

1131 **Myosoton aquaticum** (L.) Moen.
De (–) Gemeiner Wasserdarm
En (–) water chickweed
Gr (1) μυόσωτο
Hu (2) vízi csillaghúr
Ru (0) мягковолосник водяной
Sk (–) mäkkuľa vodná

1132 **Myosurus minimus** L.
Bg (1) дребна миша опашка
Cs (2) myší ocásek nejmenší
De (–) Zwerg-Mäuseschwänzchen
En (–) mousetail
Gr (1) μύοσυρος
Hu (–) egérfarkfű
Pl (2) mysiurek drobny
Ro (1) codiculă
Ru (1) мышехвостник маленький
Sh (1) sićušna malorepka; sitna malorepka
Sk (–) myšací chvostík najmenší

1133 **Myriophyllum alternifolium** DC.
Cs (0) stolístek střídavokvětý
De (–) Wechselblütiges Tausendblatt
En (–) alternate-leaved water-milfoil
Pl (1) wywłócznik skrętoległy
Ru (0) уруть очередноцветковая
Sk (–) stolístok striedavokvetý

1134 **Myriophyllum spicatum** L.
Bg (1) класовиден мириофилум
Cs (0) stolístek klasnatý
De (–) Ähriges Tausendblatt
En (–) spiked water-milfoil
Gr (1) σταχόμορφο μυριόφυλλο
Hu (1) füzéres süllőhínár
Pl (1) wywłócznik kłosowy
Ro (1) peniţă
Ru (1) уруть колосовая
Sh (1) klasasti krocanj
Sk (–) stolístok klasnatý

1135 **Myriophyllum verticillatum** L.
Bg (1) прешленест мириофилум
Cs (0) stolístek přeslenitý
De (–) Quirlblättriges Tausendblatt
En (–) whorled water-milfoil
Gr (1) σπονδυλωτό μυριόφυλλο
Hu (1) gyűrűs süllőhínár
Pl (1) wywłócznik okółkowy
Ro (1) peniţă
Ru (1) уруть мутовчатая
Sh (1) pršljenasti krocanj
Sk (–) stolístok praslenatý

N

1136 ***Najas graminea*** Delile
Bg (1) русалка
Ru (1) наяда злаковидная
Sk (–) riečňanka trávovitá

* ***Najas major*** All. → *Najas marina*

1137 ***Najas marina*** L.
Al (–) lapagrethi bregdetar
Bg (0) морска русалка
Cs (0) řečanka mořská
De (–) Großes Nixenkraut
En (–) hollyleaf naiad
Gr (1) μεγάλη ναϊδα
Hu (1) nagy tüskéshínár
Pl (1) jezierza morska
Ro (0) inariţă mare
Ru (1) наяда морская
Sh (1) morska podvodnica
Sk (–) riečňanka prímorská

1138 ***Najas minor*** L.
Al (–) lapagrethi më i vogël
Bg (1) малка русалка
Cs (0) řečanka menší
De (–) Kleines Nixenkraut
En (–) lesser naiad
Gr (1) μικρή ναϊδα
Hu (1) kis tüskéshínár
Pl (1) jezierza mniejsza
Ro (1) inariţă mare
Ru (1) наяда малая
Sh (1) mala podvodnica
Sk (–) riečňanka menšia

1139 ***Nardus stricta*** L.
Al (–) bar mali; xhufka
Bg (2) конец
Cs (3) smilka tuhá
De (–) Borstgras
En (–) mat-grass
Hu (1) szőrfű
Pl (1) bliźniczka psia trawka
Ro (3) ţăpoşică
Ru (1) белоус торчащий
Sh (1) osorka tvrdača
Sk (–) psica tuhá

* ***Nasturtium lippizense*** (Wulf.) DC. → *Rorippa lippizensis*

1140 ***Nasturtium officinale*** R Br.
Al (–) purqarku mjekësor
Bg (1) поточарка
Cs (0) potočnice lékařská
De (–) Echte Brunnenkresse
En (–) water-cress
Gr (1) νεροκάρδαμο
Hu (0) vízitorma
Pl (1) rukiew wodna
Ro (0) năsturel
Ru (0) жеруха лекарственная
Sh (1) perastolisna potočarka
Sk (–) potočnica lekárska

* ***Nasturtium sylvestre*** (L.) R Br. → *Rorippa sylvestris*

1141 ***Neotinia maculata*** (Desf.) Stearn
En (–) dense-flowered orchid
Gr (1) νεοτίνια
Sk (–) neotinea škvrnitá

1142 ***Nepeta cataria*** L.
Al (–) nepeta katare
Bg (1) коча билка
Cs (1) šanta kočičí
De (–) Echte Katzenminze
En (–) cat-mint
Hu (1) illatos macskamenta
Pl (1) kocimiętka właściwa
Ro (0) cătuşnică
Ru (1) котовник кошачий
Sk (–) kocúrnik mačací

1143 ***Nepeta nuda*** L.
De (–) Ungarische Katzenminze
Ru (–) котовник венгерский
Sh (1) glatka potplotusa
Sk (–) kocúrnik lysý

* ***Nepeta pannonica*** L. → *Nepeta nuda*

* ***Neslia apiculata*** Fisch., Meyer & Avé. → *Neslia paniculata* ssp. *thracica*

1144 ***Neslia paniculata*** (L.) Desv. **ssp. *paniculata***
Al (–) neslia në formë melthi; neslie melthore
Bg (1) метловидна дробица
Cs (3) řepinka latnatá
De (–) Rispen-Finkensame
En (–) ball-mustard
Hu (2) sömörje
Pl (2) ożędka groniasta
Ro (1) drob
Ru (1) неслия метельчатая
Sh (1) metličasta tršlja; metličasta žumanka
Sk (–) repinka metlovitá latnatá

1145 ***Neslia paniculata*** (L.) Desv. **ssp. *thracica*** (Velen.) Bornm.
Bg (1) остроплодна дробица
De (–) Rispen-Finkensame
Gr (–) νέσλια
Hu (2) sömörje
Ru (1) неслия тракийская
Sh (1) obična tršlja; ušiljena žumanka
Sk (–) repinka metlovitá trácka

1146 ***Nicandra physaloides*** (L.) Gaertn.
Al (–) luvër kopshtaqe
Bg (1) отровно биле
Cs (0) lilík mochynovitý
De (–) Giftbeere
En (–) Chinese-lantern
Hu (1) szilkesark
Pl (0) nikandra miechunkowa
Ru (0) никандра физалисовидная
Sk (–) nikandra machovkovitá

1147 ***Nigella arvensis*** L.
Al (–) nigela e arave
Bg (1) полска челебийка
Cs (1) černucha rolní
De (–) Acker-Schwarzkümmel
En (–) field nigella
Gr (1) μαυροκούκι
Hu (2) vetési katicavirág
Pl (1) czarnuszka polna
Ro (1) negrușcă
Ru (1) чернушка полевая
Sh (1) poljska crnika; poljska crnjika
Sk (–) čiernucha roľná

1148 ***Nigella damascena*** L.
Al (–) nigela e Damaskut
Bg (1) дамаска челебийка
Cs (0) černucha damašská
De (–) Jungfer im Grünen
En (–) love-in-the-mist
Gr (1) κοινό μαυροκούκκι
Hu (0) borzaskata
Pl (0) czarnuszka damasceńska
Ro (0) chica voinicului
Ru (0) чернушка дамасская
Sh (1) čupava crnika; čupava crnjika
Sk (–) čiernucha damašská

* ***Nigella latifolia*** Mill. → *Nigella arvensis*

1149 ***Nigella sativa*** L.
Bg (1) посевна челебийка
Cs (0) černucha setá
De (–) Echter Schwarzkümmel
En (–) black cumin
Gr (1) εδώδιμο μαυροκούκκι
Pl (1) czarnuszka siewna
Ro (0) negrilică
Ru (1) чернушка посевная
Sk (–) čiernucha siata

1150 ***Nigella segetalis*** Bieb.
Ru (1) чернушка пашенная

* ***Nigella tenuiflora*** Gilib. → *Nigella arvensis*

1151 ***Nonea lutea*** (Desr.) DC.
Cs (0) pipla žlutá
De (–) Gelbes Mönchskraut
En (–) yellow nonea
Ru (1) нонея бледно-жёлтая
Sk (–) ostreň žltý

1152 ***Nonea pulla*** (L.) DC.
Bg (1) тъмно шекерче
Cs (1) pipla osmahlá
De (–) Braunes Mönchskraut
En (–) common nonea
Hu (–) apácavirág
Pl (0) zapłonka brunatna
Ro (1) ochiul lupului
Ru (1) нонея тёмно-бурая
Sh (1) obični samak; tamnocrvena srdovica
Sk (–) ostreň počerný

1153 ***Nuphar lutea*** (L.) Sibth. & Sm.
Al (–) neloferi i verdhë; zambak uji i verdhë
Bg (0) жълта водна роза
Cs (2) stulík žlutý
De (–) Gelbe Teichrose
En (–) yellow water-lily
Gr (1) κίτρινο νούφαρο
Hu (0) vízitök
Pl (1) grążel żółty
Ro (0) nufăr galben
Ru (1) кубышка жёлтая
Sk (–) leknica žltá

1154 ***Nymphaea alba*** L.
Al (–) lëkue i bardhë; zambak uji i bardhë
Bg (0) бяла водна роза
Cs (1) leknín bílý
De (–) Weiße Seerose
En (–) white water-lily
Gr (1) νυμφαία
Hu (0) fehér tündérrózsa
Pl (1) grzybień biały
Ro (0) nufăr alb
Ru (0) кувшинка белая
Sk (–) lekno biele

1155 ***Nymphoides peltata*** (Gmel.) Kuntze
Cs (–) plavin lekninovitý; plavin stitnaty
De (–) Seekanne
En (–) fringed water-lily
Hu (0) tündérfátyol
Pl (0) grzybieńczyk wodny
Ro (1) plutică
Ru (1) болотоцветник щитолистный
Sh (1) obični lokvanjčić
Sk (–) leknovec štítnatý

O

* ***Oberna behen*** (L.) Ikonn. → *Silene vulgaris*

1156 ***Odontites lutea*** (L.) Clairv.
Cs (–) záhořanka žlutá
De (–) Gelber Zahntrost
En (–) yellow bartsia
Gr (1) κίτρινος δοντίτης
Hu (0) sárga fogfű
Pl (1) zagorzałek żółty
Ru (1) зубчатка осенняя
Sh (1) žuta vidoka
Sk (–) zúbkokvet žltý

* ***Odontites rubra*** Bess. → *Odontites verna*

1157 ***Odontites verna*** (Bella.) Dumort.
Cs (3) zdravínek červený
De (–) Roter Zahntrost
En (–) red bartsia
Gr (1) κόκκινος δοντίτης
Hu (1) piros fogfű
Pl (2) zagorzałek wiosenny
Ro (1) iarba dintelui
Ru (1) зубчатка весенняя
Sh (1) crvena vidoka
Sk (–) zdravienok červený jarný

1158 ***Oenanthe* spp.** *L.*
Bg (1) йонанте
De (–) Wasserfenchel
En (–) water-dropwort
Gr (1) οινάνθη
Ru (0) омежник
Sk (–) halucha

1159 ***Oenanthe aquatica*** (L.) Poir.
Bg (1) водно йонанте
Cs (0) halucha vodní
De (–) Wasserfenchel
En (–) water fennel
Hu (1) vízi mételykóró
Pl (1) kropidło wodne
Ro (1) mararaş
Sh (1) obična trbulja; vodena trbulja
Sk (–) halucha vodná

1160 ***Oenanthe banatica*** Heuff.
Bg (1) банатско йонанте
Cs (0) halucha banátská
Hu (0) bánsági borgyökér
Pl (0) kropidło banackie
Sk (–) halucha banátska

1161 ***Oenanthe fistulosa*** L.
Al (–) barujca; luledhri fyle
Bg (1) тръбесто йонанте
Cs (0) halucha dutá
De (–) Rohren-Wasserfenchel
En (–) common water-dropwort
Hu (0) bördős borgyökér
Pl (1) kropidło piszczałkowate
Ru (0) омежник дудчатый
Sk (–) halucha dutá

* ***Oenanthe media*** Griseb. → *Oenanthe silaifolia*

* ***Oenanthe meifolia*** Schloss. & Vukot. → *Oenanthe fistulosa*

* ***Oenanthe phellandrium*** Lam. → *Oenanthe aquatica*

1162 ***Oenanthe pimpinelloides*** L.
Al (–) luledhri si netërzë
Bg (1) анасоновиден воден морач
De (–) Bibernell-Rebendolde
En (–) branched water-dropwort
Ru (0) омежник бедренецовидный
Sk (–) halucha bedrovníkovitá

1163 ***Oenanthe silaifolia*** Bieb.
Al (–) luledhri me gjethe silaje

Cs (0) halucha koromáčolistá
De (–) Silgblättriger Wasserfenchel
En (–) narrow-leaved water-dropwort
Hu (0) gyűrűs borgyökér
Ru (0) омежник морковниколистный
Sh (1) mala trbulja
Sk (–) halucha siličkolistá

1164 ***Oenothera biennis*** L.
Al (–) yll i mbrëmjes
Bg (0) свещица
Cs (2) pupalka dvouletá
De (–) Gemeine Nachtkerze
En (–) common evening-primrose
Gr (1) οινόθηρα
Hu (1) parlagi ligetszépe
Pl (1) wiesiołek dwuletni
Ro (1) luminiţă
Ru (1) ослинник двулетний
Sh (1) žuti noćurak
Sk (–) púpalka dvojročná

1165 ***Oenothera muricata*** L.
Cs (–) pupalka mělkoostenná
De (–) Kleinblütige Nachtkerze
Ru (–) ослинник коротконглый
Sh (1) purpurni noćurak
Sk (–) púpalka múrová

1166 ***Oenothera parviflora*** L.
De (–) Kleinblütige Nachtkerze
En (–) small-flowered evening-primrose
Hu (1) magyar ligetszépe
Sk (–) púpalka malokvetá

1167 ***Omalotheca sylvatica*** (L.) Schultz
Al (–) gnafala e pyllit
Cs (0) protěž lesní
De (–) Wald-Ruhrkraut
En (–) heath cudweed
Hu (0) erdei gyopár
Pl (1) szarota leśna
Ru (0) сушеница лесная
Sh (1) klasati runolistak; šumski srcopuc
Sk (–) plesnivček lesný

1168 ***Onobrychis aequidentata*** (Sibth. & Sm.) D'Urv.
Al (–) esparceta dhëmbëbarabartë
Bg (1) равнозъба еспарзета
Gr (1) ισοδοντωτή ονοβρυχίδα
Sk (–) ľadníček rovnozubý

1169 ***Onobrychis caput-galli*** (L.) Lam.
Al (–) ballëbreshtëza lafshëgjeli; esparceta kokëgjeli
De (–) Hahnenkamm
Gr (1) κοινή ονοβρυχίδα
Ru (0) эспарцет петушья голова
Sk (–) ľadníček kuria hlava

* ***Onobrychis foveolata*** DC. → *Onobrychis aequidentata*

1170 ***Onobrychis viciifolia*** Scop.
Al (–) esparceta me gjethe grashine
Bg (1) фиева еспарзета
Cs (1) vičenec ligrus
De (–) Futter-Esparsette
En (–) common sainfoin
Hu (0) takarmánybalatacím
Pl (0) esparceta siewna
Ro (0) sparcetă
Ru (0) эспарцет виколистный
Sk (–) ľadníček obyčajný

1171 ***Ononis*** **spp.** L.
Bg (1) гръмотрън
De (–) Hauhechel
En (–) restharrow
Gr (1) ανονίδα
Ru (1) стальник
Sk (–) ihlica

1172 ***Ononis arvensis*** L.
Al (–) gromatër e fushave; kalmuthi i arave
Bg (1) обикновен гръмотрън
Cs (0) jehlice kozlí
De (–) Bocks-Hauhechel

En (–) field restharrow
Hu (1) nehézszagú iglice
Pl (2) wilżyna bezbronna
Ro (1) osul iepurelui
Ru (1) стальник пашенный
Sh (1) kozji gladiš; poljski zečji trn
Sk (–) ihlica roľná

* ***Ononis procurrens*** Wallr. → *Ononis repens*

1173 ***Ononis pusilla*** L.
Al (–) gromatër voglushe
Bg (1) дребен гръмотрън
Cs (0) jehlice nizounká
De (–) Zwerg-Hauhechel
Hu (0) sárga iglice
Ru (0) стальник маленький
Sk (–) ihlica nízka

1174 ***Ononis repens*** L.
Bg (1) пълзящ гръмотрън
Cs (0) jehlice plazivá
De (–) Kriechender Hauhechel
En (–) common restharrow
Pl (0) wilżyna rozłogowa
Ru (0) стальник ползучий
Sk (–) ihlica plazivá

1175 ***Ononis spinosa*** L.
Al (–) ferrëbukja; kalmuthi me gjemba
Cs (2) jehlice trnitá
De (–) Dorniger Hauhechel
En (–) spring restharrow
Hu (2) tövises iglice
Pl (1) wilżyna ciernista
Ru (1) стальник колючий
Sh (1) trnoviti gladiš; trnoviti zečji trn
Sk (–) ihlica tŕnistá

1176 ***Onopordum* spp.** L.
Bg (1) гингер
De (–) Eselsdistel
Gr (1) γαϊδουράγκαθο
Ru (0) онопордум
Sk (–) trubil

1177 ***Onopordum acanthium*** L.
Al (–) arshylli; xhixhëlla dashtër
Bg (1) обикновен гингер
Cs (2) ostropes trubil
De (–) Eselsdistel
En (–) cotton thistle
Hu (2) szamárbogáncs
Pl (1) popłoch pospolity
Ro (2) scai măgăresc
Ru (1) онопордум колючий
Sh (1) vunasti kravačac
Sk (–) ostropes trubil

1178 ***Onopordum illyricum*** L.
Al (–) xhixhëlla e Ilirisë
Bg (1) илирийски гингер

1179 ***Onopordum tauricum*** Willd.
Bg (1) кримски гингер
Ru (1) онопордум таврический

1180 ***Onosma arenaria*** Waldst. & Kit.
Al (–) çiklla e ranishteve
Bg (1) пясъчно омразниче
Cs (0) ruměnice písečná
De (–) Sand-Lotwurz
En (–) purple golden-drop
Hu (0) homoki vértő
Ro (0) otrăţel
Ru (0) оносма песчаная
Sh (1) pešćarski ožućak; piješćarski ožućak
Sk (–) rumenica piesočná

1181 ***Onosma visianii*** Clem.
Bg (1) Визианово омразниче
Cs (0) ruměnice visianiho
De (–) Visiani's Lotwurz
Hu (0) borzas vértő
Ru (0) оносма Визиани
Sk (–) rumenica Visianiho

1182 *Ophioglossum vulgatum* L.
Al (–) bar nëngjuhe; nëngjuha
Bg (1) змийска папрат
Cs (0) jazyk hadí
De (–) Gemeine Natternzunge
En (–) adder's-tongue fern
Hu (0) kígyónyelv
Pl (0) nasięźrzał pospolity
Ro (0) limba şarpelui
Ru (0) ужовник обыкновенный
Sk (–) hadí jazyk obyčajný

1183 *Ophrys* spp. L.
Bg (0) пчелица
De (–) Ragwurz
Gr (1) οφρύδα
Ru (0) офрис
Sk (–) trčník

* ***Oplismenus crus-galli*** Dumort. → *Echinochloa crus-galli*

1184 *Opopanax chironium* (L.) Koch
Bg (1) хирониум
De (–) Große Gummiwurz
En (–) opopanax

1185 *Opuntia ficus-indica* (L.) Mill.
Al (–) fik indie
De (–) Echter Feigenkaktus
En (–) prickly-pear
Gr (1) φραγκοσυκιά
Sk (–) opuncia figovníková

1186 *Orchis* spp. L.
Bg (0) салеп
De (–) Knabenkraut
Gr (1) σερνικοβότανο
Ru (1) ятрышник
Sk (–) vstavač

1187 *Orchis coriophora* L.
Al (–) salep i lëkurtë
Bg (0) дървеницов салеп
Cs (0) vstavač štěničný
De (–) Wanzen-Knabenkraut
En (–) bug-orchid
Hu (0) poloskaszagú kosbor
Pl (1) storczyk cuchnący
Ro (0) ploşniţoasă
Ru (0) ятрышник клопоносный
Sh (1) mirisavi kaćun
Sk (–) vstavač bĺší

* ***Orchis latifolia*** L. → *Dactylorhiza majalis*

1188 *Orchis militaris* L.
Bg (0) шлемовиден салеп
Cs (0) vstavač vojenský
De (–) Helm-Knabenkraut
En (–) military orchid
Hu (0) vitézkosbor
Pl (1) storczyk kukawka
Ro (0) poroinic
Ru (0) ятрышник шлемоносный
Sk (–) vstavač vojenský

1189 *Orchis morio* L.
Cs (–) vstavač kukačka
De (–) Kleines Knabenkraut
En (–) green-winged orchid
Ru (1) ятрышник малый
Sh (1) obični kaćun
Sk (–) vstavač obyčajný

1190 *Origanum heracleoticum* L.
De (–) Falscher Staudenmajoran
En (–) winter marjoram
Gr (1) ρίγανη

1191 *Origanum vulgare* L.
Al (–) çaj mali; rigoni
Bg (0) риган
Cs (0) dobromysl obecná
De (–) Dost
En (–) pot marjoram
Gr (1) αγριορίγανη

Hu (0) szurokfű
Pl (1) lebiodka pospolita
Ro (0) şovîrv
Sh (1) obični mravinac
Sk (–) pamajoran obyčajný

1192 ***Orlaya* spp.** Hoffm.
De (–) Breitsame
En (–) orlaya
Gr (1) ορλάγια
Ru (0) орлайа

1193 ***Orlaya grandiflora*** (L.) Hoffm.
Al (–) orlaja lulemadhe
Bg (1) срамниче
Cs (0) paprska velkokvětá
De (–) Großblütiger Breitsame
Hu (1) Orlay-murok
Ru (1) орлайа крупноцветковая
Sh (1) krupnocvetni stidak; krupnocvjetni stidak
Sk (–) orlaya veľkokvetá

1194 ***Ornithogalum* spp.** L.
Bg (1) гарвански лук
De (–) Milchstern
Gr (1) ορνιθόγαλο
Ru (1) птицемлечник
Sk (–) bledavka

1195 ***Ornithogalum boucheanum*** Asch.
Bg (1) бушенов гарвански лук
Cs (0) snědek nící Bouchéův
De (–) Garten-Milchstern
Hu (1) kónya madártej
Pl (0) śniadek Buchego
Ru (1) птицемлечник Буше
Sk (–) bledavka ovisnutá

1196 ***Ornithogalum comosum*** L.
Bg (1) голямокачулест гарвански лук
Cs (1) snědek chocholatý
De (–) Schopfiger Milchstern
Hu (0) üstökös sárma
Sh (1) obični vranji luk
Sk (–) bledavka chocholatá

* ***Ornithogalum nanum*** Sibth. & Sm. → *Ornithogalum sibthorpii*

1197 ***Ornithogalum narbonense*** L.
Bg (1) нарбонски гарвански лук

1198 ***Ornithogalum orthophyllum*** Ten.
Bg (1) теснолистен гарвански лук
Hu (1) pusztai madártej
Sk (–) bledavka priamolistá

1199 ***Ornithogalum pyramidale*** L.
Bg (1) пирамидален гарвански лук
Cs (1) snědek jehlancovitý
De (–) Pyramiden-Milchstern
Hu (1) nyúlánk madártej
Ro (1) băluşcă
Sk (–) bledavka ihlancovitá

1200 ***Ornithogalum pyrenaicum*** L.
Al (–) minikëla e Pirinejeve
Bg (1) пиренейски гарвански лук
Cs (0) snědek pyrenejský
De (–) Pyrenäen-Milchstern
En (–) spiked star-of-Bethlehem
Hu (1) déli madártej
Ru (0) птицемлечник пиренейский
Sk (–) bledavka pyrenejská

1201 ***Ornithogalum refractum*** Kit.
Bg (1) пречупен гарвански лук
Hu (1) csilláros madártej
Ru (1) птицемлечник преломлённый

1202 ***Ornithogalum sibthorpii*** Greut.
Bg (1) малък гарвански лук
Sk (–) bledavka sibtorpova

1203 ***Ornithogalum sphaerocarpum*** Kern.
Cs (0) snědek pyrenejský kulatoplodý
Hu (1) déli madártej
Sk (–) bledavka guľatoplodá

* ***Ornithogalum tenuifolium*** auct. → *Ornithogalum orthophyllum*

1204 ***Ornithogalum umbellatum*** L.
Al (–) lulë dalldyshe; minikëla ombrelore
Bg (1) сенников гарвански лук
Cs (0) snědek chocholičnatý
De (–) Dolden-Milchstern
En (–) common star-of-Bethlehem
Hu (1) ernyős madártej
Pl (1) śniadek baldaszkowaty
Ru (1) птицемлечник зонтичный
Sh (1) štitasti vranji luk
Sk (–) bledavka chocholíkatá

1205 ***Ornithopus compressus*** L.
Al (–) ornitopa e ngjeshur
Bg (1) сбита серадела
Cs (0) ptačí noha smáčknutá
De (–) Gelbe Serradella
Ru (0) сераделла сжатая
Sk (–) vtačia noha stlačená

1206 ***Orobanche* spp.** L.
Al (–) sparqa
Bg (1) синя китка
De (–) Sommerwurz
En (–) broomrape
Gr (2) οροβάγχη
Pl (–) zaraza
Ro (–) lupoaie
Ru (1) заразиха
Sk (–) záraza

1207 ***Orobanche aegyptica*** Pers.
Ru (1) заразиха египетская
Sk (–) záraza egyptská

1208 ***Orobanche cernua*** Loefl.
Bg (1) слънчогледова синя китка
Cs (1) záraza kumánská
De (–) Nickende Sommerwurz
Hu (2) napraforgó vajvirág
Ro (3) lupoaia florii soarelui
Ru (1) заразиха поникшая
Sh (1) suncokretova vodnjača; suncokretov volovod
Sk (–) záraza krivokvetá

* ***Orobanche cumana*** Wallr. → *Orobanche cernua*

1209 ***Orobanche elatior*** Sutt.
Al (–) sparqe e madne
Bg (1) голяма синя китка
Cs (0) záraza větší
De (–) Ginster-Sommerwurz
En (–) knapweed broomrape
Hu (1) sárga vajvirág
Sh (1) velika vodnjača
Sk (–) záraza vyššia

1210 ***Orobanche gracilis*** Sm.
Al (–) sparqe e hajthshme
Bg (1) нежна синя китка
Cs (0) záraza štíhlá
De (–) Blutrote Sommerwurz
Hu (1) vérveres vajvirág
Pl (0) zaraza krwistoczerwona
Ru (1) заразиха тонкая
Sk (–) záraza štíhla

1211 ***Orobanche lutea*** Baumg.
Al (–) sparq i jonxhës; sparqe e verdhë
Cs (0) záraza žlutá
De (–) Gelbe Sommerwurz
En (–) yellow broomrape
Hu (1) sárga vajvirág
Pl (0) zaraza czerwonawa
Ru (1) зразиха жёлтая
Sh (–) rumenkasti volovod
Sk (–) záraza žltá

1212 ***Orobanche minor*** Sm.
Al (–) sparq i tërfilit; sparqe më e vogël
Bg (0) малка синя китка
Cs (3) záraza menší
De (–) Kleine Sommerwurz

En (–) common broomrape
Hu (1) lóhere-vajvirág
Pl (0) zaraza drobnokwiatowa
Ru (1) заразиха малая
Sh (1) mala vodnjača
Sk (–) záraza menšia

* ***Orobanche mutelii*** Schultz → *Orobanche ramosa* ssp. *mutelii*

1213 ***Orobanche ramosa*** L. **ssp.** ***mutelii*** (Schultz) Cout.
Bg (2) Мутелиев воловодец
Cs (1) záraza Mutelova
Ru (1) заразиха Мутеля
Sk (–) záraza tabaková

1214 ***Orobanche ramosa*** L. **ssp.** ***ramosa***
Al (–) sparqe e degëzuar; sparq i kërpit
Bg (2) тютюнева синя китка
Cs (1) záraza větevnatá
De (–) Ästige Sommerwurz
En (–) hemp broomrape
Gr (1) λυκόχορτο
Hu (2) dohányfojtó vajvirág
Pl (2) zaraza gałęzista
Ro (3) lupoaia tutunului
Ru (2) заразиха ветвистая
Sh (1) granata vodnjača; granati volovod
Sk (–) záraza vetvistá

* ***Oryza clandestina*** A Br. → *Leersia oryzoides*

* ***Oryzopsis miliacea*** (L.) Benth. & Hook. → *Piptatherum miliaceum*

1215 ***Osmunda regalis*** L.
Cs (0) podezřeň královská
De (–) Königs-Rispenfarn
En (–) royal fern
Gr (1) οσμούνδα
Pl (0) długosz królewski
Ru (0) чистоуст величавый
Sk (–) osmunda kráľovská

1216 ***Oxalis acetosella*** L.
Al (–) uthëllnike
Bg (1) обикновено киселиче
Cs (3) šťavel kyselý
De (–) Wald-Sauerklee
En (–) wood-sorrel
Hu (0) erdei madársóska
Pl (1) szczawik zajęczy
Ro (0) macrişul iepurelui
Ru (1) кислица обыкновенная
Sh (1) batvasti cecelj; zečja soca
Sk (–) kyslička obyčajná

1217 ***Oxalis corniculata*** L.
Al (–) uthëllnike në formë briri
Bg (1) рогчесто киселиче
Cs (3) šťavel růžkatý
De (–) Horn-Sauerklee
En (–) procumbent yellow-sorrel
Hu (1) szürke madársóska
Pl (2) szczawik różkowaty
Ru (1) кислица рожковая
Sh (1) zlatna soca
Sk (–) kyslička rožkatá

* ***Oxalis fontana*** Bung. → *Oxalis stricta*

1218 ***Oxalis pes-caprae*** L.
De (–) Nickender Sauerklee
En (–) buttercup oxalis
Gr (3) οξαλίδα
Ru (1) кислица козья
Sk (–) kyslička kozia noha

1219 ***Oxalis stricta*** L.
Cs (3) šťavel tuhý
De (–) Steifer Sauerklee
En (–) upright yellow-sorrel
Hu (3) parlagi madársóska
Pl (3) szczawik żółty
Ru (2) кислица прямостоящая
Sh (1) obični cecelj
Sk (–) kyslička tuhá

1220 ***Oxytropis campestris*** (L.) DC.
Bg (1) полски окситропис
Cs (0) vlnice ladní
De (–) Gemeiner Spitzkiel
En (–) yellow milk-vetch
Pl (0) ostrołódka polna
Ru (0) остролодочник полевой
Sk (–) ostropysk poľný

1221 ***Oxytropis pilosa*** (L.) DC.
Bg (1) висок окситропис
Cs (0) vlnice chlupatá
De (–) Zottiger Spitzkiel
Hu (0) csajkavirág
Pl (1) ostrołódka kosmata
Ru (0) остролодочник волосистый
Sk (–) ostropysk chlpatý pravý

P

1222 ***Pallenis spinosa*** L.
Al (–) asterisku me gjemba
De (–) Goldauge
Gr (1) καρφόχορτο

1223 ***Panicum capillare*** L.
Al (–) mel flokëngjashëm
Cs (0) proso vláskovité
De (–) Haarstiel-Hirse
En (–) witchgrass
Hu (3) cérnaköles
Pl (0) proso włosowate
Ro (1) mei mărunt
Ru (1) просо волосовидное
Sh (1) vlasasto proso
Sk (–) proso vláskovité

* ***Panicum coarctatum*** Stev. → *Echinochloa crus-galli*

* ***Panicum crus-galii*** L. → *Echinochloa crus-galli*

1224 ***Panicum dichotomiflorum*** Michx
De (–) Gabelblütige Hirse
En (–) autumn witchgrass
Ru (1) просо ветвистометельчатое
Sk (–) proso vidličkokveté

* ***Panicum eruciforme*** Sibth. & Sm. → *Brachiaria eruciformis*

* ***Panicum glaucum*** L. → *Setaria glauca*

* ***Panicum hostii*** Bieb. → *Echinochloa crus-galli*

1225 ***Panicum miliaceum*** L.
Al (–) meli
Bg (0) просо
Cs (–) proso seté
De (–) Echte Hirse
En (–) wild proso-millet
Gr (1) κοινό πάνικο
Hu (2) termesztett köles
Pl (0) proso zwyczajne
Ro (0) mei
Ru (1) просо посевное
Sk (–) proso siate

* ***Panicum oryzicola*** Vsain. → *Echinochloa oryzoides*

1226 ***Panicum repens*** L.
En (–) torpedograss
Gr (1) πολυετές πάνικο
Sk (–) proso plazivé

* ***Panicum sanguinale*** L. → *Digitaria sanguinalis*

* ***Panicum verticillatum*** L. → *Setaria verticillata*

* ***Panicum viride*** L. → *Setaria viridis*

1227 ***Papaver arenarium*** Bieb.
Ru (1) мак песчаный
Sk (–) mak piesočný

1228 ***Papaver argemone*** L.
Al (–) sygjarpëri
Cs (2) mák polní
De (–) Sand-Mohn
En (–) prickly poppy
Gr (1) τριχωτή παπαρούνα
Hu (1) ördög mák
Pl (3) mak piaskowy
Ru (1) мак аргемона
Sk (–) mak poľný

1229 ***Papaver dubium*** L.
Al (–) lulëkuqe e dyshimtë
Bg (2) продълговатоплоден мак
Cs (2) mák pochybný

De (–) Saat-Mohn
En (–) long-headed poppy
Gr (1) μακρόκαρπη παπαρούνα
Hu (1) bujdosó mák
Pl (3) mak wątpliwy
Ro (2) mac de cîmp
Ru (2) мак сомнительный
Sh (1) bela bulka; turčinak mak
Sk (–) mak pochybný

1230 ***Papaver hybridum*** L.
Al (–) lulëkuqe hibride
Bg (0) хибриден мак
Cs (0) mák zvrhlý
De (–) Bastard-Mohn
En (–) rough poppy
Hu (1) korcs mák
Ru (1) мак гибридный
Sk (–) mak zvrhlý

1231 ***Papaver macrostomum*** Boiss.
Ru (1) мак крупнокоробочный

1232 ***Papaver rhoeas*** L.
Al (–) lulëkuqe
Bg (2) див мак
Cs (3) mák vlčí
De (–) Klatsch-Mohn
En (–) common poppy
Gr (3) κοινή παπαρούνα
Hu (3) pipacs
Pl (3) mak polny
Ro (3) mac roşu
Ru (3) мак самосейка
Sh (2) divlji mak; poljska bulka
Sk (–) mak vlčí

1233 ***Parietaria cretica*** L.
Gr (1) κρητικό περδικούλι
Sk (–) múrovník krétsky

1234 ***Parietaria diffusa*** Mert. & Koch
Al (–) karafileigjen i hapët
Bg (0) разклонена разваленка
De (–) Ausgebreitetes Glaskraut
Gr (1) περδικούλι

* ***Parietaria judaica*** L. → *Parietaria officinalis*

1235 ***Parietaria officinalis*** L.
Al (–) bar muresh; karafileigjen mjekësor
Cs (0) drnavec lékařský
De (–) Glaskraut
En (–) wall pellitory
Gr (1) φαρμακευτικό περδικούλι
Hu (1) falgyom
Pl (0) parietaria lekarska
Ro (1) paracherniţă
Ru (1) постенница лекарственная
Sh (1) velika crkvina
Sk (–) múrovník lekársky

1236 ***Paris quadrifolia*** L.
Al (–) shtarca katërgjethëshe
Bg (0) вранско око
Cs (1) vraní oko čtyřlisté
De (–) Einbeere
En (–) herb-Paris
Hu (0) farkasszőlő
Pl (1) czterolist pospolity
Ro (0) dalac
Ru (1) вороний глаз обыкновенный
Sk (–) vranie oko štvorlisté

1237 ***Parnassia palustris*** L.
Al (–) parnasia e kënetës
Bg (1) блатна росица
Cs (0) tolije bahenní
De (–) Sumpf-Herzblatt
En (–) grass-of-Parnassus
Hu (0) fehérmájvirág
Pl (1) dziewięciornik błotny
Ro (0) şopîrliţă albă
Ru (0) белозор болотный
Sh (1) obična talija
Sk (–) parnasia močiarna

1238 ***Paspalum dilatatum*** Poir.
En (–) Dallis grass
Gr (1) πάσπαλος
Sk (–) paspal roztiahnutý

1239 ***Paspalum paspalodes*** (Michx) Scrib.
Gr (2) νεραγριάδα
Ru (1) гречка двуколосая
Sk (–) paspal dvojradý

* ***Pastinaca fleischmannii*** Hlad. → *Pastinaca sativa*

* ***Pastinaca opaca*** Bernh. → *Pastinaca sativa*

1240 ***Pastinaca sativa*** L.
Al (–) pastinaka
Bg (1) посевен пащърнак
Cs (2) pastiňák setý
De (–) Gewöhnlicher Pastinak
En (–) wild parsnip
Gr (1) παστινάκα
Hu (1) pasztinák
Pl (1) pasternak zwyczajny
Ro (1) păstîrnac
Ru (1) пастернак посевной
Sh (1) obični pastrnak; obični pastrnjak
Sk (–) paštrnák siaty

* ***Pastinaca sylvestris*** Mill. → *Pastinaca sativa*

1241 ***Pedicularis comosa*** L.
Al (–) pedikularia me baluka
Cs (0) všivec chocholatý
En (–) crested lousewort
Ru (0) мытник Кауфмана
Sh (1) dlakavi ušljivac
Sk (–) všivec chocholatý

1242 ***Pedicularis palustris*** L.
Cs (–) všivec bahenní
De (–) Sumpf-Läusekraut
En (–) marsh lousewort
Pl (1) gnidosz błotny
Ru (0) мытник болотный
Sh (1) močvarni ušljivac
Sk (–) všivec močiarny

1243 ***Peganum harmala*** L.
En (–) Turkish rue
Ru (1) гармала обыкновенная
Sh (1) mali sočivac; obični zrneš
Sk (–) harmala mnohodielna

* ***Peplis portula*** L. → *Peucedanum officinale*

1244 ***Periploca graeca*** L.
Al (–) shtalpra
Bg (1) гърбач
De (–) Baumschlinge
En (–) Grecian silk-vine
Hu (0) görögtekercs
Ru (0) обвойник греческий
Sk (–) svidna grécka

* ***Persicaria amphibia*** (L.) Gray → *Polygonum amphibium*

* ***Persicaria argyrocoleon*** Steud. → *Polygonum argyrocoleum*

* ***Persicaria hydropiper*** (L.) Opiz → *Polygonum hydropiper*

* ***Persicaria lapathifolia*** (L.) Gray → *Polygonum lapathifolium*

* ***Persicaria minor*** Opiz → *Polygonum minus*

* ***Persicaria vulgaris*** Samp. → *Polygonum persicaria*

1245 ***Petasites albus*** (L.) Gaertn.
Bg (0) бяла чобанка
Cs (–) devětsil bílý
De (–) Weiße Pestwurz
En (–) white butterbur
Hu (0) fehér keserűlapu
Pl (1) lepiężnik biały
Ro (0) cucuruz
Ru (0) белокопытник белый
Sh (1) beli lopuh; bijeli lopuh
Sk (–) kvemenina biela

1246 ***Petasites hybridus*** (L.) Gaertn., Meyer & Scherb.
Al (–) lëpusha; rapuha
Cs (3) devětsil lékařský
De (–) Gemeine Pestwurz
En (–) common butterbur
Gr (1) πετασίτης
Hu (0) vörös keserűlapu
Pl (1) lepiężnik różowy
Ro (0) captalan
Ru (0) белокопытник гибридный
Sh (1) obični lopuh
Sk (–) kvemenina hybridná

1247 ***Petasites spurius*** (Retz.) Reichenb.
De (–) Filzige Pestwurz
En (–) false butterbur
Pl (1) lepiężnik kutnerowaty
Ru (0) белокопытник ложный
Sk (–) kvemenina mäkkovlasá

1248 ***Peucedanum carvifolia*** Vill.
Bg (1) кимионолистен пеуцеданум
Cs (0) smldník kmínolistý
De (–) Kümmelblättriger Haarstrang
Hu (0) köménylevelű kocsord
Pl (0) gorysz podolski
Ru (–) белокопытник тминолистный
Sk (–) smldník rascolistý

1249 ***Peucedanum cervaria*** (L.) Lapeyr.
Al (–) paucedanumi si qarë
Cs (0) smldník jelení
De (–) Hirschwurz
En (–) broad-leaved spignel
Hu (0) szarvas kocsord
Pl (1) gorysz siny
Ru (0) горичник олений
Sk (–) smldník jelený

1250 ***Peucedanum officinale*** L.
Al (–) kopër thiu; paucedanumi mjekësor
Bg (0) лечебен пеуцеданум
Cs (0) smldník lékařský
De (–) Echter Haarstrang
En (–) hog's fennel
Hu (0) sziki kocsord
Ru (0) горичник аптечный
Sh (1) obična siljevina
Sk (–) smldník lekársky

1251 ***Peucedanum oreoselinum*** (L.) Moen.
Al (–) majdanoz mali
Cs (0) smldník olešníkovitý
De (–) Berg-Haarstrang
Hu (0) citromkocsord
Pl (1) gorysz pagórkowy
Ro (0) chimionul porcului
Ru (0) горичник горный
Sh (1) šumska siljevina
Sk (–) smldník olšovníkovitý

1252 ***Peucedanum palustre*** (L.) Moen.
Al (–) majdanoz kënete
Cs (0) olešníkovec bahenní
De (–) Sumpf-Haarstrang
En (–) milk-parsley
Hu (0) mocsári kocsord
Pl (1) gorysz błotny
Ru (0) горичник болотный
Sk (–) smldník močiarny

1253 ***Phacelia tanacetifolia*** Benth.
Cs (2) svazenka vratičolistá
De (–) Büschelschön
En (–) tansy phacelia

Pl (0) facelia błękitna
Ru (0) фацелия рябинолистная
Sk (–) sväzenka vratičolistá

1254 ***Phalaris arundinacea*** L.
Al (–) kokëmadhe si kallam
Bg (1) тръстиковидно тръстичина
Cs (2) chrastice rákosovitá
De (–) Rohr-Glanzgras
En (–) reed canary-grass
Hu (1) pántlikafű
Pl (1) mozga trzcinowata
Ro (1) ierbăluţă
Ru (1) двукисточник тростниковидный
Sh (1) trskolika svetlice; trskolika svjetlica
Sk (–) chrastnica trsťovitá

1255 ***Phalaris brachystachys*** Link
Al (–) kokëmadhe kallishkurtër
Cs (–) lesknice krátkoklasá
De (–) Kurzähriges Glanzgras
En (–) short-eared canary-grass
Sk (–) lesknica krátkoklasá

* ***Phalaris bulbosus*** L. → *Phleum subulatum*

1256 ***Phalaris coerulescens*** Desf.
Al (–) kokëmadhe e kaltër
Cs (0) lesknice namodralá
Gr (1) βολβοφάλαρη
Sk (–) lesknica modravá

1257 ***Phalaris minor*** Retz.
De (–) Kleines Glanzgras
En (–) lesser canary-grass
Gr (2) μικρόκαρπη φάλαρη
Ru (0) канареечник малый
Sk (–) lesknica menšia

1258 ***Phalaris paradoxa*** L.
Al (–) kokëmadhe paradoksale
Cs (0) lesknice podivná
De (–) Seltsames Glanzgras
En (–) awned canary-grass
Gr (2) παράδοξη φάλαρη
Ru (1) канареечник странный
Sk (–) lesknica podivná

* ***Phalaris semiverticillata*** Forsk. → *Polypogon viridis*

* ***Phelypaea ramosa*** (L.) Meyer → *Orobanche ramosa* ssp. *mutelii*

1259 ***Phleum* spp.** L.
Bg (1) тимотейка
De (–) Lieschgras
Gr (1) φλεός
Ru (1) тимофеевка
Sk (–) bojínek

* ***Phleum boehmeri*** Wib. → *Phleum phleoides*

1260 ***Phleum paniculatum*** Huds.
Cs (0) bojínek latnatý
De (–) Rauhes Lieschgras
Hu (0) érdes komócsin
Ru (0) тимофеевка метельчатая
Sk (–) bojínek latnatý

1261 ***Phleum phleoides*** (L.) Karst.
Al (–) bishtmi i stepeve; fleumi i Bohmerit
Bg (1) гръцка тимотейка
Cs (1) bojínek Boehmerův
De (–) Glanz-Lieschgras
En (–) purple cat's-tail
Hu (0) sima komócsin
Pl (–) tymotka Boehmera
Ru (1) тимофеевка степная
Sk (–) bojínek stepný

1262 ***Phleum pratense*** L.
Al (–) bishtmi i livadheve; fleumi i livadheve
Bg (1) ливадна тимотейка

Cs (2) bojínek luční
De (–) Wiesen-Lieschgras
En (–) timothy
Hu (0) mezei komócsin
Pl (2) tymotka łąkowa
Ro (0) timoftică
Ru (1) тимофеевка луговая
Sk (–) timotejka lúčna

1263 ***Phleum subulatum*** (Savi) Asch. & Graeb.
Al (–) fleumi i hollë
Bg (1) грудеста тимотейка
Cs (0) bojínek šidlovitý
En (–) sand cat's-tail
Ru (0) тимофеевка шиловидная
Sk (–) bojínek šidlovitý

* ***Phleum tenue*** Schrad. → *Phleum subulatum*

1264 ***Phlomis fruticosa*** L.
Al (–) bezga; sfakë e rëndomtë
En (–) Jerusalem sage
Gr (1) ασφάκα
Ru (0) зопник кустарниковидный
Sk (–) sápa obyčajná

1265 ***Phlomis herba-venti*** L.
Bg (1) бодлива руница
Ro (0) scorogoi
Ru (0) зопник колючий
Sk (–) sápa veternicovitá

* ***Phlomis pungens*** Willd. → *Phlomis herba-venti*

1266 ***Phlomis tuberosa*** L.
Al (–) sfakë zhardhoke
Bg (1) грудеста руница
Cs (0) sápa hlíznatá
De (–) Knollen-Brandkraut
Hu (0) macskahere
Pl (0) żeleźniak bulwiasty
Ro (1) şolovîrfiţă
Ru (0) зопник клубненосный
Sk (–) sápa hľuzovitá

1267 ***Phragmites australis*** (Cav.) Trin.
Al (–) kallmishtja; plishi
Bg (1) тръстика
Cs (3) rákos obecný
De (–) Schilfrohr
En (–) common reed
Gr (2) νεροκάλαμο
Hu (3) nád
Pl (2) trzcina pospolita
Ro (3) trestie
Ru (3) тростник обыкновенный
Sh (1) obična trska
Sk (–) trsť obyčajná

* ***Phragmites communis*** Trin. → *Phragmites australis*

* ***Phragmites phragmites*** Karst. → *Phragmites australis*

1268 ***Phyllitis scolopendrium*** (L.) Newm.
Cs (–) jelení jazyk celolistý
De (–) Gemeine Hirschzunge
En (–) hart's-tongue fern
Ru (0) листовик обыкновенный
Sh (1) obični jelenak
Sk (–) jelení jazyk

1269 ***Physalis alkekengi*** L.
Al (–) fshikakuqi
Bg (1) мехунка
Cs (1) mochyně židovská třešeň
De (–) Judenkirsche
En (–) Cape gooseberry
Gr (1) φυσαλίδα
Hu (–) zsidócseresznye
Pl (0) miechunka rozdęta
Ro (1) păpălău
Ru (1) физалис обыкновенный

Sh (1) obična mjehurica; obični ljoskovac
Sk (–) machovka židovska; čerešňa židovska

* **Physalis ixiocarpa** Brot. → *Physalis philadelphica*

1270 Physalis philadelphica Lam.
Ru (1) физалис клейкоплодный
Sk (–) machovka dužnoplodá

1271 Physalis pubescens L.
Ru (1) физалис пушистый
Sk (–) machovka vlasatá

1272 Physospermum cornubiense (L.) DC.
Ru (0) даная голостебельная
Sh (1) primorska mukavica

1273 Phytolacca americana L.
Al (–) çapëza
Bg (1) американски лаконос
Cs (0) líčidlo americké
De (–) Amerikanische Kermesbeere
En (–) common pokeweed
Gr (1) αγριοσταφίδα
Hu (–) alkörmös
Ro (0) cîrmîz
Ru (0) лаконос американский
Sh (1) američka vinoboja; grozdasti kermes
Sk (–) líčidlo americké

1274 Picnomon acarna (L.) Cass.
De (–) Sperrdistel
En (–) soldier thistle
Gr (1) πίκνομο
Ru (0) пикномон колючий
Sh (1) primorska sikalina

1275 Picris spp. L.
Al (–) cifuri
Bg (1) горчивка
De (–) Bitterkraut
En (–) oxtongue
Gr (1) χοιροβότανο
Pl (–) goryczel
Ro (1) iarba găii
Ru (1) горчак
Sk (–) horčík

1276 Picris echioides L.
Al (–) cifuri si shlligë; zgjebe
Bg (1) устойчива горчивка
Cs (0) draslavec hadincovitý
De (–) Wurmlattich
En (–) bristly oxtongue
Hu (2) vándorvirág
Pl (1) goryczel żmijowcowy
Ru (1) горлюха румянковидная
Sh (1) ježasta grkuša
Sk (–) draslavec hadincovitý

1277 Picris hieracioides L.
Al (–) cifuri këmashen
Bg (1) обикновена горчивка
Cs (0) hořčík jestřábníkovitý
De (–) Gemeines Bitterkraut
En (–) hawkweed oxtongue
Hu (–) keserűgyökér
Pl (1) goryczel jastrzębcowaty
Ru (1) горчак жёлтый
Sh (1) ostriješni jagušac; runjolika grkuša
Sk (–) horčík jastrabníkovitý

* **Pimpinella alpestris** (Spreng.) Schultes → *Pimpinella saxifraga*

1278 Pimpinella anisum L.
Al (–) anasoni
Bg (0) анасон
Cs (1) bedrník anýz
De (–) Anis
En (–) anise
Gr (1) γλυκάνισσο
Hu (0) illatos ánizs
Pl (1) biedrzeniec anyż
Ro (0) anason

Ru (0) анис обыкновенный
Sk (–) aníz voňavý

1279 ***Pimpinella peregrina*** L.
Al (–) pimpinela shtegëtare
Bg (0) чуждоземен анасон
De (–) Fremde Bibernelle
Gr (1) τριχόκαρπη πιμπινέλλα
Sk (–) bedrovník cudzí

1280 ***Pimpinella saxifraga*** L.
Al (–) pimpinela riqësë
Cs (2) bedrník obecný
De (–) Kleine Bibernelle
En (–) common burnet-saxifrage
Gr (1) λειόκαρπη πιμπινέλλα
Hu (1) hasznos földitömjén
Pl (2) biedrzeniec mniejszy
Ro (0) pètrinjel de cîmp
Ru (1) бедренец камнеломковый
Sh (1) obična bedrinka; obični bedrinac
Sk (–) bedrovník lomikameňový

1281 ***Piptatherum miliaceum*** (L.) Cosson
Al (–) orizopsisi si mel
Bg (1) осилесто просо
De (–) Grannenhirse
En (–) milo
Gr (2) γρήλαρη
Sk (–) ryžovka prosovitá

* ***Pisum arvense*** L. → *Pisum sativum*

* ***Pisum elatius*** Bieb. → *Pisum sativum*

1282 ***Pisum sativum*** L.
Al (–) bizelja
Bg (1) полски грах
Cs (0) hrách setý
De (–) Garten-Erbse
En (–) field pea
Hu (0) takarmányborsó
Pl (–) groch polny; peluszka
Ro (1) mazăre sălbatică
Ru (1) горох посевной
Sk (–) hrach siaty

* ***Pisum tuberosum*** L. → *Lathyrus tuberosus*

1283 ***Plantago* spp.** L.
Bg (1) жиловлек
De (–) Wegerich
En (–) plantain
Gr (1) πεντάνευρο
Ru (2) подорожник
Sk (–) skorocel

1284 ***Plantago altissima*** L.
Al (–) gjethedelli shumë i lartë
Cs (0) jitrocel nejvyšší
De (–) Hoher Wegerich
Hu (1) magas útifű
Ru (1) подорожник высочайший
Sk (–) skorocel najvyšši

1285 ***Plantago arenaria*** Waldst. & Kit.
Al (–) gjethedelli i ranishteve
Bg (1) пясъчен жиловлек
Cs (0) jitrocel indický
De (–) Sand-Wegerich
En (–) branched plantain
Hu (1) homoki útifű
Pl (1) babka gałęzista
Ro (1) ochiul lupului
Ru (1) подорожник песчаный
Sk (–) skorocel piesočný

1286 ***Plantago bellardii*** All.
Al (–) gjethedelli i Belardit
Bg (1) Белардиев жиловлек
En (–) silky plantain
Ru (1) подорожник Белларда
Sk (–) skorocel vlasatý

1287 ***Plantago coronopus*** L.
Al (–) gjethedelli këmbësore
Bg (1) перестолистен жиловлек

Cs (1) jitrocel vranožka
De (–) Schlitz-Wegerich
En (–) buck's-horn plantain
Pl (1) babka pierzasta
Ru (1) подорожник перистолопастный
Sk (–) skorocel vrania nôžka

* **Plantago indica** L. → *Plantago arenaria*

* **Plantago intermedia** DC. → *Plantago major*

1288 **Plantago lagopus** L.
Bg (1) мъхнатоцветен жиловлек
De (–) Zottiger Wegerich
En (–) hare's-foot plantain
Ru (0) подорожник заячий
Sk (–) skorocel vločkovitý

1289 **Plantago lanceolata** L.
Al (–) gjethedelli heshtor
Bg (1) теснолистен хиловлек
Cs (3) jitrocel kopinatý
De (–) Spitz-Wegerich
En (–) ribwort-plantain
Hu (2) lándzsás útifű
Pl (3) babka lancetowata
Ru (2) подорожник ланцетный
Sh (2) uskolisna bokvica; uskolisni trputac
Sk (–) skorocel kopijovitý

1290 **Plantago major** L.
Al (–) dejza; gjethedelli i madh
Bg (1) едролистен жиловлек
Cs (3) jitrocel větší
De (–) Großer Wegerich
En (–) greater plantain
Hu (3) nagy útifű
Pl (3) babka zwyczajna
Ro (1) pătlagină mare
Ru (2) подорожник большой
Sh (2) velika bokvica; veliki trputac
Sk (–) skorocel väčší

1291 **Plantago media** L.
Al (–) gjethedelli i ndërmjem
Bg (1) среден жиловлек
Cs (3) jitrocel prostřední
De (–) Mittlerer Wegerich
En (–) hoary plantain
Hu (2) réti útifű
Pl (3) babka średnia
Ru (1) подорожник средний
Sh (1) dugoklasni trputac; srednja bokvica
Sk (–) skorocel prostredný

* **Plantago ramosa** Asch. → *Plantago arenaria*

* **Plantago stricta** Schous. → *Plantago arenaria*

1292 **Platanthera bifolia** (L.) Richard
Al (–) plantantera dygjethëshe
Bg (1) двулистна платантера
Cs (0) vemeník dvoulistý
De (–) Zweiblättrige Kuckucksblume
En (–) lesser butterfly-orchid
Hu (0) kétlevelű sarkavirág
Pl (0) podkolan biały
Ro (2) stupiniţă
Ru (1) любка двулистная
Sh (1) bijeli vimenjak; dvolisni vimenjak
Sk (–) vemenník dvojlistný biely

* **Pleconax conica** L. → *Silene conica*

* **Pleuropterus cuspidatus** (Sieb. & Zucc.) Gross → *Reynoutria japonica*

1293 **Plumbago europaea** L.
Bg (1) европейски саркофай
Cs (0) olověnec evropský
De (–) Bleiwurz
En (–) European plumbago
Gr (1) λεπιδόχορτο
Pl (–) zawciąg

Ru (0) плюмбаго европейский
Sk (–) olovník európsky

1294 ***Poa* spp.** L.
Bg (1) ливадина
De (–) Rispengras
En (–) meadow-grass
Gr (1) πόα
Ru (2) мятлик
Sk (–) lipnica

1295 ***Poa alpina*** L.
Al (–) flokësa e Alpeve
Bg (1) алпийска ливадина
Cs (0) lipnice alpská
De (–) Alpen-Rispengras
En (–) alpine meadow-grass
Pl (0) wiechlina alpejska
Ro (0) firuşca şopîrlelor
Ru (0) мятлик альпийский
Sk (–) lipnica alpínska

1296 ***Poa annua*** L.
Al (–) flokësa njëvjeçare
Bg (1) едногодишна ливадина
Cs (3) lipnice roční
De (–) Einjähriges Rispengras
En (–) annual meadow-grass
Gr (2) κοινή πόα
Hu (2) egynyári perje
Pl (3) wiechlina roczna
Ro (1) hiruşor
Ru (2) мятлик однолетний
Sh (2) jednogodišnja livadarka
Sk (–) lipnica ročná

1297 ***Poa bulbosa*** L.
Al (–) flokësa me qepujkë
Bg (1) луковична ливадина
Cs (2) lipnice cibulkatá
De (–) Knolliges Rispengras
En (–) bulbous meadow-grass
Gr (1) βολβοπόα
Hu (–) gumós perje
Pl (1) wiechlina cebulkowata
Ro (0) firicea
Ru (1) мятлик луковичный
Sk (–) lipnica cibuľkovitá

1298 ***Poa chaixii*** Vill.
Bg (1) Шексиева ливадина
Cs (0) lipnice Chaixova
De (–) Wald-Rispengras
En (–) broad-leaved meadow-grass
Pl (1) wiechlina Chaixa
Ru (0) мятлик Шэ
Sk (–) lipnica širokolistá

1299 ***Poa compressa*** L.
Al (–) flokësa e rrasur
Bg (1) сплесната ливадина
Cs (2) lipnice smáčknutá
De (–) Flaches Rispengras
En (–) flattened meadow-grass
Hu (2) laposszárú perje
Pl (1) wiechlina spłaszczona
Ru (1) мятлик сплюснутый
Sk (–) lipnica stisnutá

* ***Poa dura*** (L.) Beauv. → *Sclerochloa dura*

1300 ***Poa nemoralis*** L.
Al (–) flokës pyjesh
Bg (0) горска ливадина
Cs (2) lipnice hajní
De (–) Hain-Rispengras
En (–) wood meadow-grass
Hu (0) ligeti perje
Pl (1) wiechlina gajowa
Ro (0) iarbă deasă
Ru (0) мятлик боровый
Sk (–) lipnica hájna

1301 ***Poa palustris*** L.
Al (–) flokës kënetore
Bg (1) мочурна ливадина
Cs (1) lipnice bahenní

De (–) Sumpf-Rispengras
En (–) swamp meadow-grass
Hu (0) mocsári perje
Pl (1) wiechlina błotna
Ru (1) мятлик болотный
Sk (–) lipnica močiarna

* **Poa pilosa** L. → *Eragrostis pilosa*

1302 **Poa pratensis** L.
Al (–) flokësa e livadheve
Bg (1) ливадна ливадина
Cs (2) lipnice luční
De (–) Wiesen-Rispengras
En (–) smooth meadow-grass
Hu (2) réti perje
Pl (2) wiechlina łąkowa
Ro (0) firuţă
Ru (2) мятлик луговой
Sk (–) lipnica lúčna

* **Poa sylvicola** Guss. → *Poa trivialis*

* **Poa triangularis** Gilib. → *Poa annua*

1303 **Poa trivialis** L.
Al (–) flokësa e zakonshme
Bg (1) обикновена ливадина
Cs (3) lipnice obecná
De (–) Gemeines Rispengras
En (–) rough meadow-grass
Gr (2) τραχεία λειβαδοπόα
Hu (3) sovány perje
Pl (2) wiechlina zwyczajna
Ru (2) мятлик обыкновенный
Sh (2) obična livadarka; obična vlasnjača
Sk (–) lipnica pospolitá

* **Podospermum laciniatum** (L.) DC. → *Scorzonera laciniata*

1304 **Polycarpon tetraphyllum** (L.) L.
Al (–) polikarpa me katër gjethe
Cs (0) kýlatka čtýrlistá
De (–) Vierblättriges Nagelkraut
En (–) four-leaved allseed
Gr (1) πολύκαρπο
Ru (1) многоплодник четырёхлистный
Sh (1) četverolisna pljevuša
Sk (–) stosemä štvorlisté

1305 **Polycnemum arvense** L.
Al (–) poliknema e fushës
Bg (1) полска хрупенка
Cs (0) chruplavník rolní
De (–) Acker-Knorpelkraut
Gr (1) μικρό πολύκνημο
Hu (1) közönséges torzon
Pl (0) chrząstkowiec polny
Ro (1) scîrţîitoare
Ru (1) хруплявник полевой
Sh (1) njivska jelica
Sk (–) chrumkavec roľný

1306 **Polycnemum majus** Braun
Al (–) poliknema më e madhe
Bg (1) едра хрупенка
Cs (0) chruplavník větší
De (–) Großes Knorpelkraut
Gr (1) μεγάλο πολύκνημο
Hu (1) közönséges torzon
Pl (0) chrząstkowiec większy
Ru (1) хруплявник крупный
Sk (–) chrumkavec väčší

* **Polycnemum minus** Kit. → *Polycnemum arvense*

1307 **Polygala spp.** L.
Bg (1) телчарка
De (–) Kreuzblume
En (–) milkwort
Gr (1) πολύγαλο
Ru (0) истод
Sk (–) vitod

1308 **Polygala comosa** Schk.
Al (–) poligala me baluke

Bg (1) качулеста телчарка
Cs (0) vitod chocholatý
De (–) Schopfige Kreuzblume
En (–) tufted milkwort
Hu (0) üstökös pacsirtafű
Pl (1) krzyżownica czubata
Ro (0) amăreală
Ru (0) истод хохлатый
Sh (1) dlakavi krestušac
Sk (–) vitod chocholatý

1309 ***Polygala nicaeensis*** Risso
Al (–) poligala Niceas
Bg (1) ниценска телчарка
De (–) Nizza-Kreuzblume
Ro (0) iarba laptoasa
Ru (1) истод инееватый

1310 ***Polygala vulgaris*** L.
Al (–) poligala e zakonshme
Cs (1) vítod obecný
De (–) Gemeine Kreuzblume
En (–) common milkwort
Hu (0) hegyi pacsirtafű
Pl (1) krzyżownica zwyczajna
Ro (0) şopîrliţă
Ru (1) истод обыкновенный
Sh (1) obični krestušac
Sk (–) vitod obyčajný

1311 ***Polygonatum latifolium*** (Jacq.) Desf.
Bg (1) широколистна момкова сълза
Cs (0) kokořík širolistý
De (–) Breitblättrige Weißwurz
En (–) broad-leaved Solomon's-seal
Hu (0) széleslevelű salamonpecsét
Pl (0) kokoryczka szerokolistna
Ru (0) купена широколистная
Sk (–) kokorík širokolistý

1312 ***Polygonatum verticillatum*** (L.) All.
Bg (1) прешленеста момкова сълза
Cs (0) kokořík přeslenitý
De (–) Quirlblättrige Weißwurz
En (–) whorled Solomon's-seal
Hu (0) pávafarkú salamonpecsét
Pl (1) kokoryczka okółkowa
Ru (0) купена мутовчатая
Sh (1) pršljenasta pokosnica
Sk (–) kokorík praslenatý

1313 ***Polygonum* spp.** L.
Bg (1) пиперче
De (–) Knöterich
En (–) polygonums
Gr (1) πολύγωνο
Ru (2) горец

* ***Polygonum aequale*** Lindm. → *Polygonum arenastrum*

1314 ***Polygonum amphibium*** L.
Al (–) nejcë amfibe
Bg (1) земеводно пиперче
Cs (3) rdesno obojživelné
De (–) Wasser-Knöterich
En (–) amphibious bistort
Hu (3) vidra keserűfű
Pl (2) rdest ziemno-wodny
Ro (1) sălcuţă
Ru (1) горец земноводный
Sh (1) barski dvornik
Sk (–) stavikrv obojživelný

1315 ***Polygonum arenarium*** Waldst. & Kit.
Al (–) nejca e ranishteve
Bg (2) красива пача трева
Cs (0) rdesno písečné
Hu (1) homoki keserűfű
Ro (0) troscot de nisipuri
Ru (1) горец песчаный
Sh (1) pešćarski dvornik; pješćarski dvornik
Sk (–) stavikrv piesočný

1316 ***Polygonum arenastrum*** Boreau
Bg (2) пача трева
De (–) Gleichblättriger Vogel-Knöterich

Hu (3) porcsin keserűfű
Sk (–) stavikrv pobrežný

1317 ***Polygonum argyrocoleum*** Steud.
Cs (–) rdesno stříbrné
De (–) Silberscheidiger Knöterich
Ru (1) горец серебристый
Sk (–) stavikrv strieborný

1318 ***Polygonum aviculare*** L.
Al (–) bar pate; bari i oborreve
Bg (2) гъшеница
Cs (3) truskavec ptačí
De (–) Vogel-Knöterich
En (–) knotgrass
Gr (3) πολυκόμπι
Hu (3) porcsin keserűfű
Pl (3) rdest ptasi
Ro (2) troscot
Ru (3) горец птичий
Sh (3) ptičji dvornik
Sk (–) stavikrv vtáčí

* ***Polygonum binicola*** Sutulov →
Polygonum lapathifolium

1319 ***Polygonum bistorta*** L.
Al (–) nejca e përdredhur
Cs (2) rdesno hadí kořen
De (–) Schlangen-Knöterich
En (–) common bistort
Hu (0) kígyógyökerű keserűfű
Pl (2) rdest wężownik
Ro (0) răculeț
Ru (1) горец змеиный
Sh (1) uvinuti dvornik
Sk (–) hadí koreň väčší

* ***Polygonum convolvulus*** L. →
Bilderdykia convolvulus

* ***Polygonum cuspidatum*** Sieb. & Zucc.
→ *Reynoutria japonica*

* ***Polygonum dumetorum*** L. →
Bilderdykia dumetorum

1320 ***Polygonum equisetiforme*** Sibth. & Sm.
Bg (1) хвощоподобна пача трева
Ru (0) горец хвощевидный
Sk (–) stavikrv prasličkovitý

* ***Polygonum geniculatum*** Poir. →
Polygonum aviculare

1321 ***Polygonum hydropiper*** L.
Al (–) nejca piper uji
Bg (1) водно пиперче
Cs (2) rdesno peprník
De (–) Wasserpfeffer
En (–) water-pepper
Gr (1) νεροπιπεριά
Hu (1) borsos keserűfű
Pl (2) rdest ostrogorzki
Ro (1) piperul bălții
Ru (1) горец перечный
Sh (1) papreni dvornik
Sk (–) stavikrv korenistý

* ***Polygonum kitaibelianum*** Sadl. →
Polygonum patulum

1322 ***Polygonum lapathifolium*** L.
Al (–) nejca gjethelëpjete
Bg (2) лападоволистно пиперче
Cs (3) rdesno blešník
De (–) Ampfer-Knöterich
En (–) pale persicaria
Gr (2) λαπάτσα
Hu (3) lapulevelű keserűfű
Pl (3) rdest kolankowaty
Ro (2) iarbă roșie
Ru (3) горец узловатый
Sh (2) uzlati dvornik; veliki lisac
Sk (–) stavikrv plsťnatý pravý

1323 ***Polygonum minus*** Huds.
Al (–) hejda e vogël

Bg (1) малко пиперче
Cs (0) rdesno menší
De (–) Kleiner Knöterich
En (–) small water-pepper
Hu (1) keskenylevelű keserűfű
Pl (2) rdest mniejszy
Ru (1) горец малый
Sh (1) uskolisni dvornik
Sk (–) stavikrv menší

1324 ***Polygonum mite*** Schrank
Cs (0) rdesno řídkokvěté
De (–) Milder Knöterich
En (–) tasteless water-papper
Hu (1) szelíd keserűfű
Pl (1) rdest łagodny
Ru (1) горец мягкий
Sh (1) srednji dvornik
Sk (–) stavikrv riedkokvetý

1325 ***Polygonum orientale*** L.
Al (–) hejda e Lindjes
Cs (0) rdesno východní
De (–) Orient-Knöterich
Hu (0) karmazsinvirág
Pl (0) rdest wschodni
Ro (0) moţul curcanului
Ru (1) горец восточный
Sk (–) stavikrv východný

1326 ***Polygonum oxyspermum*** Meyer & Bun.
Bg (1) райева пача трева
De (–) Strand-Knöterich
En (–) Ray's knotgrass
Pl (1) rdest nadbrzeżny
Ru (1) горец остроплодный

1327 ***Polygonum patulum*** Bieb.
Bg (1) разперена пача трева
Cs (0) rdesno rozkladité
Hu (1) kitaibel keserűfű
Pl (0) rdest rozłożysty
Ru (1) горец отклонённый
Sh (1) široki dvornik
Sk (–) stavikrv rozložitý

1328 ***Polygonum persicaria*** L.
Al (–) nejca gjethepjeshk
Bg (1) прасковолистно пиперче
Cs (3) rdesno červivec
De (–) Floh-Knöterich
En (–) redshank
Gr (2) αγριοπιπεριά
Hu (3) baracklevelű keserűfű
Pl (3) rdest plamisty
Ro (1) iarbă roşie
Ru (2) горец почечуйный
Sh (2) mali lisac; obični dvornik
Sk (–) horčiak obyčajný

* ***Polygonum pulchellum*** Loisel. → *Polygonum arenarium*

* ***Polygonum pusillum*** Lam. → *Polygonum minus*

* ***Polygonum raii*** Bab. → *Polygonum oxyspermum*

* ***Polygonum tomentosum*** Schrank → *Polygonum lapathifolium*

1329 ***Polypodium vulgare*** L.
Cs (–) osladič obecný
De (–) Gemeiner Tüpfelfarn
En (–) common polypody
Ru (0) многоножка обыкновенная
Sh (1) obični oslad; slatka paprat
Sk (–) sladič obyčajný

1330 ***Polypogon maritimus*** Willd.
Al (–) polipoga bregdetare
Bg (1) приморски полипогон
Ru (0) многобородник морской
Sk (–) bradatec prímorský

1331 ***Polypogon monspeliensis*** (L.) Desf.
Al (–) polipoga e Monspelies
Cs (0) vousatec středozemský
De (–) Gemeines Bürstengras
En (–) annual beard-grass
Ru (1) многобородник монпельинский
Sk (–) bradatec stredozemský

* ***Polypogon semiverticillatus*** (Forsk.) Hyl. → *Polypogon viridis*

1332 ***Polypogon viridis*** (Gouan) Breistr.
En (–) green beard-grass
Ro (0) păiuş
Ru (1) многобородник полумутовчатый
Sk (–) bradatec zelený

1333 ***Polystichum aculeatum*** (L.) Roth
De (–) Borstiger Schildfarn
En (–) hard shield-fern
Ru (0) многорядник шиповатый
Sh (1) krpasta papratnjača
Sk (–) papradina laločnatá

* ***Porrum rotundum*** Riechenb. → *Allium scorodoprasum* ssp. *rotundum*

* ***Portulaca officinarum*** Cran. → *Portulaca oleracea*

1334 ***Portulaca oleracea*** L.
Al (–) burdullaku
Bg (2) тученица
Cs (2) šrucha zelná
De (–) Portulak
En (–) purslane
Gr (3) αντράκλα
Hu (3) kövér porcsin
Pl (2) portulaka pospolita
Ro (3) graşiţă
Ru (2) портулак огородный
Sh (2) obični portulak; obični tušt
Sk (–) šrucha kapustná

* ***Portulaca parvifolia*** Haw. → *Portulaca oleracea*

1335 ***Potamogeton spp.*** L.
Bg (1) ръждавец
De (–) Laichkraut
En (–) pondweed
Gr (1) ποταμόχορτο
Ru (1) рдест
Sk (–) červenavec

1336 ***Potamogeton alpinus*** Balbis
Cs (0) rdest alpský
De (–) Alpen-Laichkraut
En (–) red pondweed
Pl (1) rdestnica alpejska
Ru (0) рдест альпийский
Sk (–) červenavec alpínsky

1337 ***Potamogeton crispus*** L.
Al (–) kalli njëkaçurrel; potamogeta kaçurrele
Bg (1) къдраволистен ръждавец
Cs (0) rdest kadeřavý
De (–) Krauses Laichkraut
En (–) curled pondweed
Hu (1) bodros békaszőlő
Pl (1) rdestnica kędzierzawa
Ro (1) paşă
Ru (1) рдест курчавый
Sh (1) kovrdjavi mresnjak; kudrasti mrijesnjak
Sk (–) červenavec kučeravý

1338 ***Potamogeton gramineus*** L.
Al (–) potamogeta gramore
Cs (0) rdest trávovitý
De (–) Grasartiges Laichkraut
En (–) various-leaved pondweed
Hu (1) felemáslevelű békaszőlő
Pl (1) rdestnica trawiasta
Ru (1) рдест разнолистный
Sk (–) červenavec trávolistý

1339 *Potamogeton lucens* L.
Al (–) potamogeta e ndritshme
Cs (0) rdest světlý
De (–) Glänzendes Laichkraut
En (–) shining pondweed
Hu (1) üveglevelű békaszőlő
Pl (1) rdestnica połyskująca
Ro (0) broasca apei
Ru (0) рдест блестящий
Sk (–) červenavec lesklý

1340 *Potamogeton natans* L.
Al (–) gjuhujësja; potamogeta notuese
Bg (1) плуващ ръждавец
Cs (0) rdest plovoucí
De (–) Schwimmendes Laichkraut
En (–) broad-leaved pondweed
Hu (1) úszó békaszőlő
Pl (1) rdestnica pływająca
Ro (1) broscariţă
Ru (1) рдест плавающий
Sk (–) červenavec plávajúci

1341 *Potamogeton nodosus* Poir.
Cs (0) rdest vzplývavý
De (–) Flutendes Laichkraut
En (–) Loddon pondweed
Hu (0) imbolygó békaszőlő
Pl (1) rdestnica nawodna
Ru (0) рдест узловатый
Sh (1) čvorasti mrijesnjak; rečni mresnjak
Sk (–) červenavec riečny

1342 *Potamogeton pectinatus* L.
Al (–) kalli uji kreshtak; potamogeta si krehër
Bg (1) гребенест ръждавец
Cs (0) rdest hřebenitý
De (–) Kamm-Laichkraut
En (–) fennel-leaved pondweed
Hu (1) fésüs békaszőlő
Pl (1) rdestnica grzebieniasta
Ru (0) рдест гребенчатый
Sk (–) červenavec hrebenitý

1343 *Potamogeton perfoliatus* Pour.
Al (–) kalli uji gjethembështjellës; potamogeta me gjethe pushtuese
Bg (1) стъблообхващащ ръждавец
Cs (0) rdest prorostlý
De (–) Durchwachsenes Laichkraut
En (–) perfoliate pondweed
Hu (1) hínáros békaszőlő
Pl (1) rdestnica przeszyta
Ro (0) moţ
Ru (1) рдест стеблеобъемлющий
Sk (–) červenavec prerastaný

1344 *Potamogeton polygonifolius* Pour.
Bg (1) листен ръждавец
De (–) Knöterich-Laichkraut
En (–) bog pondweed
Pl (0) rdestnica podługowata
Ru (0) рдест гречиховидный
Sk (–) červenavec stavikrvolistý

1345 *Potamogeton pusillus* L.
Al (–) kalli uji i vogël; potamogeta e vogël
Bg (1) малък ръждавец
Cs (0) rdest maličký
De (–) Kleines Laichkraut
En (–) lesser pondweed
Hu (1) apró békaszőlő
Pl (1) rdestnica drobna
Ru (1) рдест Брехтольда
Sk (–) červenavec maličký pravý

1346 *Potentilla* spp. L.
Bg (1) очиболец
De (–) Fingerkraut
Gr (1) ποτεντίλλα
Ru (1) лапчатка
Sk (–) nátržník

1347 *Potentilla alba* L.
Bg (0) бял очиболец
Cs (0) mochna bílá
De (–) Weißes Fingerkraut
En (–) white cinquefoil

Hu (0) fehér pimpó
Pl (1) pięciornik biały
Ru (–) лапчатка белая
Sk (–) nátržník biely

1348 ***Potentilla anserina*** L.
Bg (1) патицов очиболец
Cs (3) mochna husí
De (–) Gänse-Fingerkraut
En (–) silverweed
Hu (1) liba pimpó
Pl (2) pięciornik gęsi
Ro (1) coada racului
Ru (1) лапчатка гусиная
Sh (2) gušćija petoprsnica; plazeća petoprsta
Sk (–) nátržník husí

1349 ***Potentilla argentea*** L.
Al (–) pesëfoja e argjentë; zorrëca e argjentë
Bg (1) сребролистен очиболец
Cs (2) mochna stříbrná
De (–) Silber-Fingerkraut
En (–) hoary cinquefoil
Hu (1) ezüstös pimpó
Pl (2) pięciornik srebrny
Ro (1) scrîntitoare
Ru (1) лапчатка серебристая
Sh (1) srebrna petoprsnica
Sk (–) nátržník strieborný

1350 ***Potentilla bifurca*** L.
Ru (1) лапчатка вильчатая

1351 ***Potentilla erecta*** (L.) Räusch.
Al (–) zorrëca e përpjetë
Bg (1) планински очиболец
Cs (1) mochna nátržník
De (–) Aufrechtes Fingerkraut
En (–) tormentil
Hu (0) vérontófű
Pl (2) pięciornik kurze ziele
Ro (0) sclipeţi
Ru (1) лапчатка прямостоящая
Sh (1) osovna petoprsta; velika petoprsnica
Sk (–) nátržník rovný

1352 ***Potentilla hirta*** L.
Al (–) zorrëca leshatake
Bg (1) космат очиболец
Cs (0) mochna chlupatá
En (–) hairy cinquefoil
Sk (–) nátržník chlpatý

* ***Potentilla orientalis*** Juz. → *Potentilla bifurca*

1353 ***Potentilla recta*** L.
Al (–) zorrëca e drejtë
Bg (1) изправен очиболец
Cs (0) mochna přímá
De (–) Hohes Fingerkraut
En (–) sulphur cinquefoil
Hu (0) egyenes pimpó
Pl (1) pięciornik wyprostowany
Ro (1) buruiană cu cinci degete
Ru (1) лапчатка прямая
Sh (1) uspravna petoprsta
Sk (–) nátržník vzpriamený

1354 ***Potentilla reptans*** L.
Al (–) zorrëca zvaranike
Bg (0) пълзящ очиболец
Cs (3) mochna pětilístek
De (–) Kriechendes Fingerkraut
En (–) creeping cinquefoil
Hu (1) indás pimpó
Pl (2) pięciornik rozłogowy
Ro (1) cinci degete
Ru (1) лапчатка ползучая
Sh (2) puzava petoprsta
Sk (–) nátržník plazivý

1355 ***Potentilla supina*** L.
Bg (0) лежащ очиболец
Cs (0) mochna poléhavá

De (–) Niedriges Fingerkraut
Hu (2) henye pimpó
Pl (1) pięciornik niski
Ru (1) лапчатка низкая
Sk (–) nátržník poliehavý

* ***Potentilla sylvestris*** Neck. → *Potentilla erecta*

* ***Poterium polygamum*** Waldst. & Kit. → *Sanguisorba minor*

* ***Poterium sanguisorba*** L. → *Sanguisorba minor*

* ***Poterium spinosum*** L. → *Sarcopterium spinosum*

1356 ***Primula*** **spp.** L.
Bg (1) иглика
De (–) Schlüsselblume
Gr (1) πριμούλα
Ru (0) первоцвет
Sk (–) prvosienka

1357 ***Primula elatior*** (L.) Hill
Al (–) aguliçja e ngritur
Bg (1) висока иглика
Cs (0) prvosenka vyšší
De (–) Hohe Schlüsselblume
En (–) oxlip
Hu (0) sugár kankalin
Pl (1) pierwiosnek wyniosły
Ro (1) țîța vacii
Ru (0) первоцвет высокий
Sk (–) prvosienka bledožltá

1358 ***Primula veris*** L.
Al (–) aguliçe e vërtetë; aguliçe me erë
Cs (2) prvosenka jarní
De (–) Wiesen-Schlüsselblume
En (–) cowslip
Hu (0) tavaszi kankalin
Ro (1) ciuboțica cucului
Ru (0) первоцвет весенний
Sh (1) ljekoviti jaglac; prolećna jagorčevina
Sk (–) prvosienka jarná pravá

1359 ***Prunella grandiflora*** (L.) Scholl.
Al (–) prunela lulemadhe
Cs (–) černohlávek velkokvětý
De (–) Großblutige Braunelle
En (–) large self-heal
Gr (1) μεγάλο βουτυρόχορτο
Hu (0) nagyvirágú gyíkfű
Pl (1) głowienka wielkokwiatowa
Ru (1) черноголовка крупноцветковая
Sh (1) krupnocvetni crnjevac; krupnocvjetna celinščica
Sk (–) čiernohlávok veľkokvetý

1360 ***Prunella laciniata*** (L.) L.
Al (–) prunela e rrypzuar
Cs (–) černohlávek dřípeny
De (–) Weiße Braunelle
En (–) cut-leaved self-heal
Gr (1) λευκό βουτυρόχορτο
Hu (0) fehér gyíkfű
Pl (0) głowienka kremowa
Ru (0) черноголовка разрезная
Sh (1) žućkasti crnjevac
Sk (–) čiernohlávok driapatý

1361 ***Prunella vulgaris*** L.
Al (–) kokëza; prunela e rëndomtë
Bg (1) обикновено усойниче
Cs (3) černohlávek obecný
De (–) Gemeine Braunelle
En (–) common self-heal
Gr (1) κοινό βουτυρόχορτο
Hu (1) közönséges gyíkfű
Pl (2) głowienka pospolita
Ro (1) busuioc de cîmp
Ru (1) черноголовка обыкновенная
Sh (1) obična celinščica; obični crnjevac
Sk (–) čiernohlávok obyčajný

* **Psammophilliella muralis** (L.) Ikonn. → *Gypsophila muralis*

1362 **Psoralea bituminosa** L.
Al (–) psorala katranore
Bg (1) смолиста псоралея
De (–) Asphaltklee
En (–) pitch trefoil
Gr (1) καλοσύκι
Ru (0) псоралея смолистая
Sk (–) psoralea asfaltová

1363 **Pteridium aquilinum** (L.) Kuhn
Al (–) fier shqiponje
Bg (2) орлова папрат
Cs (0) hasivka orličí
De (–) Adlerfarn
En (–) bracken
Gr (2) φτέρη
Hu (1) saspáfrány
Pl (1) orlica pospolita
Ro (2) ţolul lupului
Ru (1) орляк обыкновенный
Sh (2) obična bujad
Sk (–) papradie orličie

* **Pteridium tauricum** (Presl) Krecz. → *Pteridium aquilinum*

* **Pteris aquilina** L. → *Pteridium aquilinum*

* **Pterocephalus brachiatus** Spreng. → *Tremastelma palaestinum*

1364 **Pulicaria dysenterica** (L.) Bernh.
Al (–) brengeçë; plenëra
Bg (1) дизентерийно шумкавче
Cs (0) blešník úplavičný
De (–) Ruhrwurz
En (–) common fleabane
Gr (1) σκυλόχορτο
Hu (1) réti bolhafű
Pl (1) płesznik czerwonkowy
Ro (1) tătăiş
Ru (1) блошница дизентерийная
Sk (–) blšník úplavičný

1365 **Pulicaria odora** (L.) Reichenb.
Gr (1) μυρωδάτη πουλικάρια
Ru (1) блошница душистая
Sk (–) blšník voňavý

1366 **Pulicaria vulgaris** Gaertn.
Bg (1) обикновено шумкавче
Cs (0) blešník obecný
De (–) Kleines Flohkraut
En (–) small fleabane
Gr (1) κοινή πουλικάρια
Hu (1) parlagi bolhafű
Pl (1) płesznik zwyczajny
Ro (1) puricariţă
Ru (1) блошница простёртая
Sh (2) obični srčun; pahuljavi businja
Sk (–) blšník obyčajný

1367 **Pulmonaria angustifolia** L.
Cs (0) plicník úzkolistý
De (–) Schmalblättriges Lungenkraut
Hu (0) keskenylevelű tüdőfű
Pl (1) miodunka wąskolistna
Ru (0) медуница узколистная
Sk (–) pľúcnik úzkolistý

1368 **Pulmonaria officinalis** L.
Al (–) bar i mushkërisë; pulmonaria mjekësore
Bg (1) лечебна медуница
Cs (1) plicník lékařský
De (–) Echtes Lungenkraut
En (–) common lungwort
Hu (0) bársonyos tüdőfű
Pl (0) miodunka plamista
Ro (0) cuscrişor
Ru (0) медуница лекарственная
Sh (1) ljekoviti plućnjak; obični plućnjak
Sk (–) pľúcnik lekársky

1369 *Pulsatilla pratensis* (L.) Mill.
Bg (1) полско котенце
Cs (1) koniklec luční
De (–) Wiesen-Küchenschelle
En (–) small Pasque-flower
Pl (1) sasanka łąkowa
Ru (0) прострел луговой
Sk (–) poniklec lúčny

1370 *Pulsatilla vulgaris* Mill.
Bg (1) обикновено котенце
Cs (0) koniklec obecný
De (–) Gewöhnliche Küchenschelle
En (–) Pasque-flower
Pl (1) sasanka zwyczajna
Ro (1) dediţei
Ru (1) прострел обыкновенный
Sh (1) blijeda kučika; obična sasa
Sk (–) poniklec obyčajný

R

1371 *Ranunculus* spp. L.
Bg (1) лютиче
De (–) Hahnenfuß
En (–) buttercup
Gr (1) ρανούγκουλος
Ru (2) лютик
Sk (–) izkiernik

* ***Ranunculus acer*** auct. → *Ranunculus acris*

1372 *Ranunculus aconitifolius* L.
Al (–) zhabinë e verdhë
Cs (1) pryskyřník omějolistý
De (–) Eisenhutblättriger Hahnenfuß
En (–) white buttercup
Sk (–) iskiernik prilbicolistý

1373 *Ranunculus acris* L.
Al (–) zhabinë e mprehtë
Bg (1) обикновено лютиче
Cs (3) pryskyřník prudký
De (–) Scharfer Hahnenfuß
En (–) meadow buttercup
Hu (1) réti boglárka
Pl (3) jaskier ostry
Ro (1) floare broştească
Ru (2) лютик едкий
Sh (1) oštri ljutić; žabnjak ljutić
Sk (–) iskiernik prudký

1374 *Ranunculus aquatilis* L.
Al (–) zhabinë e ujit
Bg (1) водно лютиче
Cs (1) lakušník vodní
De (–) Wasser-Hahnenfuß
En (–) common water-crowfoot
Hu (1) nagy víziboglárka
Ru (0) шелковник обыкновенный
Sk (–) mociarka vodná

1375 *Ranunculus arvensis* L.
Al (–) zhabinë e arës
Bg (2) полско лютиче
Cs (3) pryskyřník rolní
De (–) Acker-Hahnenfuß
En (–) corn buttercup
Hu (3) vetési boglárka
Pl (2) jaskier polny
Ro (2) piciorul cocoşului de semănături
Ru (2) лютик полевой
Sh (2) njivski ljutić; poljski žabnjak
Sk (–) iskiernik roľný

1376 *Ranunculus auricomus* L.
Bg (1) златисто лютиче
Cs (1) pryskyřník zlatožlutý
De (–) Goldberger Hahnenfuß
En (–) goldilocks buttercup
Hu (0) változó boglárka
Pl (1) jaskier różnolistny
Ru (0) лютик золотистый
Sk (–) iskiernik zlatožltý

1377 *Ranunculus bulbosus* L.
Al (–) zhabinë e egër
Bg (1) луковично лютиче
Cs (2) pryskyřník hlíznatý
De (–) Knolliger Hahnenfuß
En (–) bulbous buttercup
Hu (1) hagymás boglárka
Pl (1) jaskier bulwkowy
Ru (1) лютик клубненосный
Sk (–) iskiernik hľuznatý

1378 *Ranunculus circinatus* Sibth.
De (–) Spreizender Hahnenfuß
En (–) fan-leaved water-crowfoot
Pl (1) włosienicznik krążkolistny
Ru (0) шелковник жестколистный
Sk (–) močiarka okrúhla

* ***Ranunculus echinatus*** Cran. → *Ranunculus arvensis*

* ***Ranunculus falcatus*** L. → *Ceratocephalus falcatus*

1379 ***Ranunculus ficaria*** L.
Al (–) lulari
Bg (1) жълтурче
Cs (3) orsej jarní
De (–) Frühlings-Scharbockskraut
En (–) lesser celandine
Hu (1) saláta boglárka
Pl (1) pszonka ziarnopłon
Ro (0) untişor
Ru (1) чистяк весенний
Sh (1) ledinjak žabnjak; prolećni ledinjak
Sk (–) blyskač jarný

1380 ***Ranunculus flammula*** L.
Al (–) zhabina flakë
Bg (0) огненоцветно лютиче
Cs (1) pryskyřník plamének
De (–) Brennender Hahnenfuß
En (–) lesser spearwort
Hu (1) béka boglárka
Pl (1) jaskier płomiennik
Ru (1) лютик жгучий
Sk (–) iskiernik plamenný

1381 ***Ranunculus illyricus*** L.
Bg (1) илирийско лютиче
Cs (0) pryskyřník illyrský
De (–) Illyrischer Hahnenfuß
Pl (0) jaskier ilyryjski
Ro (0) trînjoaică
Ru (0) лютик иллирийский
Sk (–) iskiernik ilýrsky

1382 ***Ranunculus lanuginosus*** L.
Al (–) zhabinë pushtake
Bg (1) вълнесто лютиче
Cs (1) pryskyřník kosmatý
De (–) Wolliger Hahnenfuß
En (–) woolly buttercup
Hu (0) gyapjas boglárka
Pl (1) jaskier kosmaty
Ru (1) лютик шерстистый
Sk (–) iskiernik páperistý

1383 ***Ranunculus lingua*** L.
Al (–) zhabinë gjuhë
Bg (1) езичесто лютиче
Cs (0) pryskyřník veliký
De (–) Großer Hahnenfuß
En (–) greater spearwort
Hu (0) nádi boglárka
Pl (1) jaskier wielki
Ru (0) лютик длиннолистный
Sk (–) iskiernik jazykovitý

1384 ***Ranunculus montanus*** Willd.
Bg (1) планинско лютиче
Cs (0) pryskyřník horský
De (–) Berg-Hahnenfuß
En (–) mountain buttercup
Pl (1) jaskier górski
Ru (0) лютик горный
Sk (–) iskiernik horský

1385 ***Ranunculus muricatus*** L.
Bg (1) шипоплодно лютиче
De (–) Stachelfrüchtiger Hahnenfuß
En (–) roughseed buttercup
Hu (3) vetési boglárka
Ru (2) лютик мягкоигольчатый
Sk (–) iskiernik mäkkoostenný

1386 ***Ranunculus neapolitanus*** Ten.
Bg (1) неаполско лютиче
Ru (0) лютик неаполитанский
Sk (–) iskiernik neapolský

1387 ***Ranunculus nemorosus*** DC.
Bg (1) горско лютиче
Cs (0) pryskyřník Breynův
De (–) Wald-Hahnenfuß
Pl (1) jaskier gajowy
Ro (0) gălbenele de pădure
Ru (0) лютик дубравный
Sk (–) iskiernik hájny

1388 ***Ranunculus ophioglossifolius*** Vill.
Al (–) zhabinë me gjethe si gjuhëgjarpëri
Bg (1) змиоезичестолистно лютиче
En (–) adder's-tongue spearwort
Ru (0) лютик ужовниколистный
Sk (–) iskiernik hadojazyčnolistý

* ***Ranunculus orthoceras*** Benth. & Hook. → *Ranunculus sardous*

1389 ***Ranunculus polyanthemos*** L.
Al (–) zhabinë me shumë lule
Bg (1) многоцветно лютиче
Cs (0) pryskyřník mnohokvětý
De (–) Vielblütiger Hahnenfuß
Hu (1) sokvirágú boglárka
Pl (1) jaskier wielkokwiatowy
Ro (1) gălbenele
Ru (1) лютик многоцветковый
Sh (1) mnogocvetni ljutić; mnogocvjetni žabnjak
Sk (–) iskiernik mnohokvetý

1390 ***Ranunculus repens*** L.
Al (–) zhabinë zvaranike
Bg (1) пълзящо лютиче
Cs (3) pryskyřník plazivý
De (–) Kriechender Hahnenfuß
En (–) creeping buttercup
Hu (1) kúszó boglárka
Pl (2) jaskier rozłogowy
Ro (1) floare de leac
Ru (2) лютик ползучий
Sh (2) puzavi žabnjak; vrezasti ljutić
Sk (–) iskiernik plazivý

1391 ***Ranunculus sardous*** Cran.
Al (–) zhabinë e Sardenjës
Bg (1) сардинско лютиче
Cs (0) pryskyřník sardinský
De (–) Sardinischer Hahnenfuß
En (–) hairy buttercup
Hu (2) buborcs boglárka
Pl (1) jaskier sardyński
Ro (2) piciorul cocoşului
Ru (1) лютик сардинский
Sh (1) obični ljutić; sardinijski žabnjak
Sk (–) iskiernik sardínsky

1392 ***Ranunculus sceleratus*** L.
Al (–) zhabinë e moçaleve
Bg (1) отровно лютиче
Cs (2) pryskyřník lítý
De (–) Gift-Hahnenfuß
En (–) celery-leaved buttercup
Hu (1) torzsika boglárka
Ro (2) boglari
Ru (1) лютик ядовитый
Sh (1) otrovni ljutić; otrovni žabnjak
Sk (–) iskiernik jedovatý

* ***Ranunculus stevenii*** auct. → *Ranunculus acris*

1393 ***Ranunculus trichophyllus*** Chaix
Bg (1) нишколистно лютиче
Cs (–) lakušník chudoprašný
De (–) Haarblättriger Hahnenfuß
En (–) thread-leaved water-crowfoot
Hu (1) hínáros víziboglárka
Ru (0) шелковник волосолистный
Sk (–) močiarka vlásočnolistá

1394 ***Ranunculus velutinus*** Ten.
Al (–) zhabinë si kadife
Bg (1) кадифенолистно лютиче
Sk (–) močiarka zamatová

* ***Raphanistrum segetum*** Baumg. → *Raphanus raphanistrum*

* ***Raphanus landra*** Moret. → *Raphanus raphanistrum*

1395 ***Raphanus raphanistrum*** L.
Al (–) rrapanidhë repë e egër; rrepan i egër

Bg (2) дива ряпа
Cs (3) ředkev ohnice
De (–) Hederich
En (–) wild radish
Gr (2) ραπανίδα
Hu (3) repcsényretek
Pl (3) rzodkiew świrzepa
Ro (3) ridichioară
Ru (3) редька дикая
Sh (2) divlja rotkva
Sk (–) ohnica poľná

1396 *Rapistrum* spp. Cran.
Bg (1) репник
De (–) Rapsdotter
Gr (1) ράπιστρο
Ru (–) репник
Sk (–) repča

1397 *Rapistrum perenne* (L.) All.
Bg (1) многогодишен репник
Cs (1) řepovník vytrvalý
De (–) Ausdauernder Rapsdotter
En (–) steppe cabbage
Hu (2) rekenyő
Pl (1) świrzepa trwała
Ro (2) ciurlan alb
Ru (1) репник многолетний
Sh (1) trajna poveska
Sk (–) repča trváca

* ***Rapistrum perfoliatum*** Berg. → *Myagrum perfoliatum*

1398 *Rapistrum rugosum* (L.) All.
Al (–) rrepicë
Bg (1) едногодишен репник
Cs (0) řepovník svraskalý
De (–) Runzliger Rapsdotter
En (–) bastard cabbage
Pl (0) świrzepa pomarszczona
Ru (0) репник морщинистый
Sk (–) repča vráskavá

1399 *Reseda alba* L.
Al (–) djegës i bardhë; shukulli i bardhë
Cs (0) rýt bílý
De (–) Weißer Wau
En (–) white mignonette
Gr (1) άσπρη ρεζεντά
Ru (0) резеда белая
Sk (–) rezeda biela

1400 *Reseda lutea* L.
Al (–) shukulli i verdhë
Bg (1) жълта резеда
Cs (2) rýt žlutý
De (–) Gelbe Resede
En (–) wild mignonette
Gr (1) ώχρα
Hu (3) vadrezeda
Pl (1) rezeda żółta
Ro (2) rechie
Ru (2) резеда жёлтая
Sh (1) obična rezeda; žuti katanac
Sk (–) rezeda žltá

1401 *Reseda luteola* L.
Al (–) djegës verbues
Bg (1) жълтеникава резеда
Cs (0) rýt barvířský
De (–) Farber-Resede
En (–) weld
Hu (1) festő rezeda
Pl (0) rezeda żółtawa
Ro (0) gaude
Ru (1) резеда жёлтенькая
Sh (1) žućkasta rezeda
Sk (–) rezeda žltavá

1402 *Reseda phyteuma* L.
Al (–) djegës i egër
Cs (0) rýt velkokališní
De (–) Wilde Resede
En (–) rampion mignonette
Hu (1) terpedt rezeda
Pl (0) rezeda mała
Sk (–) rezeda veľkokališná

1403 ***Reynoutria japonica*** Houtt.
Cs (0) křídlatka hrotolistá
De (–) Zugespitzter Knöterich
En (–) Japanese knotweed
Hu (1) japánkóró
Pl (0) rdest ostrokończysty
Sk (–) krídlatka končistolistá

1404 ***Rhagadiolus stellatus*** (L.) Gaertn.
Al (–) ragadiola yllake
Cs (0) kosatka hvězdovitá
De (–) Sichelsalat
En (–) star hawkbit
Ru (1) рагадиолюс звездчатый
Sk (–) kosatka hviezdovitá

* ***Rhinanthus aestivalis*** (Zing.) Schisk. & Serg. → *Rhinanthus angustifolius*

1405 ***Rhinanthus alectorolophus*** (Scop.) Pollich
Bg (2) шушнярка
Cs (2) kokrhel luštinec
De (–) Zottiger Klappertopf
Hu (1) borzas kakascímer
Pl (2) szelężnik kosmaty
Ru (2) погремок большой
Sk (–) štrkáč kohútik

1406 ***Rhinanthus angustifolius*** Gmel.
Al (–) lafshë gjeli e madhe; taklinia
Bg (2) едра клопачка
Cs (2) kokrhel pozdní
De (–) Großer Klappertopf
En (–) great yellow-rattle
Hu (1) nagyvirágú kakascímer
Pl (3) szelężnik większy
Ru (2) погремок остистый
Sh (2) uskolisni šuškavac; veliki šuškavac
Sk (–) štrkáč úzkolistý

* ***Rhinanthus apterus*** (Fries) Ostenf. → *Rhinanthus angustifolius*

* ***Rhinanthus crista-galli*** L. var. ***major*** → *Rhinanthus angustifolius*

* ***Rhinanthus crista-galli*** L. var. ***minor*** → *Rhinanthus minor*

* ***Rhinanthus major*** L. → *Rhinanthus alectorolophus*

1407 ***Rhinanthus minor*** L.
Al (–) lafshë gjeli e vogël; taklinia më e vogël
Bg (1) малка клопачка
Cs (2) kokrhel menší
De (–) Kleiner Klappertopf
En (–) lesser yellow-rattle
Hu (1) csörgő kakascímer
Pl (1) szelężnik mniejszy
Ru (1) погремок малый
Sh (1) mali šuškavac
Sk (–) štrkáč menší

1408 ***Rhinanthus rumelicus*** Velen.
Bg (2) румелийска клопачка
Cs (0) kokrhel rumelský
De (–) Drüsiger Klappertopf
Hu (1) mirigyes kakascímer
Ro (2) clocotici
Ru (1) погремок эзельский
Sh (1) rumelijski šuškavac
Sk (–) štrkáč rumelský

* ***Rhinanthus serotinus*** (Schön.) Obor. → *Rhinanthus angustifolius*

1409 ***Roemeria hybrida*** (L.) DC.
Ru (1) рёмерия гибридная

1410 ***Rorippa* spp.** Scop.
Bg (–) пореч
De (–) Sumpfkresse
En (–) yellow-cress
Gr (1) ρορίππα

Ru (1) жерушник
Sk (–) roripa

1411 ***Rorippa amphibia*** (L.) Bess.
Bg (1) воден пореч
Cs (0) rukev obojživelná
De (–) Wasser-Sumpfkresse
En (–) great yellow-cress
Hu (1) vízi kányafű
Pl (1) rzepicha ziemnowodna
Ro (1) gălbenea
Ru (1) жерушник земноводный
Sk (–) roripa obojživelná

1412 ***Rorippa austriaca*** (Cran.) Bess.
Bg (1) австрийски пореч
Cs (0) rukev rakouská
De (–) Österreichische Sumpfkresse
En (–) Austrian yellow-cress
Hu (2) osztrák kányafű
Pl (0) rzepicha austriacka
Ro (1) gălbenea
Ru (1) жерушник австрийский
Sh (1) austrijski grbak; austrijski ugaz
Sk (–) roripa rakúska

1413 ***Rorippa islandica*** (Oed.) Borb.
Bg (1) блатен пореч
Cs (0) rukev islandská
De (–) Sumpfkresse
En (–) marsh yellow-cress
Hu (1) mocsári kányafű
Ro (1) rapiţă sălbatică
Ru (1) жерушник болотный
Sk (–) roripa islandská

1414 ***Rorippa lippizensis*** (Wulf.) Reichenb.
Al (–) roripa e Lipices
Bg (1) липицензов пореч

* ***Rorippa palustris*** (L.) Bess. → *Rorippa islandica*

1415 ***Rorippa pyrenaica*** (Lam.) Reichenb.
Al (–) roripa e Pirinejeve
Bg (1) пиренейски пореч
Cs (0) rukev pyrenejská
De (–) Pyrenäen-Sumpfkresse
Pl (0) rzepicha pirenejska
Ro (0) gălbenea
Ru (0) жерушник пиренейский
Sk (–) roripa pyrenejská

1416 ***Rorippa sylvestris*** (L.) Bess.
Al (–) roripa e pyllit
Bg (1) горски пореч
Cs (3) rukev lesní
De (–) Wald-Sumpfkresse
En (–) creeping yellow-cress
Hu (2) iszapkányafű
Pl (3) rzepicha lesna
Ru (2) жерушник лесной
Sh (2) obični grbak; žuti ugaz
Sk (–) roripa lesná

1417 ***Rosa* spp.** L.
Bg (1) роза
De (–) Rose
En (–) rose
Gr (1) αγριοτριανταφυλλιά
Ru (1) роза
Sk (–) ruža

1418 ***Rosa arvensis*** Huds.
Cs (–) růže rolní
De (–) Feld-Rose
En (–) field rose
Ru (0) роза пашенная
Sh (1) poljska ruža
Sk (–) ruža roľná

1419 ***Rosa canina*** L.
Al (–) trëndafil i egër
Bg (1) кучешка роза
Cs (1) růže šipková
De (–) Hunds-Rose
En (–) dog-rose

Hu (0) gyepűrózsa
Pl (1) róża dzika
Ro (1) măceş
Ru (1) роза собачья
Sh (1) obična ruža; pasja ruža
Sk (–) ruža šípová

1420 ***Rosa corymbifera*** Borkh.
De (–) Busch-Rose
Sh (1) divlja ruža

1421 ***Rosa gallica*** L.
Al (–) trëndafil i Francës; trëndafil i kuq
Bg (1) французка роза
Cs (0) růže galská
De (–) Essig-Rose
En (–) Provence rose
Hu (1) parlagi rózsa
Pl (0) róża francuska
Ro (1) răsură
Ru (0) роза французская
Sh (1) mesečarka ruža; srcolisna ruža
Sk (–) ruža galská

1422 ***Rosa glutinosa*** Sibth. & Sm.
Al (–) trëndafil veshtullak
Bg (1) лепкава роза
Ru (0) роза клейкая
Sk (–) ruža lepkavá

1423 ***Rosa micrantha*** Borr.
Al (–) trëndafil lulevogël
Bg (1) дребноцветна роза
Cs (0) růže drobnokvětá
De (–) Kleinblütige Rose
En (–) lesser sweet-briar
Hu (0) kisvirágú rózsa
Pl (1) róża drobnokwiatowa
Ru (0) роза мелкоцветковая
Sk (–) ruža malokvetá

1424 ***Rosa pendulina*** L.
Al (–) trëndafil mali
Bg (1) алпийска роза
Cs (0) růže alpská
De (–) Alpen-Rose
Hu (0) havasalji rózsa
Pl (1) róża alpejska
Ro (0) trandafir de munte
Sk (–) ruža alpská

1425 ***Rosa pimpinellifolia*** L.
Cs (–) růže bedrníkolistá
De (–) Dünen-Rose
En (–) burnet rose
Sh (1) bodljikava ruža
Sk (–) ruža tŕňová

* ***Rosa spinosissima*** L. → *Rosa pimpinellifolia*

1426 ***Rubia tinctorum*** L.
Al (–) rrëza ngjyruese; rrëzore ngjyruese
Bg (1) брош
Cs (0) mořena barvířská
De (–) Echte Färberröte
En (–) common madder
Hu (0) festő buzér
Ro (1) roibă
Ru (1) марена красильная
Sh (1) obični broć; pravi broć
Sk (–) morena farbiarska

1427 ***Rubus spp.*** L.
Bg (1) калина
De (–) Brombeere
Gr (1) βάτος
Ru (1) ежевика
Sk (–) ostružina

1428 ***Rubus caesius*** L.
Al (–) manaferë e kaltërme; mjedhër e vogël
Bg (1) полска калина
Cs (3) ostružiník ježiník
De (–) Acker-Brombeere
En (–) dewberry
Hu (3) hamvas szeder

Pl (2) jeżyna popielica
Ro (2) rugi de mirişte
Ru (2) ежевика сизая
Sh (2) divlja kupina
Sk (–) ostružina sivobelasá

1429 ***Rubus canescens*** DC.
De (–) Filzige Brombeere
Sh (1) dlakava kupina

1430 ***Rubus fruticosus*** L.
Cs (1) ostružiník křovitý
De (–) Echte Brombeere
En (–) bramble
Gr (2) βάτος
Ru (0) ежевика несская
Sk (–) černica

1431 ***Rubus hirtus*** Waldst. & Kit.
Bg (1) косматa калина
De (–) Borstige Brombeere
Pl (1) jeżyna gruczołowata
Ru (1) ежевика щетинистая
Sh (1) čekinjasta kupina
Sk (–) ostružina srstnatá

1432 ***Rubus idaeus*** L.
Al (–) mjedhëra
Bg (0) малина
Cs (2) ostružiník maliník
De (–) Himbeere
En (–) wild raspberry
Hu (0) málna
Pl (1) malina właściwa
Ro (0) smeur
Ru (0) малина обыкновенная
Sh (1) gajena kupina
Sk (–) malina

* ***Rubus tomentosus*** Borkh. → *Rubus canescens*

1433 ***Rudbeckia laciniata*** L.
Bg (1) златна топка
Cs (0) třapatka dřípatá
De (–) Schlitzblättriger Sonnenhut
En (–) cutleaf coneflower
Hu (1) magas kúpvirág
Pl (0) rudbekia naga
Ro (0) rujă galbenă
Ru (0) рудбекия рассечённая
Sk (–) rudbekia driapatý

1434 ***Rumex acetosa*** L.
Al (–) lëpjetë e thartë; lëpjetë livadhi
Bg (1) киселец
Cs (3) šťovík kyselý
De (–) Großer Ampfer
En (–) common sorrel
Gr (1) ξινολάπαθο
Hu (3) mezei sóska
Pl (2) szczaw zwyczajny
Ro (1) măcriş
Ru (1) щавель обыкновенный
Sh (1) livadski štavelj; obična kiselica
Sk (–) šťavel kyslý

1435 ***Rumex acetosella*** L.
Al (–) ufullaçkë; uthëllnike
Bg (1) козя брада
Cs (3) kyselka obecná; šťovík menší
De (–) Kleiner Ampfer
En (–) sheep's sorrel
Gr (1) ξινάκι
Hu (3) juhsóska
Pl (3) szczaw polny
Ro (2) măcriş mărunt
Ru (3) щавель воробьиный
Sh (1) mala kiselica; mali kiseljak
Sk (–) štavíček obyčajný

1436 ***Rumex alpinus*** L.
Al (–) lëpjetë e Alpeve
Bg (1) алпийски лапад
Cs (3) šťovík alpský
De (–) Alpen-Ampfer
En (–) monk's-rhubarb
Pl (1) szczaw alpejski

Ru (1) щавель альпийский
Sk (–) šťavel alpínsky

1437 ***Rumex aquaticus*** L.
Al (–) lëpjetë e ujit
Bg (1) воден лапад
Cs (0) šťovík vodní
De (–) Wasser-Ampfer
En (–) Scottish dock
Gr (1) νερολάπαθο
Hu (0) vízi lórom
Pl (1) szczaw wodny
Ro (0) ştevie de baltă
Ru (0) щавель водяной
Sk (–) šťavel vodný

1438 ***Rumex bucephalophorus*** L.
De (–) Ochsenkopf-Ampfer
En (–) horned dock
Gr (1) μικρό λάπαθο

1439 ***Rumex confertus*** Willd.
Cs (–) šťovík stěsňaný
Ru (1) щавель конский
Sk (–) šťavel stlačený

1440 ***Rumex conglomeratus*** Murray
Al (–) lëpjetë e dendur
Bg (1) кичурест лапад
Cs (1) šťovík klubkatý
De (–) Knäuelblütiger Ampfer
En (–) clustered dock
Gr (1) κοπανολάπαθο
Hu (1) murvás lórom
Pl (1) szczaw skupiony
Ro (1) măcrişul calului
Ru (1) щавель клубковатый
Sh (1) nabita kiselica; zbijeni štavel
Sk (–) šťavel kĺbkovitý

1441 ***Rumex crispus*** L.
Bg (2) обикновен лапад
Cs (3) šťovík kadeřavý
De (–) Krauser Ampfer
En (–) curled dock
Gr (1) λάπαθο
Hu (2) fodros lórom
Pl (2) szczaw kędzierzawy
Ro (1) dragavei
Ru (1) щавель курчавый
Sh (2) kudrava kiselica; obični štavel
Sk (–) šťavel kučeravý

* ***Rumex domesticus*** Hartm. → *Rumex longifolius*

1442 ***Rumex hydrolapathum*** Huds.
Bg (0) речен лапад
Cs (1) šťovík koňský
De (–) Fluß-Ampfer
Hu (0) tavi lórom
Pl (1) szczaw lancetowaty
Ru (1) щавель воднощавелевый
Sh (1) velika kiselica; veliki štavelj
Sk (–) šťavel konský

* ***Rumex infestus*** Salisb. → *Rumex acetosella*

1443 ***Rumex longifolius*** DC.
Cs (0) šťovík domácí
De (–) Gemüse-Ampfer
En (–) northern dock
Ru (1) щавель домашний
Sk (–) šťavel domáci

1444 ***Rumex maritimus*** L.
Al (–) lëpjetë deti
Cs (0) šťovík přímořský
De (–) Strand-Ampfer
En (–) golden dock
Hu (1) tengeri lórom
Pl (1) szczaw nadmorski
Ru (1) щавель морской
Sh (1) morski štavelj; ružičasta kiselica
Sk (–) šťavel prímorský

1445 ***Rumex obtusifolius*** L.
Al (–) lëpjetë e rëndomtë; lëpjetë me gjethe shytë
Bg (1) тъполистен лапад
Cs (3) šťovík tupolistý
De (–) Stumpfblättriger Ampfer
En (–) broad-leaved dock
Gr (1) μεγάλο λάπαθο
Hu (3) réti lórom
Pl (3) szczaw tępolistny
Ru (2) щавель туполистный
Sh (2) štavelj konjštak; tupolisna kiselica
Sk (–) šťavel tupolistý

1446 ***Rumex palustris*** Sm.
Bg (1) блатен лапад
Cs (–) šťovík mokradní
De (–) Sumpf-Ampfer
En (–) marsh dock
Gr (1) υγρολάπαθο
Hu (1) mocsári lórom
Pl (1) szczaw błotny
Ru (1) щавель болотный
Sh (1) močvarna kiselica; močvarni štavelj
Sk (–) šťavel močiarny

1447 ***Rumex patientia*** L.
Al (–) lëpjetë e kopshtit
Bg (1) спанаков лапад
Cs (1) šťovík zahradní
De (–) Englischer Spinat
En (–) patience dock
Gr (1) σιδερολάπαθο
Hu (1) paréj lórom
Pl (1) szczaw żółty
Ro (0) ştevie de grădină
Ru (1) шавель шпинатный
Sk (–) šťavel záhradný

1448 ***Rumex pulcher*** L.
Al (–) lëpjetë e bukur
Bg (1) красив лапад
De (–) Schöner Ampfer
En (–) fiddle dock
Hu (1) csinos lórom
Ru (1) щавель красивый
Sh (1) lepi štavelj; razastrta kiselica
Sk (–) šťavel pekný

1449 ***Rumex sanguineus*** L.
Bg (1) кървавочервен лапад
Cs (0) šťovík krvavý
De (–) Hain-Ampfer
En (–) wood dock
Hu (1) erdei lórom
Pl (1) szczaw gajowy
Ro (1) dragavei
Ru (1) щавель кровяной
Sh (1) cjelovita kiselica; crveni štavelj
Sk (–) šťavel krvavý

* ***Rumex sylvestris*** Wallr. → *Rumex obtusifolius*

1450 ***Rumex tenuifolius*** (Wallr.) Löve
De (–) Schmalblättriger Ampfer
En (–) fine-leaved sorrel
Sh (1) tankolisni štavelj
Sk (–) šťavel tenkolistý

1451 ***Rumex thyrsiflorus*** Fingerh.
Ru (1) щавель пирамидальный
Sk (–) šťavel mnohokvetý

1452 ***Rumex tuberosus*** L.
Bg (1) грудест лапад
Ru (0) щавель клубневой
Sk (–) šťavel hľuzovitý

1453 ***Ruscus aculeatus*** L.
Al (–) rrushkulli shpues
Bg (0) миши чимшир
De (–) Stechender Mäusedorn
En (–) butcher's-broom
Gr (1) λαγομηλιά
Hu (0) szúrós csodabogyó
Ro (0) ghimpe

Ru (0) иглица колючая
Sh (1) bodljikava veprina; obična kostrika
Sk (–) listnatec čipkový

1454 ***Ruscus hypoglossum*** L.
Al (–) rrushkulli nëngjuhës
Bg (0) миши трън
Cs (–) listnatec čípkový
Hu (0) lónyelvű csodabogyó
Ru (0) иглица подъязычная
Sh (1) jezičasta veprina; širokolisna kostrika
Sk (–) listnatec podjazyčný

1455 ***Ruta chalepensis*** L.
Cs (–) routa vonna
En (–) fringed rue
Gr (1) απήγανος

1456 ***Ruta graveolens*** L.
De (–) Wein-Raute
Gr (1) απήγανος
Pl (0) ruta zwyczajna
Ru (0) рута душистая
Sk (–) ruta obyčajná

S

1457 *Saccharum ravennae* (L.) Murray
Al (–) kuqëla e Ravenës
Bg (1) равенска кънора
Gr (1) καλαμίδι
Ru (1) эриантус равенский
Sk (–) cukrová trstina

1458 *Sagina apetala* Ard.
Al (–) sagina pa petale
Bg (0) дребна мъховка
Cs (0) úrazník bezplátečný
De (–) Kronblattloses Mastkraut
En (–) annual pearlwort
Gr (1) απέταλη σαγήνη
Hu (1) sziromtalan zöldhúr
Pl (1) karmnik bezpłatkowy
Ru (1) мшанка безлепестная
Sk (–) úrazník bezlístočný

1459 *Sagina maritima* Don
Bg (0) приморска мъховка
De (–) Strand-Mastkraut
En (–) sea pearlwort
Pl (1) karmnik nadmorski
Sk (–) úrazník prímorský

1460 *Sagina nodosa* (L.) Fenzl
Cs (0) úrazník uzlovitý
De (–) Knotiges Mastkraut
En (–) knotted pearlwort
Hu (0) csomós zöldhúr
Pl (3) karmnik kolankowaty
Ru (1) мшанка узловатая
Sk (–) úrazník uzlovatý

1461 *Sagina procumbens* L.
Al (–) sagina e shtrirë
Bg (1) полегнала мъховка
Cs (0) úrazník poléhavý
De (–) Niederliegendes Mastkraut
En (–) procumbent pearlwort
Gr (1) πλαγιαστή σαγήνη
Hu (1) heverő zöldhúr
Pl (2) karmnik rozesłany
Ro (1) grăşătoare
Ru (1) мшанка лежачая
Sh (1) polegla čvorika; polegla mokrica
Sk (–) úrazník poliehavý

1462 *Sagina subulata* (Swartz) Presl
Cs (0) úrazník šidlovitý
De (–) Pfriemen-Mastkraut
En (–) heath pearlwort
Hu (1) szálkás zöldhúr
Pl (1) karmnik ościsty
Sk (–) úrazník šidlovitý

1463 *Sagittaria sagittifolia* L.
Al (–) sygomari
Bg (1) стрелолист
Cs (0) šípatka střelolistá
De (–) Pfeilkraut
En (–) arrowhead
Gr (1) σαγκιτάρια
Hu (1) nyílfű
Pl (2) strzałka wodna
Ro (0) săgeata apei
Ru (1) стрелолист стрелолистный
Sh (1) obična keka; strelicolisna strelica
Sk (–) šípovka vodná

1464 *Salix* spp. L.
Al (–) shelgu
Bg (1) върба
Cs (1) vrba
De (–) Weide
En (–) willow
Gr (1) ιτιά
Pl (–) wierzba
Ro (–) salcie
Ru (1) ива
Sk (–) vŕba

1465 ***Salsola kali*** L.
Al (–) këmbëkuqja
Bg (1) руско вълмо
Cs (0) slanobýl draselný
De (–) Kali-Salzkraut
En (–) prickly saltwort
Gr (1) αλμυρίδι
Hu (3) homoki ballangó
Pl (–) solanka kolczysta
Ro (1) ciurlan
Ru (1) солянка калийная
Sh (1) bodljikava solnjača; obična solnjača
Sk (–) slaná byľ draselná

* ***Salsola ruthenica*** Ilj. → *Salsola kali*

1466 ***Salsola collina*** Pallas
Ru (1) солянка холмовая
Sk (–) slaná byľ kopcová

1467 ***Salsola soda*** L.
De (–) Sodakraut
Ru (1) солянка содоносная
Sh (1) slanica solnjača

1468 ***Salvia* spp.** L.
Bg (1) конски босилек
De (–) Salbei
Gr (1) σάλβια
Ru (1) шалфей
Sk (–) šalvia

1469 ***Salvia aethiopis*** L.
Al (–) sherbela e Etiopisë
Bg (1) етиопски конски босилек
Cs (0) šalvěj habešská
De (–) Mohren-Salbei
En (–) woolly clary
Hu (1) magyar zsálya
Ro (1) şerlai
Ru (0) шалфей эфиопский
Sh (1) bela žalfija; češljasta kadulja
Sk (–) šalvia habešská

1470 ***Salvia austriaca*** Jacq.
Bg (1) австрийски конски босилек
Cs (0) šalvěj rakouská
De (–) Pannonischer Salbei
Hu (1) osztrák zsálya
Ru (1) шалфей австрийский
Sh (1) austrijska žalfija; žutkasta kadulja
Sk (–) šalvia rakúska

1471 ***Salvia glutinosa*** L.
Al (–) sherbelë mali; sherbela veshtullake
Bg (1) жлезист конски босилек
Cs (0) šalvěj lepkavá
De (–) Klebriger Salbei
Hu (0) enyves zsálya
Pl (1) szałwia lepka
Ro (0) cinsteţ
Ru (0) шалфей железистый
Sk (–) šalvia lepkavá

1472 ***Salvia grandiflora*** Etlin.
Bg (1) едроцвтен конски босилек
Ru (0) шалфей крупноцветковый
Sk (–) šalvia veľkokvetá

1473 ***Salvia nemorosa*** L.
Al (–) sherbelë e egër
Bg (1) горски конски босилек
Cs (0) šalvěj hajní
De (–) Wald-Salbei
En (–) wild sage
Hu (1) ligeti zsálya
Pl (1) szałwia omszona
Ro (1) jaleş de cîmp
Ru (1) шалфей дубравный
Sh (1) šumska kadulja; šumska žalfija
Sk (–) šalvia hájna

1474 ***Salvia officinalis*** L.
Al (–) sherbela mjekësore
Bg (1) лечебен конски босилек
Cs (1) šalvěj lékařská

De (–) Echter Salbei
En (–) garden sage
Hu (0) kerti zsálya
Pl (0) szałwia lekarska
Ro (0) jaleş de grădină
Ru (0) шалфей лекарственный
Sk (–) šalvia lekárska

1475 ***Salvia pratensis*** L.
Al (–) sherbelë e livadheve
Bg (1) ливаден конски босилек
Cs (2) šalvěj luční
De (–) Wiesen-Salbei
En (–) meadow clary
Hu (0) mezei zsálya
Pl (1) szałwia łąkowa
Ro (1) salvie de cîmp
Ru (1) шалфей луговой
Sh (1) livadna kadulja; livadarka žalfija
Sk (–) šalvia lúčna

1476 ***Salvia sclarea*** L.
Al (–) shëngjini
Bg (1) мускатен конски босилек
Cs (0) šalvěj muškátová
De (–) Muskat-Salbei
En (–) garden clary
Ro (0) iarba Sfîntului Ioan
Ru (0) шалфей мускатный
Sk (–) šalvia muškátová

1477 ***Salvia verbenaca*** L.
Al (–) sherbelë shporicë; sherbelë shporizëngjashme
De (–) Eisenkraut-Salbei
En (–) Guernsey clary
Ru (1) шалфей вербеновый
Sk (–) šalvia železníkovitá

1478 ***Salvia verticillata*** L.
Al (–) sherbela qerthullake
Bg (1) прешленест конски босилек
Cs (0) šalvěj přeslenitá
De (–) Quirlblütiger Salbei
En (–) whorled clary
Hu (1) ló zsálya
Pl (1) szałwia okręgowa
Ro (1) urechea porcului
Ru (1) шалфей мутовчатый
Sh (1) pršljenasta kadulja; pršljenasta žalfija
Sk (–) šalvia praslenitá

1479 ***Salvinia natans*** (L.) All.
Al (–) bar peshku
Bg (1) плаваща лейка
Cs (0) nepulkalka plovoucí
De (–) Gemeiner Schwimmfarn
Gr (1) σαλβίνια
Hu (1) rucaüröm
Pl (1) salwinia pływająca
Ro (1) peştişoară
Ru (1) сальвиния плавающая
Sk (–) salvinia plávajúca

1480 ***Sambucus ebulus*** L.
Al (–) qingla
Bg (1) тревист бъз
Cs (2) bez chebdí
De (–) Attich
En (–) danewort
Gr (1) βουζιά
Hu (2) gyalogbodza
Pl (2) bez hebd
Ro (2) boz
Ru (2) бузина травянистая
Sh (1) obični burjan; smrdljiva bazga
Sk (–) baza chabzda; chabzdie

* ***Sambucus humilis*** Lam. → *Sambucus ebulus*

1481 ***Sambucus nigra*** L.
Al (–) shtogu
Bg (1) черен бъз
Cs (3) bez černý
De (–) Schwarzer Holunder

En (–) common elder
Gr (1) κουφοξυλιά
Hu (1) fekete bodza
Pl (1) bez czarny
Ro (0) soc
Ru (1) бузина чёрная
Sk (–) baza čierna

1482 *Samolus valerandi* L.
Al (–) li ujës; samola e Velerandit
Cs (0) solenka Valerandova
De (–) Salz-Bunge
En (–) brookweed
Gr (1) σάμολος
Hu (0) árokvirág
Pl (0) jarnik solankowy
Ru (0) самолус Валеранда
Sk (–) soľanka Valerandova

1483 *Sanguisorba minor* Scop.
Al (–) dorëvata sanguisorbë; lule komishti
Bg (1) дрбна динка
Cs (2) krvavec menší
De (–) Kleiner Wiesenknopf
En (–) salad-burnet
Hu (1) csábaíre
Pl (1) krwiściąg mniejszy
Ro (1) cebarea
Ru (1) черноголовник кровохлёбковый
Sh (1) mala krvara; obična dinjica
Sk (–) krvavec menší pravý

1484 *Sanguisorba officinalis* L.
Bg (1) лечбна динка
Cs (2) krvavec toten
De (–) Großer Wiesenkopf
En (–) great burnet
Hu (0) őszi vérfű
Pl (1) krwiściąg lekarski
Ro (0) sorbestrea
Ru (1) кровохлёбка аптечная
Sh (–) lekovita dinjica; tamnorujna krvara
Sk (–) krvavec lekársky

1485 *Saponaria bellidifolia* Sm.
Al (–) sapunqyqja gjethebukur
Bg (1) красиволистно сапунче
En (–) yellow soapwort
Gr (1) σαπωνάρια
Sk (–) mydlica žltokvetá

1486 *Saponaria officinalis* L.
Al (–) sapunqyqja mjekësore; shkumbëza e rëndomtë
Bg (1) лечебно сапунче
Cs (3) mydlice lékařská
De (–) Gemeines Seifenkraut
En (–) soapwort
Gr (1) σαπουνόχορτο
Hu (1) szappanfű
Pl (2) mydlnica lekarska
Ro (1) odogaci
Ru (1) мыльнянка лекарственная
Sh (1) obična sapunika; obična sapunjača
Sk (–) mydlica lekárska

1487 *Sarcopterium spinosum* (L.) Spach
Al (–) dorëvata
En (–) thorny burnet
Gr (1) αστοιβή
Ru (0) саркоптериум колючий
Sk (–) krvavec ostnatý

* ***Sarothamnus scoparius*** (L.) Wim. → *Cytisus scoparius*

* ***Satureja acinos*** (L.) Schee. → *Acinos arvensis*

1488 *Satureja montana* L.
De (–) Winter-Bergminze
En (–) savory
Ru (0) чабер горный
Sh (1) brdski čubar
Sk (–) štubrik horský

* ***Satureja vulgaris*** (L.) Frit. → *Clinopodium vulgare*

1489 ***Saxifraga* spp.** L.
Bg (0) камеломка
De (–) Steinbrech
En (–) saxifrage
Gr (1) σαξίφράγα
Ru (0) камеломка
Sk (–) lomikameň

1490 ***Saxifraga tridactylites*** L.
Al (–) iriqëza tregishtash
Bg (0) трипръстна каменоломка
Cs (0) lomikámen trojprstý
De (–) Dreifinger-Steinbrech
En (–) rue-leaved saxifrage
Hu (0) apró kőtörőfű
Pl (2) skalnica trójpalczasta
Ru (0) каменоломка трёхпалая
Sk (–) lomikameň trojprstý

1491 ***Scabiosa* spp.** L.
Bg (1) самогризка
De (–) Grindkraut
En (–) scabious
Gr (1) σκαβιόζα
Ru (1) скабиоза
Sk (–) hlaváč

1492 ***Scabiosa canescens*** Waldst. & Kit.
Cs (0) hlaváč šedavý
De (–) Graues Grindkraut
Hu (0) szürkés ördögszem
Pl (1) driakiew wonna
Ru (0) скабиоза собачья
Sk (–) hlaváč šedavý

1493 ***Scabiosa columbaria*** L.
Al (–) bar zgjebe pëllumb; pllumbini
Bg (0) гълъбова самогризка
Cs (0) hlaváč fialový
De (–) Tauben-Grindkraut
En (–) small scabious
Hu (1) vajszínű ördögszem
Pl (1) driakiew gołębia
Ru (1) скабиоза голубиная
Sh (1) golubija udovičica; obična udovičica
Sk (–) hlaváč fialový

* ***Scabiosa integrifolia*** L. → *Knautia integrifolia*

1494 ***Scabiosa ochroleuca*** L.
Al (–) bar zgjebe në të verdhëm
Bg (1) жълтеникава самогризка
Cs (1) hlaváč bledožlutý
De (–) Gelbes Grindkraut
En (–) yellow scabious
Hu (1) vajszínű ördögszem
Pl (1) driakiew żółtawa
Ro (1) sipică
Ru (1) скабиоза бледно-жёлтая
Sh (1) blijedožuta udovičica
Sk (–) hlaváč bledožltý

* ***Scabiosa palaestina*** L. → *Tremastelma palaestinum*

* ***Scabiosa syriaca*** L. → *Cephalaria syriaca*

1495 ***Scandix australis*** L.
En (–) southern shepherd's-needle
Gr (1) σκαντζίκι
Ru (0) скандикс южный
Sk (–) česák veľkokvetý

* ***Scandix grandiflora*** L. → *Scandix australis*

1496 ***Scandix pecten-veneris*** L.
Al (–) finraspori
Bg (1) венерин гребен
Cs (0) vochlice hřebenitá
De (–) Echter Venuskamm
En (–) shepherd's-needle
Gr (1) μυρώνι
Hu (1) berzenke
Pl (1) czechrzyca grzebieniowa

Ro (1) acul Doamnei
Ru (1) скандикс гребенчатый
Sh (1) Venerina češljika
Sk (–) česák hrebenitý

* **Schlosseria chabrei** Schloss. & Vukot. → *Peucedanum carvifolia*

* **Schoenoplectus lacustris** (L.) Palla → *Scirpus lacustris*

* **Schoenoplectus litoralis** (Schrad.) Palla → *Scirpus litoralis*

* **Schoenoplectus supinum** (L.) Palla → *Scirpus supinus*

* **Schoenoplectus tabernaemontam** (Gmel.) Palla → *Scirpus lacustris*

* **Schoenoplectus triqueter** (L.) Palla → *Scirpus triqueter*

1497 Schoenus nigricans L.
Al (–) skenusi i zi
Bg (0) черен схьонус
Cs (0) šášina načernalá
De (–) Schwarzes Kopfried
En (–) bog-rush
Hu (0) kormos csáté
Pl (1) mokrzyca czarniawa
Ru (0) схенус черноватый
Sh (1) crna šiljevina
Sk (–) šašina čierna

1498 Scilla autumnalis L.
Al (–) boçka e vjeshtës
Bg (1) есенен синчец
De (–) Herbst-Blaustern
En (–) autumn squill
Gr (1) σκίλα
Hu (0) őszi csillagvirág
Ru (0) пролеска осенняя
Sk (–) scila jesenná

1499 Scilla bifolia L.
Al (–) boçka dygjethëshe
Bg (1) обикновен синчец
Cs (0) ladoňka dvoulistá
De (–) Zweiblättriger Blaustern
En (–) alpine squill
Hu (0) tavaszi csillagvirág
Pl (1) cebulica dwulistna
Ru (0) пролеска двулистная
Sk (–) scila dvojlistá

1500 Scirpus spp. L.
Bg (1) камъш
De (–) Simse
En (–) club-rush
Gr (1) σκίρπος
Ru (1) камыш

* **Scirpus affinis** Roth → *Scirpus lacustris*

1501 Scirpus holoschoenus L.
Bg (1) обикновен холосхьонус
Cs (0) kamýšek římský
De (–) Kugel-Simse
En (–) round-headed club-rush
Hu (1) szürke káka
Pl (1) hołoszeń główkowaty
Ro (0) ţipirig
Ru (1) голосхенус обыкновенный
Sh (1) obična vežljika
Sk (–) šarinka obyčajná

1502 Scirpus lacustris L.
Al (–) kryekuqi; sitkë liqeni
Bg (1) блатен камъш
Cs (0) skřípinec jezerní
De (–) Grüne Teichsimse
En (–) common bulrush
Hu (1) tavi káka
Pl (1) oczeret jeziorny
Ro (1) pipirig
Ru (1) камыш озёрный
Sh (1) obični oblić
Sk (–) škripinec jazerný

1503 ***Scirpus litoralis*** Schrad.
Bg (0) крайбрежен камъш
Hu (0) tengermelléki káka
Pl (1) sitowie przybrzeżne
Ro (2) trosnitoare
Ru (1) камыш приморский
Sk (–) šarina pobrežná

1504 ***Scirpus maritimus*** L.
Al (–) sitkë deti; shqirrë deti
Bg (2) морски камъш
Cs (0) kamýšník přímořský
De (–) Meer-Simse
En (–) sea club-rush
Hu (2) zsióka
Pl (1) sitowiec nadmorski
Ro (2) trosnitoare
Ru (1) клубнекамыш морской
Sh (1) visoka zuka
Sk (–) šarina prímorská

1505 ***Scirpus mucronatus*** L.
Bg (1) островръх камъш
De (–) Stachel-Teichsimse
En (–) roughseed club-rush
Hu (–) szúrós káka
Pl (0) oczeret sztyletowaty
Ro (2) pilă
Ru (1) камыш остроконечный
Sk (–) škripinec hrotitý

1506 ***Scirpus radicans*** Schk.
Cs (2) skřípina kořenující
De (–) Wurzelnde Binse
Ru (1) камыш укореняющийся
Sh (1) ukorjenjujuća vežljika
Sk (–) škripina koreňová

1507 ***Scirpus setaceus*** L.
Al (–) sitkë qimehollë
Bg (0) четиновиден изолепис
Cs (0) bezosetka štětinovitá
De (–) Borstige Moorbinse
En (–) bristle club-rush
Hu (0) kasika-káka
Pl (0) sitniczka szczecinowata
Ru (0) камыш щетиновидный
Sk (–) škripík štetinovitý

1508 ***Scirpus supinus*** L.
Bg (0) полегнал изолепис
Cs (0) bezosetka nízká
De (–) Niederliegende Moorbinse
Hu (1) henye káka
Sk (–) škripinec poliehavý

1509 ***Scirpus sylvaticus*** L.
Al (–) shqira e pyjeve
Bg (1) горски сцирпус
Cs (3) skřípiná lesní
De (–) Wald-Simse
En (–) wood club-rush
Hu (–) közönséges erdeikáka
Pl (1) sitowie leśne
Ru (1) камыш лесной
Sh (1) koljenčasta iglica; šumska vežljika
Sk (–) škripina lesná

1510 ***Scirpus triqueter*** L.
Al (–) heshta
Bg (1) трирьбест камъш
Cs (2) skřípinec trojhranný
De (–) Dreikantige Simse
En (–) triangular club-rush
Hu (0) háromélű káka
Pl (1) sitowie trójgraniaste
Ru (1) камыш трёхгранный
Sk (–) škripinec trojhranný

1511 ***Scleranthus annuus*** L.
Al (–) lauresha; skleranta njëvjeçare
Bg (1) едногодишна хрущялка
Cs (3) chmerek roční
De (–) Einjähriger Knäuel
En (–) annual knawel
Gr (1) ετήσιος σκλήρανθος
Hu (3) parlagi szikárka
Pl (3) czerwiec roczny

Ro (3) buruiana supărăturii
Ru (3) дивала однолетняя
Sh (1) jednogodišnja treskavica
Sk (–) chmerok ročný

* **Scleranthus campestris** Schur → *Scleranthus annuus*

1512 **Scleranthus perennis** L.
Bg (1) бяла хрущялка
Cs (0) chmerek vytrvalý
De (–) Ausdauernder Knäuel
En (–) perennial knawel
Gr (1) πολυετής σκλήρανθος
Hu (0) évelő szikárka
Pl (2) czerwiec trwały
Ro (1) sincerică
Ru (1) дивала многолетняя
Sh (1) višegodišnja treskavica
Sk (–) chmerok trváci

1513 **Sclerochloa dura** (L.) Beauv.
Bg (1) обикновена твърдокласица
Cs (0) tužanka tvrdá
De (–) Hartgras
Gr (1) σκληροχλόη
Hu (1) kőperje
Ru (1) жесткоколосница твёрдая
Sk (–) tvrdica obyčajná

1514 **Scolymus hispanicus** L.
Al (–) rëkualli
Bg (1) четка
De (–) Spanische Golddistel
En (–) golden thistle
Gr (1) ασκόλυμπρος
Ro (0) anghinare
Ru (1) сколимус испанпский
Sh (1) dlakava dragušica; španska dragušica

1515 **Scolymus maculatus** L.
Bg (0) петнист сколимус
De (–) Gefleckte Golddistel
Gr (1) χριστάγκαθο
Ru (0) сколимус пятнистый

1516 **Scorzonera spp.** L.
Bg (1) кокеш
De (–) Schwarzwurzel
Gr (1) σκορσονέρα
Ru (1) козелец
Sk (–) hadí mor

1517 **Scorzonera austriaca** Willd.
Al (–) cjapëzi i Austrisë; skorzonera e Austrisë
Cs (1) hadí mord rakouský
De (–) Österreichische Schwarzwurzel
En (–) Austrian viper's-grass
Hu (0) osztrák pozdor
Ro (0) lăptiucă
Ru (0) козелец австрийский
Sk (–) hadí mor rakúsky

* **Scorzonera candollei** Vis. → *Scorzonera humilis*

1518 **Scorzonera humilis** L.
Al (–) cjapëzi xhuxh
Cs (0) hadí mord nízký
De (–) Niedrige Schwarzwurzel
En (–) common viper's-grass
Hu (–) alacsony pozdor
Pl (1) wężymord niski
Ru (0) козелец низкий
Sk (–) hadí mor nízky

1519 **Scorzonera laciniata** L.
Al (–) skorzonera me rrypëza
Cs (0) hadí mordec dřípený
De (–) Stielsamenkraut
Hu (1) sallangos pozdor
Ru (–) козелец разрезной
Sh (1) obični politovac; rasperjani politovac
Sk (–) nohoplodec driapatý

1520 ***Scorzonera purpurea*** L.
Al (–) cjapëzi i purpurt; skorzonera e purpurt
Cs (0) hadí mord nachový
De (–) Purpur-Schwarzwurzel
En (–) purple viper's-grass
Hu (0) piros pozdor
Pl (1) wężymord stepowy
Ru (0) козелец пурпурный
Sk (–) hadí mor nachový

* ***Scorzonera rosea*** Waldst. & Kit. → *Scorzonera purpurea*

1521 ***Scrophularia* spp.** L.
Bg (1) живениче
De (–) Braunwurz
En (–) figwort
Gr (1) σκροφουλάρια
Ru (1) норичник
Sk (–) krtičník

1522 ***Scrophularia canina*** L.
Al (–) sarushta e qenit
Bg (1) кучешко живениче
Cs (0) krtičník psí
De (–) Hunds-Braunwurz
En (–) French figwort
Ru (0) норичник собачий
Sk (–) krtičník psí

1523 ***Scrophularia nodosa*** L.
Al (–) sarushta me nyje
Bg (1) възловато живениче
Cs (0) krtičník uzlovitý
De (–) Knotige Braunwurz
En (–) common figwort
Hu (1) göcsös görvélyfű
Pl (1) trędownik bulwiasty
Ro (1) buberic
Ru (1) норичник узловатый
Sh (1) gomoljasti strupnik; obični ustupnik
Sk (–) krtičník uzlovitý

1524 ***Scutellaria galericulata*** L.
Bg (1) обикновена превара
Cs (0) šišák vroubkovaný
De (–) Sumpf-Helmkraut
En (–) common skullcap
Hu (0) vízmelléki csukóka
Pl (1) tarczyca pospolita
Ro (1) mirgău
Ru (1) шлемник обыкновенный
Sk (–) šišak vrúbkovaný

1525 ***Scutellaria hastifolia*** L.
Bg (1) копиелистна превара
Cs (0) šišák hrálolistý
De (–) Spießblättriges Helmkraut
En (–) spear-leaved skullcap
Hu (1) dárdalevelű csukóka
Pl (0) tarczyca oszczepowata
Ru (1) шлемник копьелистный
Sh (1) kopljasta šiška; kopljastolisna grozničica
Sk (–) šišak oštepolistý

1526 ***Securigera securidaca*** (L.) Degen & Dorf.
Bg (1) секуригера
De (–) Beilwicke
Ru (0) секуригера мечевидная

1527 ***Sedum* spp.** L.
Bg (1) тлъстига
De (–) Fetthenne
En (–) stonecrop
Gr (1) κουκοστάφυλλο
Ru (0) очиток
Sk (–) rozchodník

1528 ***Sedum acre*** L.
Al (–) rrushqyqe e athët
Bg (0) лютива тлъстига
Cs (0) rozchodník ostrý
De (–) Scharfe Fetthenne
En (–) bitter stonecrop
Hu (1) borsos varjúháj

Pl (1) rozchodnik ostry
Ro (0) iarbă de şoaldină
Ru (1) очиток едкий
Sk (–) rozchodník prudký

1529 *Sedum album* L.
Al (–) rrushqyqe e bardhë
Bg (1) бяла тлъстига
Cs (0) rozchodník bílý
De (–) Weiße Fetthenne
En (–) white stonecrop
Hu (0) fehér varjúháj
Pl (0) rozchodnik biały
Ro (0) iarba faptului
Ru (0) очиток белый
Sk (–) rozchodník biely

* ***Sedum maximum*** (L.) Suter → *Sedum telephium*

1530 *Sedum rubens* L.
Bg (1) червеникава тлъстига
Cs (0) rozchodník červený
De (–) Rotliches Fetthenne
Ru (0) очиток краснеющий
Sk (–) tučnica červeňavá

1531 *Sedum sexangulare* L.
Al (–) krimbës gjashtëbrinjësh
Cs (0) rozchodník šestiřadý
De (–) Milde Fetthenne
Hu (1) olasz varjúháj
Pl (1) rozchodnik sześciorzędowy
Ru (–) очиток шестирядный
Sk (–) rozchodník šesťuholný

1532 *Sedum telephium* L.
Al (–) rrushqyqe e madhe
Cs (3) rozchodník veliký
De (–) Rote Fetthenne
En (–) orpine
Hu (0) bablevelű varjúháj
Ro (1) iarbă de urechi
Ru (1) очиток большой
Sk (–) rozchodníkovec najväčší

1533 *Selinum carvifolia* (L.) L.
Cs (–) olešník kminolistý
De (–) Kümmelblättrige Silge
Ru (0) гирча тминолистная
Sh (1) obična selenka
Sk (–) olšovník rascolistý

1534 *Senecio* spp. L.
Bg (1) спореж
De (–) Greiskraut
Gr (1) μαρτιάκος
Ru (1) крестовник
Sk (–) starček

1535 *Senecio aquaticus* L.
Cs (0) starček vodní
De (–) Wasser-Kreuzkraut
En (–) marsh ragwort
Hu (1) vízi aggófű
Pl (0) starzec wodny
Sk (–) starček vodný

* ***Senecio bosniacus*** Beck → *Senecio papposus*

* ***Senecio campestris*** (Retz.) DC. → *Senecio integrifolius*

1536 *Senecio doria* L.
Cs (0) starček žlatý
Hu (0) kövér aggófű
Pl (1) starzec cienisty
Sk (–) starček dorský

* ***Senecio erraticus*** Bertol. → *Senecio aquaticus*

1537 *Senecio erucifolius* L.
Bg (1) ерукoлистен спореж
Cs (0) starček roketolistý

De (–) Rautenblättriges Kreuzkraut
En (–) hoary ragwort
Hu (1) keskenylevelű aggófű
Pl (1) starzec wąskolistny
Ro (0) bătătarnică
Ru (0) крестовник эруколистный
Sk (–) starček roketolistý

1538 ***Senecio fluviatilis*** Wallr.
Bg (1) речен спореж
Cs (0) starček říční
De (–) Fluß-Greiskraut
En (–) broad-leaved ragwort
Hu (0) patakparti aggófű
Pl (1) starzec nadreczny
Ru (0) крестовник речной
Sk (–) starček poriečný

* ***Senecio fuchsii*** Gmel. → *Senecio nemorensis*

1539 ***Senecio integrifolius*** (L.) Clairv.
Cs (0) starček celolistý
De (–) Steppen-Greiskraut
En (–) field fleawort
Hu (0) mezei aggófű
Pl (1) starzec polny
Ru (0) крестовник равнинный
Sk (–) starinec celolistý

1540 ***Senecio jacobaea*** L.
Al (–) kryqës i livadheve; pulithi i Jakobit
Bg (1) ливаден спореж
Cs (3) starček přímětník
De (–) Jakobs-Kreuzkraut
En (–) common ragwort
Hu (1) Jakabnapi aggófű
Pl (2) starzec Jakubek
Ro (1) petimbroasă
Ru (1) крестовник Якоба
Sh (1) livadski kostriš
Sk (–) starček Jakubov

1541 ***Senecio nemorensis*** L.
Al (–) pulithi i pyllit
Bg (1) горски спореж
Cs (3) starček hajní
De (–) Hain-Greiskraut
Hu (0) berki aggófű
Pl (1) starzec gajowy
Ru (0) крестовник дубравный
Sk (–) starček hájny

1542 ***Senecio paludosus*** L.
Bg (1) блатен спореж
Cs (0) starček bažinný
De (–) Sumpf-Greiskraut
En (–) fen ragwort
Hu (0) mocsári aggófű
Pl (1) starzec bagienny
Ru (0) крестовник болотный
Sk (–) starček barinný

1543 ***Senecio papposus*** (Reichenb.) Less.
Cs (–) starček sírožlutý
En (–) downy fleawort
Pl (1) starzec długolistny
Ru (0) крестовник хохольковый
Sk (–) starinec páperistý bosniac' /

* ***Senecio rapistroides*** DC. → *Senecio vernalis*

* ***Senecio sarracenicus*** L. → *Senecio fluviatilis*

* ***Senecio sinuatus*** Gilib. → *Senecio vernalis*

* ***Senecio umbrosus*** Waldst. & Kit. → *Senecio doria*

1544 ***Senecio vernalis*** Waldst. & Kit.
Bg (1) пролетен спореж
Cs (1) starček jarní
De (–) Frühlings-Kreuzkraut
En (–) spring groundsel

Hu (1) tavaszi aggófű
Pl (3) starzec wiosenny
Ro (1) cruciuliţă
Ru (1) крестовник весенний
Sh (1) prolećni dragušac; proljetni kostriš
Sk (–) starček jarný

1545 *Senecio viscosus* L.
Al (–) pulithi veshtullak
Bg (1) леплив спореж
Cs (1) starček lepivý
De (–) Klebriges Kreuzkraut
En (–) sticky groundsel
Hu (1) enyves aggófű
Pl (1) starzec lepki
Ru (1) крестовник липкий
Sh (1) lepljiva krstica; ljepivi kostriš
Sk (–) starček lepkavý

1546 *Senecio vulgaris* L.
Al (–) pulithi i rëndomtë
Bg (1) обикновен спореж
Cs (3) starček obecný
De (–) Gemeines Kreuzkraut
En (–) common groundsel
Hu (2) közönséges aggófű
Pl (3) starzec zwyczajny
Ro (1) spălăcioşă
Ru (1) крестовник обыкновенный
Sh (1) obični dragušac; obični kostriš
Sk (–) starček obyčajný

1547 *Serapias cordigera* L.
Al (–) serapia me fije të trasha si litari
En (–) heart-flowered helleborine
Gr (1) καρδιόφυλλη σεράπια

* ***Serapias laxiflora*** Chaub. → *Serapias vomeracea*

1548 *Serapias lingua* L.
Al (–) serapia gjuhe
En (–) tongue helleborine
Gr (1) γλωσσόφυλλη σεράπια

1549 *Serapias vomeracea* (Burm.) Briq.
Al (–) serapia luleçlirtë
Gr (1) αραιή σεράπια
Ru (0) серапиас сошниковый

* ***Serratula arvensis*** L. → *Cirsium arvense*

1550 *Serratula radiata* (Waldst. & Kit.) Bieb.
Bg (1) перестолистен сърпец
Hu (0) sugaras zsoltina
Ru (0) серпуха лучевая
Sk (–) srpovník skrelý

1551 *Serratula tinctoria* L.
Bg (1) багрилен сърпец
Cs (0) srpice barvířská
De (–) Färber-Scharte
En (–) saw-wort
Hu (2) festő zsoltina
Pl (1) sierpik barwierski
Ro (0) şerpet
Ru (1) серпуха красильная
Sh (1) obična pilica; obični srpac
Sk (–) sŕpok farbiarsky

1552 *Seseli* spp. L.
Bg (1) сесел
De (–) Sesel
Gr (1) σέσελο
Ru (0) жабрица
Sk (–) sezel

1553 *Seseli annuum* L.
Cs (1) sesel roční
De (–) Steppenfenchel
Hu (0) homoki gurgolya
Pl (1) żebrzyca roczna
Ro (0) cosicel
Ru (0) жабрица однолетняя
Sh (1) crvenkasto devesilje
Sk (–) sezel ročný

1554 *Seseli tortuosum* L.
Al (–) seseli i përdredhur
Pl (0) żebrzyca pokręcona
Ru (0) жабрица извилистая
Sh (1) višegodišnje devesilje
Sk (–) sezel skrátený

* ***Setaria decipiens*** Schim. → *Setaria verticillata*

1555 *Setaria glauca* (L.) Beauv.
Al (–) kostrevë e rimtë
Bg (3) сива кощрява
Cs (3) bér sivý
De (–) Niedrige Borstenhirse
En (–) yellow bristle-grass
Hu (3) fakó muhar
Pl (3) włośnica sina
Ro (3) mohor
Ru (3) щетинник сизый
Sh (3) sivi muhar
Sk (–) štetinník sivý

1556 *Setaria italica* (L.) Beauv.
Al (–) çumiza; kostrevë e Italisë
Bg (1) италианско просо
Cs (–) bér vlašský
De (–) Kolbenhirse
En (–) foxtail millet
Gr (1) κεχροσετάρια
Hu (–) olasz muhar
Pl (0) włośnica ber
Ro (0) dughie
Ru (1) щетинник итальянский
Sh (1) talijanski muhar
Sk (–) štetinník vlašský

1557 *Setaria pumila* (Poir.) Schultes
Cs (2) bér sivý
De (–) Gelbe Borstenhirse
Gr (2) κίτρινη σετάρια
Ro (3) mohor
Sk (–) štetinník zakrpatený

1558 *Setaria verticillata* (L.) Beauv.
Al (–) kostrevë qerthullake
Bg (2) прешленеста кощрява
Cs (3) bér přeslenitý
De (–) Wirtel-Borstenhirse
En (–) rough bristle-grass
Gr (3) σπονδυλωτή σετάρια
Hu (3) ragadós muhar
Pl (2) włośnica okółkowa
Ro (2) mohor agăţător
Ru (2) щетинник мутовчатый
Sh (1) lepljivi muhar; pršljenasti muhar
Sk (–) štetinník praslenatý

1559 *Setaria viridis* (L.) Beauv.
Al (–) kostrevë e blertë; shtarpë
Bg (2) зелена кощрява
Cs (3) bér zelený
De (–) Grüne Borstenhirse
En (–) green bristle-grass
Gr (3) πράσινη σετάρια
Hu (3) zöld muhar
Pl (3) włośnica zielona
Ro (3) mohor verde
Ru (3) щетинник зелёный
Sh (2) zeleni muhar
Sk (–) štetinník zelený

1560 *Sherardia arvensis* L.
Al (–) sherardia e arës
Bg (1) полско венче
Cs (–) bračka rolní
De (–) Ackerröte
En (–) field madder
Gr (1) προβατόχορτο
Hu (2) csillagfű
Pl (1) rolnica polna
Ru (1) жерардия полевая
Sh (1) obični koljenac
Sk (–) bračka poľná

1561 *Sicyos angulatus* L.
Al (–) kastravec cepatak
Cs (–) libenka hranatá

De (–) Kantige Haargurke
En (–) bur cucumber
Hu (1) gyepűtök
Pl (1) harbuźnik kolczasty
Ru (1) сициос угловатый
Sk (–) ľubienka hranatá

1562 *Sideritis montana* L.
Al (–) sideriti i malit
Bg (1) планинско маясълче
Cs (–) hojník horský
De (–) Berg-Gliedkraut
Gr (1) τσάϊ του βουνού
Hu (1) sármányvirág
Pl (–) gojnik drobnokwiatowy
Ro (1) incheietoare
Ru (1) железница горная
Sh (1) brdski šućerak; trocjepni očist
Sk (–) ranohoj horský

* ***Sideritis purpurea*** Talb. → *Sideritis romana*

1563 *Sideritis romana* L.
Al (–) sideriti i purpurtë
De (–) Römisches Gliedkraut
Gr (1) τσάϊ βλάχου
Ru (0) железница румынская
Sk (–) ranohoj rímsky

* ***Sieglingia decumbens*** (L.) Bernh. → *Danthonia decumbens*

* ***Silaum flavescens*** (Bernh.) Hayek → *Silaum silaus*

1564 *Silaum silaus* (L.) Schinz. & Thell.
Cs (–) koromáč olešníkový
De (–) Gewöhnliche Wiesensilge
En (–) pepper-saxifrage
Hu (0) sárga kígyókapor
Pl (1) koniopłoch łąkowy
Ru (0) морковник луговой
Sk (–) silička úhľadná

* ***Silaus pratensis*** Bess. → *Silaum silaus*

1565 *Silene* spp. L.
Bg (1) плюскавче
De (–) Leimkraut
Gr (1) βοϊδοκράτνς
Ru (1) смолёвка
Sk (–) silenka

1566 *Silene alba* (Mill.) Krause
Al (–) melandra e bardhë; veshlepuri
Bg (1) бяло плюскавче
Cs (3) knotovka bílá; silenka bílá
De (–) Weiße Nachtnelke
En (–) white campion
Hu (3) fehér mécsvirág
Pl (3) bniec biały
Ro (1) opaiţă
Ru (1) дрема беловатая
Sh (1) beli pucavac; bijeli pucavac
Sk (–) knôtovka biela

* ***Silene anglica*** L. → *Silene gallica*

1567 *Silene armeria* L.
Al (–) grackë mizash; klokëza skorëliq
Bg (1) армериевидно плюскавче
Cs (–) silenka svazčitá
De (–) Nelken-Leimkraut
En (–) sweet-William catchfly
Hu (0) kerti habszegfű
Pl (0) lepnica baldaszkowa
Ro (0) miliţea roşie
Ru (0) смолёвка армериевидная
Sk (–) silenka sväzkovitá

1568 *Silene conica* L.
Al (–) klokëza konike
Bg (1) коническа хлопка
Cs (–) silenka kuželovitá
De (–) Kegel-Leimkraut
En (–) sand-catchfly
Hu (1) homoki habszegfű
Pl (0) lepnica smukła

Ru (1) смолёвка коническая
Sk (–) silenka kužeľovitá

* **Silene cucubalus** Wib. → *Silene vulgaris*

1569 **Silene dichotoma** Ehrh.
Bg (1) вилужно плюскавче
Cs (–) silenka vidličnatá
De (–) Gabeliges Leimkraut
En (–) forked catchfly
Hu (1) villás habszegfű
Pl (2) lepnica dwudzielna
Ru (1) смолёвка вильчатая
Sk (–) silenka vidlicovitá

1570 **Silene dioica** (L.) Clairv.
Cs (1) knotovka červená; silenka červená
De (–) Rote Nachtnelke
En (–) red campion
Hu (0) piros mécsvirág
Pl (0) bniec czerwony
Ru (0) смолёвка лесная
Sk (–) knôtovka červená

1571 **Silene gallica** L.
Al (–) klokëza e Francës
Bg (1) галско плюскавче
Cs (–) silenka galská
De (–) Französisches Leimkraut
En (–) small-flowered catchfly
Hu (1) francia habszegfű
Pl (1) lepnica francuska
Ru (1) смолёвка английская
Sk (–) silenka francúzska

* **Silene inflata** Sm. → *Silene vulgaris*

1572 **Silene italica** (L.) Pers.
De (–) Hain-Leimkraut
En (–) Italian catchfly
Ru (0) смолёвка итальянская
Sh (1) talijanski pucavac
Sk (–) silenka talianska

1573 **Silene multiflora** (Waldst. & Kit.) Pers.
Cs (–) silenka mnohokvětá
De (–) Vielblütiges Leimkraut
Ru (0) смолёвка многоцветковая
Sh (1) mnogocvetni pucavac; mnogocvjetni pucavac
Sk (–) silenka mnohokvetá

1574 **Silene noctiflora** L.
Bg (1) нощноцветно плюскавче
Cs (3) knotovka noční; silenka noční
De (–) Acker-Nachtnelke
En (–) night-flowering catchfly
Hu (2) esti mécsvirág
Pl (3) bniec dwudzielny
Ru (1) дрема ночная
Sh (1) crveni pucavac
Sk (–) silenka nočná

1575 **Silene nutans** L.
Bg (0) повиснало плюскавче
Cs (1) silenka nící
De (–) Nickendes Leimkraut
En (–) nodding catchfly
Hu (0) kónya habszegfű
Pl (1) lepnica zwisła
Ru (0) смолёвка поникшая
Sh (1) povijeni pucavac
Sk (–) silenka kývajúca

1576 **Silene pusilla** Waldst. & Kit.
Bg (1) лъчесто плюскавче
De (–) Vierzähniger Strahlensame
Sk (–) leskloplod lysý

1577 **Silene repens** Patrin
Ru (1) смолёвка ползучая
Sk (–) silenka plazivá

1578 **Silene trinervia** Sebast. & Mauri
Bg (1) трижилково плюскавче
Sk (–) silenka trojžilová

* **Silene venosa** Asch. → *Silene vulgaris*

1579 *Silene viscosa* (L.) Pers.
Bg (0) лепкаво плюскавче
Cs (1) knotovka lepivá; silenka lepivá
De (–) Klebriges Leimkraut
En (–) white sticky-catchfly
Hu (1) ragadós mécsvirág
Ru (1) дрема липкая
Sh (1) lepljivi pucavac; ljepiva pušina
Sk (–) silenka lepkavá

1580 *Silene vulgaris* (Moen.) Gar.
Al (–) kloköza e rëndomtë
Bg (1) мехуресто плюскавче
Cs (2) silenka nadmutá
De (–) Aufgeblasene Nachtnelke
En (–) bladder campion
Hu (1) hólyagos habszegfű
Pl (1) lepnica rozdęta
Ro (1) guşa porumbelului
Ru (1) смолёвка широколистная
Sh (1) obični pucavac
Sk (–) silenka nadutá

1581 *Silybum marianum* (L.) Gaertn.
Al (–) gjembgomari; thërrecka
Bg (1) бял трън
Cs (1) ostropestřec Mariánský
De (–) Mariendistel
En (–) Saint Mary's thistle
Gr (1) κουφάγκαθο
Hu (0) Mária-tövis
Pl (1) ostropest plamisty
Ro (0) armurariu
Ru (0) расторопша пятнистая
Sk (–) ostropestrec Mariansky

1582 *Sinapis alba* L.
Al (–) sinapi i bardhë
Bg (1) бял синап
Cs (–) horcice bila
De (–) Weißer Senf
En (–) white mustard
Gr (3) ήμερο σινάπι
Hu (1) fehér mustár
Pl (–) gorczyca biała
Ro (2) muştar alb
Ru (1) горчица белая
Sh (3) bela gorušica; bijela gorušica
Sk (–) horčica biela

1583 *Sinapis arvensis* L.
Al (–) sinapi i arës
Bg (3) полски синап
Cs (3) hořčice rolní
De (–) Acker-Senf
En (–) charlock
Gr (3) άγριο σινάπι
Hu (3) vadrepce
Pl (3) gorczyca polna
Ro (3) muştar sălbatic
Ru (3) горчица полевая
Sh (2) poljska gorušica
Sk (–) horčica roľná

* ***Sinapis dissecta*** Lag. → *Sinapis alba*

1584 *Sisymbrium* spp. L.
Bg (1) мъдрица
De (–) Rauke
Gr (1) σισύμπριο
Ru (1) гулявник
Sk (–) huľavník

1585 *Sisymbrium altissimum* L.
Bg (1) висока мъдрица
Cs (2) hulevník nejvyšší
De (–) Ungarische Rauke
En (–) tall rocket
Hu (1) magyar zsombor
Pl (0) stulisz pannoński
Ru (1) гулявник высокий
Sk (–) huľavník najvyšší

1586 *Sisymbrium irio* L.
Bg (1) ирова мъдрица
Cs (0) hulevník drobnokvětý
De (–) Schlaffe Rauke
En (–) London-rocket

Pl (1) stulisz gładki
Ru (0) гулявник ирио
Sk (–) hulavník cudzí

1587 ***Sisymbrium loeselii*** L.
Bg (1) льозелова мъдрица
Cs (3) hulevník Loeselův
De (–) Loesel's Rauke
En (–) false London-rocket
Hu (1) parlagi zsombor
Pl (1) stulisz Loesela
Ro (1) voinicică
Ru (1) гулявник Лезелиев
Sh (1) hrapavi oranj
Sk (–) hulavník Loeselov

* ***Sisymbrium murale*** L. → *Diplotaxis muralis*

1588 ***Sisymbrium officinale*** (L.) Scop.
Al (–) sisimbra mjekësore
Bg (1) лекарствен раноцелебник
Cs (3) hulevníkovec lékařský
De (–) Weg-Rauke
En (–) hedge mustard
Hu (1) szapora zsombor
Pl (2) stulisz lekarski
Ro (1) brîncuță
Ru (1) гулявник лекарственный
Sh (1) lekoviti oranj; obični oranj
Sk (–) hulavníkovec lekársky

1589 ***Sisymbrium orientale*** L.
Bg (1) източен раноцелебник
Cs (0) hulevník východní
De (–) Orientalische Rauke
En (–) Eastern rocket
Hu (2) hamvas zsombor
Pl (0) stulisz wschodni
Ru (1) гулявник восточный
Sh (1) istočni oranj; runjavi oranj
Sk (–) hulavník východný

* ***Sisymbrium pannonicum*** Jacq. → *Sisymbrium altissimum*

1590 ***Sisymbrium polyceratium*** L.
Bg (1) многорога мъдрица
Ru (0) гулявник густой
Sk (–) hulavník mnohostrukovitý

* ***Sisymbrium sophia*** L. → *Descurainia sophia*

1591 ***Sisymbrium strictissimum*** L.
Bg (1) изправеносбита мъдрица
Cs (0) hulevník stažený
De (–) Steife Rauke
En (–) upright rocket
Hu (1) magas zsombor
Pl (1) stulisz szływny
Ro (0) voinicică
Ru (1) гулявник сжатый
Sk (–) hulavník stiahnutý

* ***Sisymbrium thalianum*** (L.) Gay → *Arabidopsis thaliana*

* ***Sisymbrium torulosum*** Desf. → *Torularia torulosa*

* ***Sisymbrium villosum*** Moen. → *Sisymbrium orientale*

1592 ***Sisymbrium volgense*** Bieb.
Ru (1) гулявник волжский
Sk (–) hulavník povolžský

* ***Sisymbrium waltheri*** Cran. → *Sisymbrium orientale*

1593 ***Sium latifolium*** L.
Al (–) siumi gjethengushtë
Bg (0) широколистен крайводник
Cs (0) sevlák širolistý
De (–) Breitblättriger Merk
En (–) greater water-parsnip
Hu (1) széleslevelű békakorsó
Pl (1) marek szerokolistny
Ro (0) bolonică

Ru (0) поручейник широколистный
Sh (1) obični hrenak; širokolisni hrenak
Sk (–) potočník širokolistý

1594 *Smilax aspera* L.
Al (–) morenxa
De (–) Stechwinde
Gr (1) αρκουδόβατος
Ru (0) смилакс шероховатый

* ***Smilax nigra*** Willd. → *Smilax aspera*

1595 *Smyrnium olusatrum* L.
Al (–) majdanozi i zi
De (–) Gelbdolde
En (–) common alexanders
Gr (1) ιπποσέλινο
Ru (–) смирния овощная
Sk (–) smyrnium hrebenité

1596 *Smyrnium perfoliatum* L.
Al (–) smirna gjethepushtonjëse
Bg (0) прибел
Cs (2) tromín prorostlý
De (–) Stengelumfassende Gelbdolde
En (–) biennial alexanders
Gr (1) αγριοσέλινο
Hu (0) őzsaláta
Ro (0) salată de pădure
Ru (0) смирния пронзённолистная
Sk (–) smyrnium prerastané

1597 *Smyrnium rotundifolium* Mill.
En (–) round-leaved alexanders
Gr (1) χονδρολίβανο
Ru (0) смирния круглолистная
Sk (–) smyrnium okrúhlolisté

1598 *Solanum* spp. L.
Bg (1) куче грозде
De (–) Nachtschatten
Gr (1) σολανό
Ru (2) паслён
Sk (–) ľuľok

* ***Solanum alatum*** Moen. → *Solanum luteum*

1599 *Solanum cornutum* Lam.
De (–) Stachel-Nachtschatten
Ru (0) паслён рогатый
Sk (–) ľuľok zobákovitý

* ***Solanum decipiens*** Opiz → *Solanum nigrum*

1600 *Solanum dulcamara* L.
Al (–) idhnakthi i ëmbël i hidhur; idhnaqi
Bg (1) червено куче грозде
Cs (2) lilek potměchuť; potměchuť
De (–) Bittersüßer Nachtschatten
En (–) woody nightshade
Hu (1) keserny és ebszőlő
Pl (1) psianka słodkogórz
Ro (1) lăsnicior
Ru (1) паслён сладко-горький
Sh (1) gorkoslad pomoćnica; razvodnik pomoćnica
Sk (–) ľuľok sladkohorký

1601 *Solanum elaeagnifolium* Cav.
Cs (–) lilek hlošinolistý
De (–) Ölweidenblättriger Nachtschatten
En (–) silverleaf nightshade
Ru (1) паслён лохолистный
Sk (–) ľuľok hlošinolistý

* ***Solanum flavum*** Kit. → *Solanum luteum*

1602 *Solanum heterodoxum* Dunal
Ru (1) паслён разнообразный

1603 *Solanum luteum* Mill.
Cs (0) lilek žlutý
De (–) Gelber Nachtschatten
En (–) hairy nightshade
Hu (1) sárga ebszőlő

Pl (0) psianka kosmata
Ro (1) zîrnă roşie
Ru (1) паслён жёлтый
Sk (–) ľuľlok žltý pravý

1604 ***Solanum nigrum*** L.
Al (–) idhnakthi i zi
Bg (2) черно кучешко грозде
Cs (3) lilek černý
De (–) Schwarzer Nachtschatten
En (–) black nightshade
Gr (3) στίφνος
Hu (3) fekete ebszőlő
Pl (2) psianka czarna
Ro (3) zîrnă roşie
Ru (2) паслён чёрный
Sh (2) crna pomoćnica; obična pomoćnica
Sk (–) ľuľok čierny

1605 ***Solidago canadensis*** L.
Bg (0) канадски енчец
Cs (0) celík kanadský
De (–) Kanadische Goldrute
En (–) Canadian golden-rod
Hu (1) kanadai aranyvessző
Pl (2) nawłoć kanadyjska
Ro (0) sînzîiană de grădină
Ru (1) золотарник канадский
Sh (1) gustocvetna zlatnica; gustocvjetna zlatnica
Sk (–) zlatá byľ kanadská

1606 ***Solidago gigantea*** Aiton
Cs (0) celík obrovský
De (–) Riesen-Goldrute
En (–) giant golden-rod
Hu (2) magas aranyvessző
Pl (2) nawłoć późna
Sh (1) golema zlatnica
Sk (–) zlatá byľ obrovská

* ***Solidago serotina*** Aiton → *Solidago gigantea*

1607 ***Solidago virgaurea*** L.
Al (–) shuferarta; solidaga shuferartë
Bg (0) златист енчец
Cs (3) celík zlatobýl
De (–) Echte Goldrute
En (–) common golden-rod
Gr (1) χρυσόβεργα
Hu (0) közönséges aranyvessző
Pl (1) nawłoć pospolita
Ro (0) splinuţă
Ru (1) золотарник обыкновенный
Sk (–) zlatá byľ obyčajná

1608 ***Sonchus* spp.** L.
Bg (1) кострец
De (–) Gänsedistel
En (–) sow-thistle
Gr (2) ζωχός
Ru (2) осот
Sk (–) mliečь

1609 ***Sonchus arvensis*** L.
Al (–) ogrishta; rrëshyelli i arës
Bg (2) полски кострец
Cs (3) mléč rolní
De (–) Acker-Gänsedistel
En (–) perennial sow-thistle
Hu (3) mezei csorbóka
Pl (3) mlecz polny
Ro (3) susai
Ru (3) осот полевой
Sh (3) poljska gorčika; poljski ostak
Sk (–) mlieč roľný

1610 ***Sonchus asper*** (L.) Hill
Al (–) rrëshyelli i ashpër
Bg (1) грапав кострец
Cs (3) mléč drsný
De (–) Dornige Gänsedistel
En (–) prickly sow-thistle
Hu (3) szúrós csorbóka
Pl (3) mlecz kolczasty
Ro (3) susai aspru
Ru (3) осот шероховатый

Sh (3) hrapava gorčika
Sk (–) mlieč drsný

* **Sonchus ciliatus** Lam. → *Sonchus oleraceus*
* **Sonchus glaucescens** Jord. → *Sonchus asper*
* **Sonchus levis** Gars. → *Sonchus oleraceus*

1611 Sonchus oleraceus L.
Al (–) rrëshyelli perimesh
Bg (1) градински кострец
Cs (3) mléč zelinný
De (–) Kohl-Gänsedistel
En (–) smooth sow-thistle
Hu (3) szelíd csorbóka
Pl (3) mlecz zwyczajny
Ro (3) susai moale
Ru (3) осот огородный
Sh (1) obična gorčika; obični svinjak
Sk (–) mlieč zelinný

1612 Sonchus palustris L.
Cs (1) mléč bahenní
De (–) Sumpf-Gänsedistel
En (–) marsh sow-thistle
Hu (0) mocsári csorbóka
Pl (1) mlecz błotny
Ro (0) susai mare
Ru (1) осот болотный
Sh (1) velika gorčika
Sk (–) mlieč močiarny

1613 Sorghum halepense (L.) Pers.
Al (–) talla
Bg (2) балур
Cs (2) čirok halepský
De (–) Wilde Mohrenhirse
En (–) Johnson-grass
Gr (3) βέλιουρας
Hu (3) fenyércirok
Ro (3) costrei
Ru (3) гумай
Sh (3) divlji sirak
Sk (–) cirok halebský

1614 Sparganium emersum Rehm.
De (–) Einfacher Igelkolben
En (–) unbranched bur-reed
Gr (1) ορθιο σπαργάνιο
Hu (0) egyszerű békabuzogány
Ru (0) ежеголовник всплывающий
Sk (–) ježohlav jednoduchý

1615 Sparganium erectum L.
Al (–) spargana e degëzuar
Bg (1) разклонена ежова главица
Cs (0) zevar vzpřímený
De (–) Ästiger Igelkolben
En (–) common bur-reed
Gr (1) πλατύφυλλο σπάργάνιο
Hu (1) ágas békabuzogány
Pl (1) jeżogłówka gałęzista
Ro (1) buzdugan
Ru (1) ежеголовник прямой
Sh (1) uspravni ježinac
Sk (–) ježohlav vzpriamený

* **Sparganium neglectum** Beeby → *Sparganium erectum*

* **Sparganium ramosum** Huds. → *Sparganium erectum*

1616 Spartium junceum L.
Al (–) xana
Bg (0) спарциум
De (–) Spanischer Ginster
En (–) Spanish broom
Gr (1) σπάρτο
Ro (0) bucsău
Ru (0) метельник прутьевидный
Sk (–) prútnatec sitinovitý

* **Specularia hybrida** (L.) DC. → *Legousia hybrida*

* **Specularia pentagonia** (L.) DC. → *Legousia pentagonia*

* **Specularia speculum-veneris** (L.) DC. *Legousia speculum-veneris*

1617 Spergula arvensis L.
Bg (2) полска колянка
Cs (3) kolenec rolní
De (–) Feld-Spark
En (–) corn spurrey
Gr (1) κοινή σπεργούλα
Hu (2) mezei csibehúr
Pl (3) sporek polny
Ro (2) hrana vacii
Ru (3) торица полевая
Sh (1) njivska kolenika; poljska koljenika
Sk (–) kolenec roľný pravý

* **Spergula linicola** Boreau → *Spergula arvensis*

* **Spergula maxima** Weihe → *Spergula arvensis*

1618 Spergula morisonii Boreau
Bg (1) едросеменна колянка
Cs (0) kolenec jarní
De (–) Frühlings-Spark
En (–) pearlwort spurrey
Hu (0) homoki csibehúr
Pl (2) sporek wiosenny
Sk (–) kolenec jarný

1619 Spergula pentandra L.
Bg (1) петтичинкова колянка
Cs (0) kolenec pětimužný
De (–) Fünfmänniger Spark
Gr (1) πενταστήμονη σπεργούλα
Pl (1) sporek pięciopręcikowy
Ru (1) торица пятитычинковая
Sh (1) peščana kolenika; pješčana koljenika
Sk (–) kolenec pätmužný

* **Spergula praevisa** Zing. → *Spergula arvensis*

* **Spergula sativa** Boenn. → *Spergula arvensis*

* **Spergula vernalis** auct. → *Spergula morisonii*

* **Spergula vulgaris** Boenn. → *Spergula arvensis*

1620 Spergularia spp. (Pers.) Presl
De (–) Spärkling
Gr (1) σπεργουλάρια
Ru (1) торичник
Sk (–) pakolenec

* **Spergularia campestris** (L.) Asch. → *Spergularia rubra*

1621 Spergularia rubra (L.) Presl
Al (–) spergularia e kuqe
Bg (1) коленчица
Cs (1) kuřinka červená
De (–) Roter Spärkling
En (–) sand spurrey
Hu (2) piros budavirág
Pl (1) muchotrzew polny
Ru (1) торичник полевой
Sh (1) crvena koleničica; crvena koljeničica
Sk (–) pakolenec červený

1622 Spergularia segetalis (L.) Don
Cs (–) kuřinka ostnosemenná
De (–) Saat-Spärkling
En (–) autumn spurrey
Pl (1) delia polna
Sk (–) pakolenec jesenný

* **Spiraea filipendula** L. → *Filipendula vulgaris*

* ***Spiraea ulmaria*** L. → *Filipendula ulmaria*

1623 *Spiranthes aestivalis* (Poir.) Richard
Cs (0) švihlík letní
De (–) Sommer-Wendelähre
En (–) summer lady's-tresses
Gr (1) ανοιξιάτικος σπείρανθος
Hu (0) nyári füzértekercs
Pl (1) kręczynka jesienna
Sk (–) pokrut letný

1624 *Spirodela polyrhiza* (L.) Schleid.
Bg (1) многокоренна спиродела
Cs (0) závitka mnohokořenná
De (–) Teichlinse
En (–) great duckweed
Hu (1) bojtos békalencse
Pl (1) spirodela wielokorzeniowa
Ru (1) многокоренник обыкновенный
Sk (–) žabnička mnohokorenová

1625 *Stachys* spp. L.
Bg (1) ранилист
De (–) Ziest
Gr (1) στάχυς
Ru (1) чистец
Sk (–) čistec

1626 *Stachys annua* (L.) L.
Al (–) sarusha njëvjeçare
Bg (1) едногодишен ранилист
Cs (3) čistec roční
De (–) Einjähriger Ziest
En (–) annual woundwort
Hu (3) tarlótisztesfű
Pl (2) czyściec roczny
Ro (2) cinsteţ
Ru (2) чистец однолетний
Sh (1) jednogodišnji čistac
Sk (–) čistec ročný

1627 *Stachys arvensis* (L.) L.
Cs (0) čistec rolní
De (–) Acker-Ziest
En (–) field woundwort
Pl (2) czyściec polny
Ru (1) чистец полевой
Sk (–) čistec roľný

* ***Stachys bicolor*** Lutz → *Sideritis montana*

1628 *Stachys germanica* L.
Al (–) sarushë e Gjermanisë
Bg (1) германски ранилист
Cs (–) čistec německý
De (–) Deutscher Ziest
En (–) downy woundwort
Hu (1) fehér tisztesfű
Pl (1) czyściec kosmaty
Ro (1) jaleş
Ru (1) чистец германский
Sh (1) germanski čistac
Sk (–) čistec nemecký

1629 *Stachys milanii* Pet.
Bg (1) Миланов ранилист
Ru (0) чистец Милана
Sk (–) čistec milánsky

1630 *Stachys officinalis* (L.) Trev.
Al (–) sarushë mjekësore
Bg (1) лечебен ранилист
Cs (0) bukvice lékařská
De (–) Echter Ziest
En (–) common betony
Hu (0) bakfű
Pl (1) bukwica zwyczajna
Ro (0) creţişor
Ru (0) буквица лекарственная
Sh (1) ranilist čistac
Sk (–) bukvica lekárska

1631 *Stachys palustris* L.
Al (–) sarushë e kënetës
Bg (1) блатен ранилист
Cs (3) čistec bahenní

De (–) Sumpf-Ziest
En (-) marsh woundwort
Hu (2) mocsári tisztesfű
Pl (1) czyściec błotny
Ru (1) чистец болотный
Sh (2) barski čistac; močvarni čistac
Sk (–) čistec močiarny

1632 ***Stachys recta*** L.
Al (–) sarushë e drejtë
Bg (1) прав ранилист
Cs (0) čistec přímý
De (–) Aufrechter Ziest
En (–) perennial yellow-woundwort
Hu (0) hasznos tisztesfű
Pl (1) czyściec prosty
Ro (0) jaleş de cîmp
Ru (0) чистец прямой
Sh (1) osovni čistac; veliki čistac
Sk (–) čistec priamy pravý

1633 ***Stachys sylvatica*** L.
Al (–) sarushë e pyllit
Bg (1) горски ранилист
Cs (0) čistec lesní
De (–) Wald-Ziest
En (–) hedge woundwort
Hu (0) erdei tisztesfű
Pl (1) czyściec leśny
Ro (0) bălbisă
Ru (0) чистец лесной
Sh (1) šumski čistac
Sk (–) čistec lesný

1634 ***Stellaria* spp.** L.
Bg (1) звездица
De (–) Sternmiere
Gr (1) στελλάρια
Ru (2) звездчатка
Sk (–) hviezdica

* ***Stellaria dichotoma*** Geor. → *Stellaria media*

1635 ***Stellaria graminea*** L.
Al (–) stelara gramore; yllëz i graminorëve
Bg (1) бяла звездица
Cs (1) ptačinec trávovitý
De (–) Gras-Sternmiere
En (–) lesser stitchwort
Hu (2) réti csillaghúr
Pl (2) gwiazdnica trawiasta
Ru (1) звездчатка злачная
Sh (1) dugolisna mišjakinja; prugolisna mišjakinja
Sk (–) hviezdica trávovitá

1636 ***Stellaria holostea*** L.
Bg (1) ланцетолистна звездица
Cs (1) ptačinec velkokvětý
De (–) Große Sternmiere
En (–) greater stitchwort
Hu (0) olocsán csillaghúr
Pl (1) gwiazdnica wielkokwiatowa
Ro (0) iarbă moale
Ru (1) звездчатка ланцетовидная
Sh (1) visoka mišjakinja
Sk (–) hviezdica veľkokvetá

1637 ***Stellaria media*** (L.) Vill.
Al (–) stelara e ndërmjeme
Bg (2) врабчова звездица
Cs (3) ptačinec žabinec
De (–) Vogel-Sternmiere
En (–) common chickweed
Gr (3) στελλάρια
Hu (3) tyúkhúr
Pl (3) gwiazdnica pospolita
Ro (3) rocoină
Ru (3) звездчатка средняя
Sh (2) obična mišjakinja
Sk (–) hviezdica prostredná

1638 ***Stellaria nemorum*** L.
Al (–) stelara pyjesh; yllëz i pyllit
Bg (1) сенчеста звездица
Cs (2) ptačinec hajní

De (–) Hain-Sternmiere
En (–) wood stitchwort
Hu (0) erdei csillaghúr
Pl (1) gwiazdnica gajowa
Ro (0) steluţă
Ru (1) звездчатка лесная
Sk (–) hviezdica hájna

1639 ***Stellaria pallida*** (Dumort.) Piré
Bg (1) бледна звездица
De (–) Bleiche Sternmiere
En (–) lesser chickweed
Hu (3) tyúkhúr
Pl (1) gwiazdnica blada
Ru (1) звездчатка бледная
Sk (–) hviezdica bledá

1640 ***Stellaria palustris*** Retz.
Bg (1) блатна звездица
Cs (1) ptačinec bahenní
De (–) Sumpf-Sternmiere
En (–) marsh stitchwort
Hu (0) mocsári csillaghúr
Pl (1) gwiazdnica błotna
Ru (1) звездчатка болотная
Sk (–) hviezdica močiarna

* ***Stellaria viscida*** Bieb. → *Cerastium dubium*

* ***Stenactis annua*** (L.) Less. → *Erigeron annuus*

* ***Stenophragma thalianum*** (L.) Čelak. → *Arabidopsis thaliana*

1641 ***Steptorhamphus tuberosus*** (Jacq.) Grossh.
Gr (1) κρητικό αγριομάρουλο
Ru (0) степторамфус клубненосный

1642 ***Stipa* spp.** L.
Bg (1) коило
De (–) Federgras
Gr (1) στίπα
Ru (1) ковыль
Sk (–) kavyľ

1643 ***Stipa bromoides*** (L.) Dorf.
Bg (1) късоосилесто коило
En (–) slender feather-grass
Ru (1) ковыль костровидный
Sk (–) kavyľ štihlý

1644 ***Stipa capillata*** L.
Bg (1) влакновидно коило
Cs (0) kavýl vláskovitý
De (–) Haar-Federgras
En (–) hairy feather-grass
Hu (1) kunkorgó árvalányhaj
Pl (1) ostnica włosowata
Ro (1) năgară
Ru (1) ковыль волосатик
Sk (–) kavyľ vláskovitý

1645 ***Stipa pennata*** L.
Al (–) korilja; pendëkaposhi
Bg (0) Иоанова коило
Cs (0) kavýl Ivanův
De (–) Echtes Federgras
En (–) feather-grass
Hu (0) hegyi árvalányhaj
Ro (1) colilie
Ru (0) ковыль перистый
Sk (–) kavyľ perovitý

1646 ***Stratiotes aloides*** L.
Bg (0) аловиден стратиотес
Cs (0) řezan pilolistý
De (–) Krebsschere
En (–) water-soldier
Hu (0) kolokán
Pl (1) osoka aloesowata
Ro (0) foarfeca bălţii
Ru (1) телорез сабуровидный
Sh (1) krhkolisni rezac; testerasti rezac
Sk (–) rezavka aloesovitá

1647 ***Succisa pratensis*** Moen.
Al (–) suçiza e livadhit
Bg (1) ливадна синьоглавка
Cs (1) čertkus luční
De (–) Gemeiner Teufelsabbiß
En (–) devil's-bit scabious
Hu (0) ördögharaptafű
Pl (1) czarcikęs łąkowy
Ru (1) сивец луговой
Sh (1) obični preskoč; poljski preskoč
Sk (–) čertkus lúčny

1648 ***Symphytum bulbosum*** Schim.
Al (–) kufilma me qepujka
Bg (1) луковично зарасличе
De (–) Kleinblütiger Beinwell
En (–) bulbous comfrey
Gr (1) βολβόριζο στεκούλι
Sk (–) kostihoj vláknitý

1649 ***Symphytum officinale*** L.
Al (–) kufilma mjekësore
Cs (3) kostival lékařský
De (–) Gemeiner Beinwell
En (–) common comfrey
Gr (1) φαρμακευτικό στεκούλι
Hu (3) fekete nadálytő
Pl (2) żywokost lekarski
Ro (2) tătăneasă
Ru (1) окопник лекарственный
Sh (1) obični gavez
Sk (–) kostihoj lekársky

1650 ***Symphytum tuberosum*** L.
Cs (–) kostival hliznatý
De (–) Knolliger Beinwell
En (–) tuberous comfrey
Pl (0) żywokost bulwiasty
Ru (0) окопник клубневой
Sh (1) gomoljasti gavez; žuti gavez
Sk (–) kostihoj hľuznatý

T

1651 ***Taeniatherum caput-medusae*** (L.) Nev.
Bg (1) медузина глава
De (–) Medusenhaupt-Haargerste
En (–) medusa-head
Ru (0) лентоостник шероховатый
Sk (–) pýrovník medúzia hlava

1652 ***Tamarix* spp.** L.
Al (–) bruka; marina
Bg (0) ракитница
Cs (0) tamaryšek
De (–) Tamariske
En (–) tamarisk
Gr (1) αλμυρίκι
Pl (–) tamaryszek
Ro (–) cătină roşie
Ru (0) тамарикс
Sk (–) tamariška

1653 ***Tamus communis*** L.
Al (–) pejza; rrush gjarpëri
Bg (1) брей
De (–) Schmerwurz
En (–) black bryony
Gr (1) αβρονιά
Hu (0) pirítógyökér
Ro (0) fluerătoare
Ru (0) тамус обыкновенный
Sk (–) dioskurea čierna

1654 ***Tanacetum corymbosum*** (L.) Schultz
Cs (–) kopretina chocholičnatá
De (–) Straußblütige Wucherblume
Ru (0) пиретрум щитковый
Sh (1) štitasti kaloper
Sk (–) rimbaba chocholíkatá

1655 ***Tanacetum macrophyllum*** (Waldst. & Kit.) Schultz
Bg (1) едролистна маргаритка
Cs (–) kopretina velkolistá
De (–) Großblättrige Wucherblume
Ro (1) vetrice
Sk (–) rimbaba veľkolistá

1656 ***Tanacetum parthenium*** (L.) Schultz
Al (–) krizantema e virgjët
Bg (0) моминска маргаритка
Cs (0) kopretina řimbaba
De (–) Mutterkraut
En (–) feverfew
Hu (0) őszi margitvirág
Pl (1) złocień maruna
Ro (0) spilcuţe
Ru (0) пиретрум девичий
Sk (–) rimbaba obyčajná

1657 ***Tanacetum vulgare*** L.
Al (–) bar krimbash; karajpeli
Bg (1) вратига
Cs (3) vratič obecný
De (–) Rainfarn
En (–) common tansy
Gr (1) τανάκητο
Hu (1) gilisztaűző varádics
Pl (2) wrotycz pospolity
Ro (1) vetrice
Ru (1) пижма обыкновенная
Sh (1) obični kaloper; obični vratić
Sk (–) vratič obyčajný

1658 ***Taraxacum officinale*** Weber
Al (–) luleshurdha; lule qumështore
Bg (2) глухарче
Cs (3) smetanka lékařská
De (–) Gemeiner Löwenzahn
En (–) dandelion
Gr (1) αγριοράδικο
Hu (3) pongyola pitypang
Pl (3) mniszek pospolity
Ro (2) păpădie
Ru (3) одуванчик аптечный
Sh (2) obični maslačak
Sk (–) púpava lekárska

* ***Taraxacum vulgare*** (L.) Schk. → *Taraxacum officinale*

1659 ***Teesdalia nudicaulis*** (L.) R Br.
Cs (0) nahoprutka písečná
De (–) Bauernsenf
En (–) shepherd's-cress
Hu (0) rejtőke
Pl (2) chroszcz nagołodygowy
Ru (0) тисдайлия голостебельная
Sk (–) teesdálka piesočná

1660 ***Teucrium botrys*** L.
Cs (0) ožanka hroznatá
De (–) Trauben-Gamander
Hu (1) fürtös gamandor
Pl (1) ożanka pierzastosieczna
Ru (0) дубровник метельчатый
Sk (–) hrdobarka hroznovitá

1661 ***Teucrium chamaedrys*** L.
Al (–) arrsi; dushk i vogël
Cs (0) ožanka kalamandra
De (–) Echter Gamander
En (–) wall germander
Gr (1) χαμοδρυά
Hu (0) sarló gamandor
Pl (1) ożanka właściwa
Ro (0) dumbeţ
Ru (1) дубровник обыкновенный
Sh (1) hrastolisni dubačac; povaljeni dubačac
Sk (–) hrdobarka pravá

* ***Teucrium chamaepitys*** L. → *Ajuga chamaepitys*

1662 ***Teucrium polium*** L.
Al (–) bar majasili; lule zoje
De (–) Polei-Gamander
Gr (1) ασπρόχορτο
Ro (0) sugărel alb
Ru (0) дубровник беловойлочный
Sk (–) hrdobarka polejová

1663 ***Teucrium scordium*** L.
Al (–) arrsi si hudhër; bar hudhre
Cs (0) ožanka česneková
De (–) Lauch-Gamander
En (–) water germander
Hu (1) vízi gamandor
Pl (1) ożanka czosnkowa
Ro (0) iarbă usturoasă
Ru (1) дубровник скордиевидный
Sh (1) bodljasti dubačac; smrdljivi dubačac
Sk (–) hrdobarka cesnaková

1664 ***Thalictrum* spp.** L.
Bg (1) обичниче
De (–) Wiesenraute
En (–) meadow-rue
Gr (1) θάλικτρο
Ru (0) василистник
Sk (–) žltuška

1665 ***Thalictrum aquilegifolium*** L.
Al (–) pipanolla me gjethe akilegë
Bg (1) кандилковидно обичниче
Cs (0) žluťucha orlíčkolistá
De (–) Akeleiblättrige Wiesenraute
En (–) greater meadow-rue
Hu (0) erdei borkóró
Pl (1) rutewka orlikolistna
Ro (0) rutişor
Ru (0) василистник водосборолистный
Sh (1) modra očobajka; sinja kozlačica
Sk (–) žltuška orličkolistá

* ***Thalictrum elatum*** Jacq. → *Thalictrum minus*

1666 ***Thalictrum flavum*** L.
Al (–) pipanolla e verdhë
Bg (1) жълто обичниче
Cs (0) žluťucha žlutá
De (–) Gelbe Wiesenraute
En (–) common meadow-rue
Hu (0) sárga borkóró

Pl (0) rutewka żółta
Ro (0) rutişor galben
Ru (0) василистник жёлтый
Sk (–) žltuška obyčajná

* **Thalictrum galioides** Nestl. → *Thalictrum simplex*

* **Thalictrum majus** Jacq. → *Thalictrum minus*

* **Thalictrum medium** Jacq. → *Thalictrum minus*

1667 **Thalictrum minus** L.
Al (–) pipanolla e vogël
Bg (1) малко обичниче
Cs (0) žluťucha menší
De (–) Kleine Wiesenraute
En (–) lesser meadow-rue
Hu (0) közönséges borkóró
Pl (1) rutewka mniejsza
Ru (0) василистник малый
Sh (1) malena očobajka
Sk (–) žltuška menšia pravá

1668 **Thalictrum simplex** L.
Al (–) pipanolla e thjeshtë
Bg (1) обикновено обичниче
Cs (0) žluťucha jednoduchá
De (–) Schmalblättrige Wiesenraute
En (–) small meadow-rue
Hu (0) egyszerű borkóró
Pl (1) rutewka pojedyncza
Ru (0) василистник простой
Sk (–) žltuška jednoduchá

1669 **Thermopsis lanceolata** R Br.
Ru (1) термопсис ланцетолистный

1670 **Thesium spp.** L.
Bg (1) ленолист
De (–) Leinblatt
En (–) bastard-toadflax
Gr (1) θέσιο
Ru (0) ленец
Sk (–) ľanolistník

1671 **Thesium alpinum** L.
Bg (1) алпийски ленолист
Cs (0) lněnka alpská
De (–) Alpen-Leinblatt
En (–) alpine bastard-toadflax
Pl (1) leniec alpejski
Ru (0) ленец альпийский
Sk (–) ľanolistník alpínsky

1672 **Thesium arvense** Horv.
Al (–) armira e degëzuar
Bg (1) полски ленолист
Cs (0) lněnka větevnatá
Hu (1) homoki zsellérke
Pl (0) leniec polny
Ru (1) ленец полевой
Sh (1) livadska lanika
Sk (–) ľanolistník lúčny

1673 **Thesium divaricatum** Jan
Bg (0) разклонен ленолист
De (–) Sparriger Bergflachs
Ru (0) ленец распростёртый
Sh (1) metličasta lanika; raskrenuti lanak
Sk (–) ľanolistník roztiahnutý

1674 **Thesium humile** Vahl
Bg (1) долинеров ленолист
Gr (1) αγριολίναρο
Hu (1) ugari zsellérke
Ru (1) ленец низкий
Sk (–) ľanolistník nízky

* **Thesium ramosum** Hayne → *Thesium arvense*

1675 **Thlaspi spp.** L.
Bg (1) старчец
De (–) Hellerkraut
En (–) pennycress

Gr (1) θλάσπι
Ru (1) ярутка
Sk (–) peniažtek

1676 ***Thlaspi alliaceum*** L.
Bg (1) луковиден старчец
Cs (1) penízek prorostlý
De (–) Lauch-Hellerkraut
En (–) garlic pennycress
Hu (1) hagymaszagú tarsóka
Pl (0) tobołki czosnkowe
Ru (0) ярутка чесночная
Sh (1) smrdolisna čestika; vonjavi mošnjak
Sk (–) peniažtek cesnakovitý

1677 ***Thlaspi arvense*** L.
Bg (1) полски старчец
Cs (3) penízek rolní
De (–) Acker-Hellerkraut
En (–) field pennycress
Hu (2) mezei tarsóka
Pl (3) tobołki polne
Ro (3) punguliță
Ru (2) ярутка полевая
Sh (1) kravlja stisnuša; poljski mošnjak
Sk (–) peniažtek roľný

1678 ***Thlaspi perfoliatum*** L.
Al (–) tllaspi i përgjethur
Bg (1) прорастолистен старчец
Cs (1) penízek prorostlý
De (–) Stengelumfassendes Hellerkraut
En (–) perfoliate pennycress
Hu (1) galléros tarsóka
Pl (3) tobołki przerosłe
Ru (1) ярутка пронзённая
Sh (1) probušeni mošnjak; srcolisna stisnuša
Sk (–) peniažtek prerastaný

1679 ***Thlaspi praecox*** Wulf.
Al (–) tllaspi i hershëm
Bg (1) ранна попова лъжичка
De (–) Frühblühendes Hellerkraut
Ru (0) ярутка ранняя
Sk (–) peniažtek skorý

1680 ***Thymelaea passerina*** (L.) Cosson & Germ.
Al (–) ligia trumcakësh
Bg (1) обикновена лидия
Cs (0) vrabečnice úpolní
De (–) Gewöhnlicher Vogelkopf
Hu (1) cicó
Pl (0) wilczypieprz roczny
Ro (1) limba vrăbiei
Sh (1) jednogodišnja vrebina; obična ligija
Sk (–) vrabcovník úpolný

1681 ***Thymus capitatus*** (L.) Hoffmans. & Link
Gr (1) θυμάρι
Sk (–) materina dúška hlavatá

1682 ***Thymus pulegioides*** L.
Bg (1) мащерка
Cs (0) mateřídouška polejovitá
De (–) Arznei-Thymian
En (–) large thyme
Hu (0) vékony kakukkfű
Pl (1) macierzanka zwyczajna
Ro (0) lămîioară
Ru (0) тимьян блошиный
Sk (–) materina dúška vajcovitá

1683 ***Thymus serpyllum*** L.
Al (–) zhumbrica
Cs (2) mateřídouška úzkolistá
De (–) Feld-Thymian
En (–) creeping thyme
Hu (0) keskenylevelű kakukkfű
Pl (2) macierzanka piaskowa
Ro (0) cimbru
Ru (1) тимьян ползучий
Sh (1) obična majčina dušica; prava majčina dušica
Sk (–) materina dúška pravá

* ***Tiniaria japonica*** (Houtt.) Hedb. → *Reynoutria japonica*

1684 ***Tordylium apulum*** L.
Al (–) tordilia e Puljes
Bg (0) апулийски тордилиум
De (–) Apulische Zirmet
Gr (1) καυκαλίθρα

1685 ***Tordylium maximum*** L.
Al (–) tordilia e madhe
Bg (0) едър тордилиум
Cs (0) zapalička největší
De (–) Große Zirmet
En (–) greater hartwort
Gr (1) μεγάλη καυκαλίθρα
Hu (1) szarvasgyökér
Pl (0) wszawiec wielki
Ru (0) тордилиум крупный
Sk (–) zápalička veliká

1686 ***Tordylium officinale*** L.
Al (–) tordilia mjekësore
De (–) Gebräuchliche Zirmet
En (–) common hartwort
Gr (1) φαρμακευτική καυκαλίθρα
Sk (–) zápalička lekárska

1687 ***Torilis* spp.** Adans.
Bg (1) торилис
De (–) Klettenkerbel
En (–) hedge-parsley
Gr (1) τοριλίδα
Ru (0) торилис
Sk (–) torica

* ***Torilis anthriscus*** (L.) Gmel. → *Torilis japonica*

1688 ***Torilis arvensis*** (Huds.) Link
Al (–) torila e arave
Bg (1) пренебрегнат торилис
Cs (1) tořice rolní
De (–) Acker-Klettenkerbel
En (–) spreading hedge-parsley
Hu (1) parlagi tüskemag
Ro (2) haţmaţuchiul măgarului
Ru (1) торилис полевой
Sh (1) poljska čekinjavka; poljska krpeljica
Sk (–) torica roľná

* ***Torilis heterophylla*** Guss. → *Torilis arvensis*

1689 ***Torilis japonica*** (Houtt.) DC.
Al (–) torila e Japonisë
Bg (1) азмацуковиден торилис
Cs (0) tořice japonská
De (–) Gemeiner Klettenkerbel
En (–) upright hedge-parsley
Hu (1) bojtorjános tüskemag
Pl (2) kłobuczka pospolita
Ru (1) торилис японский
Sh (1) crljenkasta čekinjavka; velika krpeljica
Sk (–) torica japonská

1690 ***Torilis leptophylla*** (L.) Reichenb.
Al (–) torila gjethehollë
De (–) Feinblättriger Klettenkerbel
Ru (1) торилис тонколистный
Sk (–) torica tenkolistá

* ***Torilis neglecta*** (L.) Schultes → *Torilis arvensis*

1691 ***Torilis nodosa*** (L.) Gaertn.
Al (–) torila me nyje
Bg (1) възловат торилис
Cs (0) tořice uzlovitá
De (–) Knäuel-Klettenkerbel
En (–) knotted hedge-parsley
Ru (0) торилис узковатый
Sk (–) torica uzlovitá

1692 ***Torularia torulosa*** (Desf.) Schulz
Ru (1) чёточник бугорчатый

1693 ***Tragopogon* spp.** L.
Bg (1) козя брада
De (–) Bocksbart
En (–) goat's-beard
Gr (1) τραγοπώγονας
Ru (0) козлобородник
Sk (–) kozia brada

1694 ***Tragopogon dubius*** Scop.
Bg (1) едра козя брода
Cs (0) kozí brada pochybná
De (–) Großer Bocksbart
En (–) western goat's-beard
Hu (1) nagy bakszakáll
Pl (0) kozibród wielki
Ro (1) barba caprei
Ru (1) козлобородник большой
Sh (1) velika kozja brada; veliki kozobrad
Sk (–) kozia brada pochybná

* ***Tragopogon eriospermus*** Ten. → *Tragopogon porrifolius*

* ***Tragopogon major*** Jacq. → *Tragopogon dubius*

1695 ***Tragopogon porrifolius*** L.
Al (–) bar cjap
Bg (1) празолистна козя брада
Cs (0) kozí brada fialová
De (–) Haferwurz
En (–) common salsify
Ro (0) barba caprei
Ru (0) козлобородник пореелистный
Sk (–) kozia brada fialová

1696 ***Tragopogon pratensis*** L.
Al (–) bar cjap i livadheve; lulbrigje livadhësh
Bg (1) ливадна козя брада
Cs (2) kozí brada luční
De (–) Wiesen-Bocksbart
En (–) meadow goat's-beard
Hu (1) közönséges bakszakáll
Pl (1) kozibród łąkowy
Ro (0) tîţa caprei
Ru (1) козлобородник луговой
Sh (1) livadna kozja brada; livadski kozobrad
Sk (–) kozia brada lúčna

* ***Tragopogon sinuatus*** Avé. → *Tragopogon porrifolius*

1697 ***Tragus racemosus*** (L.) All.
Al (–) tragëza me vile
Bg (1) ягло
Cs (0) bodloplev hroznatý
De (–) Klettengras
En (–) small bur-grass
Hu (2) tövisperje
Ro (1) iarbă scăioasă
Ru (1) козлец кистистый
Sh (1) grozdasta toričina
Sk (–) ostroplev strapcovitý

1698 ***Trapa natans*** L.
Al (–) gështenjë ujëse
Bg (1) дяволски орех
Cs (1) kotvice plovoucí
De (–) Wassernuß
En (–) water-chestnut
Gr (1) τριβόλι του νερού
Hu (0) sulyom
Pl (0) kotewka orzech wodny
Ro (0) cornaci
Ru (0) водяной орех плавающий
Sk (–) kotvica plávajúca

1699 ***Tremastelma palaestinum*** (L.) Janch.
Gr (1) τρεμάστελμα
Sk (–) hlaváč vetvistý

1700 ***Tribulus terrestris*** L.
Al (–) tribula tokësore
Bg (1) бабини зъби
Cs (0) kotvičník zemní

De (–) Erd-Bürzeldorn
En (–) puncture-vine
Gr (1) τριβόλι
Hu (2) királydinnye
Ro (2) colții babei
Ru (1) якорцы стелющиеся
Sh (1) žuti babin zub
Sk (–) kotvičník zemný

1701 ***Trifolium* spp.** L.
Bg (1) детелина
De (–) Klee
En (–) clover
Gr (1) αγριοτρίφυλλο
Ru (2) клевер
Sk (–) ďatelina

1702 ***Trifolium alpestre*** L.
Al (–) tërfili i Alpeve
Bg (0) алпийска детелина
Cs (1) jetel alpinský
De (–) Hügel-Klee
En (–) mountain zigzag-clover
Hu (0) bérci here
Pl (1) koniczyna dwukłosowa
Ru (0) клевер альпийский
Sh (1) gorska detelina; gorska djetelina
Sk (–) ďatelina horná

1703 ***Trifolium arvense*** L.
Al (–) tërfili i arave
Bg (1) котенце детелина
Cs (2) jetel rolní
De (–) Hasen-Klee
En (–) hare's-foot clover
Hu (2) tarlóhere
Pl (3) koniczyna polna
Ro (2) papanași
Ru (2) клевер пашенный
Sh (1) ugarna djetelina; zečja detelina
Sk (–) ďatelina roľná

1704 ***Trifolium aureum*** Pollich
Cs (1) jetel zlatý
De (–) Gold-Klee
En (–) large hop-trefoil
Pl (3) koniczyna złocistożółta
Ru (2) клевер золотистый
Sk (–) ďatelina zlatá

* ***Trifolium bicorne*** Forsk. → *Trifolium resupinatum*

* ***Trifolium bonannii*** Presl → *Trifolium fragiferum*

1705 ***Trifolium campestre*** Schreb.
Al (–) tërfili i fushës
Bg (1) полска детелина
Cs (1) jetel ladní
De (–) Gelber Acker-Klee
En (–) common hop-trefoil
Hu (1) mezei here
Pl (2) koniczyna różnoogonkowa
Ro (0) trifoiaș
Ru (1) клевер полевой
Sh (1) engleska detelina; poljska djetelina
Sk (–) ďatelina ladná

1706 ***Trifolium dubium*** Sibth.
Al (–) tërfili i hollë
Bg (1) съмнителна детелина
Cs (2) jetel pochybný
De (–) Zwerg-Klee
En (–) lesser yellow-trefoil
Hu (0) apró here
Pl (2) koniczyna drobnogłówkowa
Ru (0) клевер сомнительный
Sk (–) ďatelina pochybná

1707 ***Trifolium echinatum*** Bieb.
Al (–) tërfili i kreshpët
Bg (1) бодлива детелина
Ru (0) клевер иглистый
Sk (–) ďatelina bodľavá

* ***Trifolium eriocephalum*** Ledeb. → *Trifolium aureum*

1708 ***Trifolium fragiferum*** L.
Al (–) tërfili luleshtrydhëbartës
Bg (1) крехка детелина
Cs (1) jetel jahodnatý
De (–) Erdbeer-Klee
En (–) strawberry clover
Hu (1) eperhere
Pl (1) koniczyna rozdęta
Ro (1) trifoi frăguț
Ru (1) клевер земляничный
Sk (–) ďatelina jahodnatá

1709 ***Trifolium glomeratum*** L.
Bg (1) главеста детелина
En (–) clustered clover
Ru (0) клевер скученный
Sk (–) ďatelina kĺbkovitá

1710 ***Trifolium hybridum*** L.
Al (–) tërfili hibrid
Bg (1) хибридна детелина
Cs (2) jetel zvrhlý
De (–) Bastard-Klee
En (–) alsike clover
Hu (1) korcs here
Pl (3) koniczyna białoróżowa
Ru (1) клевер розовый
Sh (1) hibridna djetelina; švedska detelina
Sk (–) ďatelina zvrhlá

1711 ***Trifolium incarnatum*** L.
Al (–) tërfili ngjyrëmishi
Bg (1) инкарнатка детелина
Cs (0) jetel růžák
De (–) Inkarnatklee
En (–) crimson clover
Hu (0) bíborhere
Pl (1) koniczyna inkarnatka
Ru (0) клевер пунцовый
Sk (–) ďatelina inkarnatá

1712 ***Trifolium lappaceum*** L.
Bg (1) репейна детелина
Cs (–) jetel lopuchový
De (–) Kletten-Klee
Ru (0) клевер репейниковый
Sk (–) ďatelina lopúchovitá

1713 ***Trifolium leucanthum*** Bieb.
Al (–) tërfili lulebardhë
Bg (1) бяловлакнеста детелина
Ru (0) клевер бледноцветный
Sk (–) ďatelina bielokvetá

* ***Trifolium maritimum*** Huds. → *Trifolium squamosum*

1714 ***Trifolium medium*** L.
Al (–) tërfili i ndërmjetmë
Bg (1) междинна детелина
Cs (0) jetel prostřední
De (–) Zickzack-Klee
En (–) zigzag-clover
Hu (0) erdei here
Pl (1) koniczyna pogięta
Ru (0) клевер средний
Sk (–) ďatelina prostredná

1715 ***Trifolium michelianum*** Savi
Bg (1) Мишелианова детелина
Ru (–) клевер Михеляна
Sk (–) ďatelina Michelianova

* ***Trifolium minus*** Sm. → *Trifolium dubium*

* ***Trifolium molineri*** Balbis → *Trifolium incarnatum*

1716 ***Trifolium montanum*** L.
Al (–) tërfili malor
Bg (1) планинска детелина
Cs (1) jetel horský
De (–) Berg-Klee
En (–) mountain clover
Hu (0) hegyi here
Pl (1) koniczyna pagórkowa
Ru (0) клевер горный

Sh (1) brdska detelina; brdska djetelina
Sk (–) ďatelina horská

1717 ***Trifolium nigrescens*** Viv.
Al (–) tërfili zëzërrëm
Bg (1) възчерна детелина
En (–) ball clover
Sk (–) ďatelina černejúca

1718 ***Trifolium ochroleucon*** Huds.
Al (–) tërfili bardhevel; tërfili verdhak
Bg (1) бледожълта детелина
Cs (0) jetel bledožlutý
De (–) Blaßgelber Klee
En (–) sulphur clover
Hu (0) vajszín here
Pl (1) kniczyna żółtobiała
Ru (0) клевер бледно-жёлтый
Sh (1) bledožuta detelina; žutikava djetelina
Sk (–) ďatelina bledožltá

1719 ***Trifolium pallidum*** Waldst. & Kit.
Al (–) tërfili i zbehtë
Bg (1) бледна детелина
Cs (0) jetel bledý
Hu (0) halvány here
Sk (–) ďatelina bledá

1720 ***Trifolium pannonicum*** Jacq.
Al (–) tërfili hungarez
Bg (1) панонска детелина
Cs (0) jetel panonský
De (–) Ungarischer Klee
En (–) Hungarian clover
Hu (0) magyar here
Ru (0) клевер паннонский
Sk (–) ďatelina panónska

* ***Trifolium parviflorum*** Ehrh. → *Trifolium retusum*

1721 ***Trifolium patens*** Schreb.
Al (–) tërfili i hapur
Bg (1) жълта детелина
Cs (0) jetel otevřený
De (–) Spreiz-Klee
Hu (0) terpedő here
Pl (0) koniczyna odstająca
Sh (1) žuta detelina; žuta djetelina
Sk (–) ďatelina otvorená

1722 ***Trifolium pratense*** L.
Al (–) tërfili i livadheve; tërfoja
Bg (1) червена детелина
Cs (2) jetel luční
De (–) Rot-Klee
En (–) red clover
Hu (0) vörös here
Pl (2) koniczyna łąkowa
Ro (1) trifoiul roşu
Ru (1) клевер луговой
Sk (–) ďatelina lúčna

1723 ***Trifolium purpureum*** Loisel.
Al (–) tërfili i purpurte
Bg (1) пурпурна детелина
De (–) Purpur-Klee
En (–) purple clover
Ru (–) клевер пурпурный
Sk (–) ďatelina nachová

1724 ***Trifolium repens*** L.
Al (–) tërfili i bardhë; tërfili zvaranik
Bg (1) бяла детелина
Cs (3) jetel plazivý
De (–) Weiß-Klee
En (–) white clover
Hu (2) fehér here
Pl (3) koniczyna biała
Ro (1) trifoi alb
Ru (3) клевер ползучий
Sh (1) bela detelina; bijela djetelina
Sk (–) ďatelina plazivá

1725 ***Trifolium resupinatum*** L.
Al (–) tërfili i përmbystë; tërfili persian
Bg (1) обърната детелина

Cs (0) jetel zvrácený
De (–) Persischer Klee
En (–) Persian clover
Hu (0) fonák here
Ru (1) клевер опрокинутый
Sk (–) ďatelina perzská

1726 ***Trifolium retusum*** L.
Bg (1) тъполистна детелина
Cs (0) jetel otupený
De (–) Kleinblütiger Klee
Hu (0) pusztai here
Sk (–) ďatelina otupená

1727 ***Trifolium spadiceum*** L.
Bg (1) кестенява детелина
Cs (1) jetel kaštanový
De (–) Moor-Klee
Pl (0) koniczyna kasztanowata
Ru (0) клевер тёмно-каштановый
Sk (–) ďatelina gaštanová

1728 ***Trifolium squamosum*** L.
Al (–) tërfili bregdetar
Bg (1) късозъба детелина
Cs (0) jetel šupinatý
En (–) sea clover
Sk (–) ďatelina šupinatá

1729 ***Trifolium squarrosum*** L.
Bg (1) разперенозъба детелина
Ru (–) клевер лохматый
Sk (–) ďatelina kostrbatá

* ***Trifolium strepens*** Cran. → *Trifolium arvense*

1730 ***Trifolium striatum*** L.
Al (–) tërfili i vijosur
Bg (1) ивичеста детелина
Cs (0) jetel žíhaný
De (–) Streifen-Klee
En (–) knotted clover
Hu (0) sávos here
Pl (1) koniczyna kreskowana
Ru (0) клевер ребристый
Sk (–) ďatelina pruhovaná

* ***Trifolium suaveolens*** Willd. → *Trifolium resupinatum*

* ***Trifolium supinum*** Savi → *Trifolium echinatum*

1731 ***Trifolium vesiculosum*** Savi
Al (–) tërfili si fshikës
Bg (1) четинеста детелина
Hu (0) hólyagos here
Ru (0) клевер пузырчатый
Sk (–) ďatelina mnohopruhovaná

1732 ***Triglochin palustris*** L.
Al (–) triglohinë kënetash
Bg (1) блатен триглохин
Cs (0) bařička bahenní
De (–) Sumpf-Dreizack
En (–) marsh arrow-grass
Hu (0) mocsári kígyófű
Pl (1) świbka błotna
Ro (0) broscariţă
Ru (0) триостренник болотный
Sh (1) barski trozubac; močvarna brula
Sk (–) barička močiarna

1733 ***Trigonella* spp.** L.
Bg (1) сминдух
De (–) Schabzigerklee
En (–) fenugreek
Gr (1) τριγωνέλλα
Ru (–) пажитник
Sk (–) senovka

* ***Trigonella besserana*** Ser. → *Trigonella procumbens*

1734 ***Trigonella caerulea*** (L.) Ser.
Al (–) makthi i kaltër
Bg (1) син сминдух

Cs (0) pískavice modrá
De (–) Schabzigerklee
En (–) blue fenugreek
Hu (1) sziki lepkeszeg
Pl (1) kozieradka błękitna
Ro (1) molotru albastru
Ru (1) пажитник голубой
Sk (–) senovka belasá

1735 ***Trigonella corniculata*** (L.) L.
Al (–) trëndelina
Bg (1) рогоплоден сминдух
De (–) Gehörnter Bockshornklee
En (–) sickle-fruited fenugreek
Sk (–) senovka rohatá

1736 ***Trigonella foenum-graecum*** L.
Al (–) bar grek; trëndelina kopër Greqie
Bg (1) гръцки сминдух
Cs (0) pískavice řecké seno
De (–) Bockshornklee; Griechischer Schabzigerklee
En (–) common fenugreek
Hu (0) görögszéna
Pl (–) kozieradka
Ro (0) schinduf
Ru (0) пажитник сенной
Sk (–) senovka grécka

1737 ***Trigonella monspeliaca*** L.
Bg (1) монспелийски сминдух
Cs (0) pískavice thessalská
De (–) Kleiner Bockshornklee
En (–) star-fruited fenugreek
Hu (0) francia lepkeszeg
Ro (0) molotru pitic
Ru (0) пажитник монпельевский
Sk (–) senovka tesálska

1738 ***Trigonella procumbens*** (Bess.) Reichenb.
Bg (0) пълзящ сминдух
Cs (0) piskavice modrá poléhavá
En (–) prostrate fenugreek
Hu (1) sziki lepkeszeg
Ro (1) molotru
Ru (1) пажитник простёртый
Sh (1) polegla smiljkita
Sk (–) senovka rozložená

* ***Trinia elatior*** Gaudin → *Trinia glauca*

1739 ***Trinia glauca*** (L.) Dumort.
Al (–) nëtërza e rimtë
Bg (1) синкава триния
Cs (0) bezobalka sivá
De (–) Seegrüner Faserschirm
En (–) honewort
Hu (0) szürke nyúlkapor
Pl (0) trinia
Ru (0) триния жёлтая
Sh (1) obična srdiska; smedja srdiska
Sk (–) bezobalka sivá

* ***Trinia vulgaris*** DC. → *Trinia glauca*

* ***Tripleurospermum inodorum*** Schultz → *Matricaria perforata*

* ***Tripleurospermum maritimum*** (L.) Koch → *Matricaria maritima*

* ***Tripleurospermum tenuifolium*** (Kit.) Freyn → *Matricaria trichophylla*

1740 ***Trisetum flavescens*** (L.) Beauv.
Al (–) triseta e verdhëme
Bg (1) ливаден заешки овес
Cs (1) trojštět žlutavý
De (–) Wiesen-Goldhafer
En (–) yellow oat-grass
Hu (0) aranyzab
Pl (2) konietlica łąkowa
Ro (0) ovăscior auriu
Ru (0) трищетинник желтоватый
Sk (–) trojštet žltavý

* ***Triticum aegilopoides*** Balan. → *Triticum baeoticum*

1741 ***Triticum baeoticum*** Boiss.
Bg (1) двуосилеста едрозърненка
En (–) wild einkorn
Ru (0) пшеница беотийская
Sk (–) pšenica beótska

1742 ***Triticum monococcum*** L.
Al (–) gruri tepa
Bg (1) еднозърнест лимец
Cs (0) pšenice jednozrnka
De (–) Einkorn
En (–) einkorn
Pl (1) pszenica płaskurka
Ro (–) alac
Ru (0) пшеница однозернянка
Sk (–) pšenica jednozrnná

* ***Triticum ovatum*** (L.) Gren. & Godr. → *Aegilops geniculata*

* ***Triticum ramosum*** Trin. → *Leymus ramosus*

* ***Triticum repens*** L. → *Elymus repens*

* ***Triticum villosum*** (L.) Bieb. → *Dasypyrum villosum*

1743 ***Trollius europaeus*** L.
Al (–) troli i Evropës
Bg (0) европейски аблен
Cs (2) úpolín evropský
De (–) Europäische Trollblume
En (–) globeflower
Hu (0) zergeboglár
Pl (1) pełnik europejski
Ro (0) bulbuci
Ru (0) купальница европейская
Sk (–) žltohlav európsky

1744 ***Tuberaria guttata*** (L.) Fourr.
De (–) Drüsiges Sandröschen
Gr (1) ετήσια τουμπεράρια
Sk (–) slnečný kvet škvrnitý

1745 ***Tuberaria lignosa*** (Sweet) Samp.
Gr (1) πολυετής τουμπεράρια
Sk (–) slnečný kvet obyčajný

* ***Tuberaria vulgaris*** Willk. → *Tuberaria lignosa*

1746 ***Turgenia latifolia*** (L.) Hoffm.
Al (–) kripësht gjethegjërë
Bg (1) широколистен бабинец
Cs (0) štětináč širolistý
De (–) Breitblättrige Haftdolde
En (–) greater bur-parsley
Hu (1) nagy ördögbocskor
Ru (1) тургеневия широколистная
Sh (1) obična sudlanica; širokolisna podlanica
Sk (–) štetináč širokolistý

* ***Turritis glabra*** L. → *Arabis glabra*

1747 ***Tussilago farfara*** L.
Al (–) thundërmushka; thundërza
Bg (1) подбел
Cs (3) podběl obecný
De (–) Huflattich
En (–) coltsfoot
Gr (1) χαμολεύκα
Hu (2) martilapu
Pl (2) podbiał pospolity
Ro (2) podbal
Ru (2) мать-и-мачеха обыкновенная
Sh (1) obični podbel; obični podbjel
Sk (–) deväťsil obyčajný

1748 ***Typha angustifolia*** L.
Al (–) shavari gjethengushtë
Bg (1) теснолистен папур
Cs (2) orobinec úzkolistý
De (–) Schmalblättriger Rohrkolben
En (–) lesser bulrush

Gr (1) βρακτειοφόρο ψαθί
Hu (1) keskenylevelű gyékény
Pl (2) pałka wąskolistna
Ro (1) papură îngustă
Ru (1) рогоз узколистный
Sh (1) uskolisni rogoz
Sk (–) pálka úzkolistá

1749 *Typha latifolia* L.
Al (–) shavari gjethegjërë
Bg (1) широколистен папур
Cs (2) orobinec širolistý
De (–) Breitblättriger Rohrkolben
En (–) great reedmace
Gr (2) πλατύφυλλο ψαθί
Hu (1) széleslevelű gyékény
Pl (1) pałka szerokolistna
Ro (1) papură lătă
Ru (1) рогоз широколистный
Sh (1) širokolisni rogoz
Sk (–) pálka širokolistá

1750 *Typha laxmannii* Lep.
Gr (1) στενόφυλλο ψαθί
Hu (1) liliom gyékény
Ru (1) рогоз Лаксмана
Sk (–) pálka sitinovitá

1751 *Typha minima* Funck
Al (–) shavari i vogël
Cs (0) orobinec nejmenší
De (–) Kleinster Rohrkolben
En (–) dwarf reedmace
Hu (0) apró gyékény
Ru (1) рогоз малый
Sk (–) pálka najmenšia

* ***Typha stenophylla*** Fisch. & Meyer → *Typha laxmannii*

* ***Typhoides arundinacea*** (L.) Moen. → *Phalaris arundinacea*

U

1752 ***Urospermum picroides*** (L.) Scop.
Al (–) safaqi si i lidhur
De (–) Bitterkrautartiger Schwefelsame
Gr (1) ουρόσπερμο
Ru (0) хвостосемянник горчаковидный

1753 ***Urtica cannabina*** L.
Ru (1) крапива коноплевая
Sk (–) žihľava konopná

1754 ***Urtica dioica*** L.
Al (–) hithri
Bg (1) обикновена коприва
Cs (3) kopřiva dvoudomá
De (–) Große Brennessel
En (–) common nettle
Gr (1) πολυετής τσουκνίδα
Hu (2) nagy csalán
Pl (2) pokrzywa zwyczajna
Ro (1) urzică
Ru (2) крапива двудомная
Sh (1) obična kopriva
Sk (–) žihľava dvojdomá

1755 ***Urtica pilulifera*** L.
Al (–) hithri
Cs (0) kopřiva kulkonosná
De (–) Pillen-Brennessel
En (–) Roman nettle
Gr (1) μεγάλη τσουκνίδα
Ru (0) крапива шариконосная
Sk (–) žihľava guľkonosná

1756 ***Urtica urens*** L.
Al (–) hithri i rëndomtë
Bg (1) гръцка коприва
Cs (3) kopřiva žahavka
De (–) Kleine Brennessel
En (–) annual nettle
Gr (1) μικρή τσουκνίδα
Hu (2) árvacsalán; kis csalán
Pl (3) pokrzywa żegawka
Ro (1) urzică mică
Ru (1) крапива жгучая
Sh (1) mala kopriva
Sk (–) žihľava malá

1757 ***Utricularia vulgaris*** L.
Al (–) utrikularia e rëndomtë
Bg (1) обикновена мехурка
Cs (0) bublinatka obecná
De (–) Gemeiner Wasserschlauch
En (–) greater bladderwort
Hu (1) közönséges rence
Pl (1) pływacz zwyczajny
Ro (1) otrățel
Ru (1) пузырчатка обыкновенная
Sk (–) bublinatka obyčajná

V

* ***Vaccaria hispanica*** (Mill.) Rausch. → *Vaccaria pyramidata*

* ***Vaccaria parviflora*** Moen. → *Vaccaria pyramidata*

1758 *Vaccaria pyramidata* Med.
Al (–) lule vishnje
Bg (1) калугерка
Cs (1) kravinec jehlancovitý
De (–) Saat-Kuhkraut
En (–) cowherb
Gr (1) βακκάρια
Hu (2) tinóöröm; tinószegfű
Pl (1) krowiziół zbożowy
Ro (1) floarea călugărului
Ru (1) тысячеголов посевной
Sh (1) obični kravajac; obični puknez
Sk (–) kraviarka obilná

* ***Vaccaria segetalis*** Gar. → *Vaccaria pyramidata*

* ***Vaccaria vulgaris*** Host → *Vaccaria pyramidata*

1759 *Vaccinium myrtillus* L.
Al (–) boronica; thrashegra mërsinës
Bg (1) черна боровинка
Cs (2) brusnice borůvka
De (–) Heidelbeere
En (–) bilberry
Hu (0) fekete áfonya
Pl (1) borówka czarna
Ro (0) afin
Ru (!) черника
Sh (1) obična borovnica
Sk (–) čučoriedka čierna; čučoriedka obyčajná

1760 *Vaccinium uliginosum* L.
Al (–) boronica e qenit; thrashegra ligatinësh
Bg (1) синя боровинка
Cs (0) brusnice vlochyné
De (–) Rauschbeere
En (–) bog bilberry
Pl (1) borówka bagienna
Ru (0) голубика
Sk (–) čučoriedka barinná

1761 *Vaccinium vitis-idaea* L.
Al (–) boronicë e kuqe; thrashegra si hardhi
Bg (1) червена боровинка
Cs (0) brusnice brusinka
De (–) Preiselbeere
En (–) cowberry
Hu (0) vörös áfonya
Pl (1) borówka brusznica
Ro (0) merişor de munte
Ru (0) брусника
Sh (1) brusnica borovnica; crvenozrna borovnica
Sk (–) brusnica obyčajná

1762 *Valeriana* spp. L.
Bg (0) дилянка
De (–) Baldrian
En (–) valerian
Gr (1) βαλεριανή
Ru (0) валериана
Sk (–) kozlík

1763 *Valeriana dioica* L.
Cs (0) kozlík dvoudomý
De (–) Kleiner Baldrian
En (–) marsh valerian
Hu (0) kétlaki macskagyökér
Pl (1) kozłek dwupienny
Ru (0) валериана двудомная
Sh (1) dvodomi odoljen; obični odoljen
Sk (–) kozlík dvojdomý

1764 ***Valeriana officinalis*** L.
Al (–) haraqinë mjekësore
Cs (1) kozlík lékařský
De (–) Arznei-Baldrian
En (–) common valerian
Hu (0) mezei macskagyökér
Pl (–) kozłek lekarski
Ro (0) odolean
Ru (0) валериана лекарственная
Sk (–) kozlík lekársky

1765 ***Valerianella* spp.** Mill.
Bg (1) мотовилка
De (–) Feldsalat
En (–) cornsalad
Gr (1) βαλεριανέλλα
Ru (–) валерианелла
Sk (–) kozlíček

* ***Valerianella auricula*** DC. → *Verbena officinalis*

1766 ***Valerianella carinata*** Loisel.
Bg (1) ръбеста мотовилка
Cs (0) kozlíček kýlnatý
De (–) Kiel-Feldsalat
En (–) keeled-fruited cornsalad
Hu (1) hasábos galambbegy
Ru (0) валерианелла килеватая
Sk (–) kozlíček kýlový

1767 ***Valerianella coronata*** (L.) DC.
Al (–) haraqinëz e kurorëzuar
Bg (1) коронеста мотовилка
Cs (0) kozlíček korunkatý
De (–) Krönchen-Feldsalat
En (–) crowned cornsalad
Hu (0) koronás galambbegy
Pl (1) roszpunka ostrogrzbiecista
Ru (0) валерианелла венценосная
Sk (–) kozlíček korunkový

1768 ***Valerianella costata*** (Stev.) Betc.
Bg (1) ребреста мотовилка
Ru (0) валерианелла ребристая
Sk (–) kozlíček rebristý

1769 ***Valerianella dentata*** (L.) Pollich
Al (–) haraqinëz e dhëmbëzuar
Bg (1) зъбчеста мотовилка
Cs (2) kozlíček zubatý
De (–) Gezähnter Feldsalat
En (–) narrow-fruited cornsalad
Hu (1) fogas galambbegy
Pl (1) roszpunka ząbkowana
Ru (1) валерианелла зубчатая
Sh (1) zubčasti matovilac
Sk (–) kozlíček zubatý

* ***Valerianella gibbosa*** (Guss.) DC. → *Valerianella costata*

1770 ***Valerianella locusta*** (L.) Laterr.
Al (–) haraqinëz lokusta; marule lepuri
Bg (1) полска мотовилка
Cs (2) kozlíček polníček
De (–) Gemeiner Feldsalat
En (–) common cornsalad
Hu (1) saláta galambbegy
Pl (2) roszpunka warzywna
Ru (1) валерианелла колосковидная
Sk (–) kozlíček locíkový

* ***Valerianella morisonii*** (Spreng.) DC. → *Valerianella dentata*

* ***Valerianella olitoria*** (L.) Pollich → *Valerianella locusta*

1771 ***Valerianella pumila*** (L.) DC.
Bg (1) малка мотовилка
De (–) Kleiner Feldsalat
Hu (0) hártyás galambbegy
Ru (1) валерианелла маленькая
Sk (–) kozlíček nízky

1772 ***Valerianella rimosa*** Bast.
Al (–) haraqinëz rimoza

Bg (1) уховиднoплодна мотовилка
Cs (0) kozlíček štěrbinatý
De (–) Geährter Feldsalat
En (–) broad-fruited cornsalad
Hu (2) füles galambbegy
Pl (2) roszpunka bruzdowana
Ru (1) валерианелла щелистая
Sk (–) kozlíček štrbinatý

1773 ***Vallisneria spiralis*** L.
Al (–) valisneria spirale
De (–) Gemeine Wasserschraube
En (–) eelweed
Ru (0) валлиснерия спиральная
Ro (1) sîrmuliţă

1774 ***Ventenata dubia*** (Leers) Schultz
Cs (0) ovsiřík pochybný
De (–) Grannenhafer
Hu (1) vékonyzab
Pl (0) wentenata zwodnicza
Ru (0) вентената сомнительная
Sh (1) obična ventenata; tankovita ventenata
Sk (–) ovsec pochybný

1775 ***Veratrum album*** L.
Al (–) shtarë e bardhë
Cs (3) kýchavice bílá
De (–) Weißer Germer
En (–) white hellebore
Gr (1) άσπρο βέρατρο
Hu (1) fehér zászpa
Pl (1) ciemiężyca biała
Ro (2) ştirigoaie
Ru (1) чемерица Лобеля
Sh (1) bela čemerika; bijela čemerika
Sk (–) kýchavica biela

1776 ***Veratrum nigrum*** L.
Al (–) shtarë e zezë
Bg (1) черна чемерика
Cs (0) kýchavice černá
De (–) Schwarzer Germer
En (–) false helleborine
Gr (1) μαύρο βέρατρο
Hu (1) fekete zászpa
Pl (1) ciemiężyca czarna
Ru (1) чемерица чёрная
Sh (1) crna čemerika
Sk (–) kýchavica čierna

1777 ***Verbascum*** **spp.** L.
Bg (1) лопен
De (–) Königskerze
En (–) mullein
Gr (1) βερμπάσκο
Ru (1) коровяк
Sk (–) divozel

1778 ***Verbascum blattaria*** L.
Al (–) bar peshku moleje
Cs (0) divizna švábová
De (–) Schaben-Königskerze
En (–) moth mullein
Hu (1) molyűző ökörfarkkóró
Pl (1) dziewanna rdzawa
Ru (1) коровяк тараканий
Sk (–) divozel červíkový

1779 ***Verbascum densiflorum*** Bertol.
Al (–) lëpushkë
Bg (1) висок лопен
Cs (1) divizna velkokvětá
De (–) Großblutige Königskerze
Hu (1) dúsvirágú ökörfarkkóró
Pl (2) dziewanna wielkokwiatowa
Ro (1) lumînărică
Ru (1) коровяк высокий
Sk (–) divozel hustokvetý

1780 ***Verbascum longifolium*** Ten.
Al (–) bar peshku gjethegjatë
Bg (1) дълголистен лопен
Sk (–) divozel dlholistý

1781 ***Verbascum lychnitis*** L.
Cs (–) divizna knotovitá

De (–) Mehlige Königskerze
En (–) white mullein
Ru (1) коровяк мучнистый
Sh (1) gorska divizma
Sk (–) divozel knôtovkovitý

1782 ***Verbascum nigrum*** L.
Al (–) bar peshku i zi
Bg (0) черен лопен
Cs (0) divizna černá
De (–) Schwarze Königskerze
En (–) dark mullein
Hu (0) fekete ökörfarkkóró
Pl (2) dziewanna pospolita
Ro (1) somnoroasă
Ru (1) коровяк чёрный
Sk (–) divozel čierny

1783 ***Verbascum phlomoides*** L.
Al (–) bar peshku si flom
Bg (1) лечебен лопен
Cs (0) divizna sápovitá
De (–) Windblumen-Königskerze
En (–) orange mullein
Hu (2) szöszös ökörfarkkóró
Pl (1) dziewanna kutnerowata
Ro (2) lumînărică
Ru (1) коровяк лекарственный
Sk (–) divozel sápovitý

1784 ***Verbascum phoeniceum*** L.
Al (–) bar peshku vjellcë-ngjyrë qielli
Bg (1) виолетов лопен
Cs (0) divizna fialová
De (–) Violette Königskerze
En (–) purple mullein
Hu (1) lila ökörfarkkóró
Pl (1) dziewanna fioletowa
Ro (1) coada calului
Ru (1) коровяк фиолетовый
Sh (1) modrocvetna divizma; modrocvjetna divizma
Sk (–) divozel nachový

1785 ***Verbascum pulverulentum*** Vill.
Al (–) bar peshku i pluhurosur
De (–) Flockige Königskerze
En (–) hoary mullein
Hu (1) pamutos ökörfarkkóró
Sh (1) pepeljasta divizma
Sk (–) divozel prachovitý

1786 ***Verbascum sinuatum*** L.
Al (–) bar peshku gjirak
Bg (1) врязанолистен лопен
Ru (1) коровяк выемчатый
Sk (–) divozel skrútený

1787 ***Verbascum speciosum*** Schrad.
Bg (1) красив лопен
Cs (0) divizna ozdobná
De (–) Prächtige Königskerze
Hu (1) pompás ökörfarkkóró
Ru (0) коровяк великолепный
Sk (–) divozel ozdobný

* ***Verbascum thapsiforme*** Schrad. → *Verbascum densiflorum*

1788 ***Verbascum thapsus*** L.
Al (–) buzë gomari
Cs (2) divizna malokvětá
De (–) Kleinblütige Königskerze
En (–) great mullein
Hu (1) molyhos ökörfarkkóró
Pl (2) dziewanna drobnokwiatowa
Ro (1) corovatic
Ru (1) коровяк обыкновенный
Sh (1) obična divizma
Sk (–) divozel málokvetý

1789 ***Verbena officinalis*** L.
Al (–) sporisi mjekësor
Bg (1) лечебна върбинка
Cs (0) sporýš lékařský
De (–) Eisenkraut
En (–) common vervain
Hu (2) közönséges vasfű

Pl (1) werbena pospolita
Ro (1) sporiş
Ru (1) вербена аптечная
Sh (1) obična vrbena
Sk (–) železník lekársky

1790 ***Veronica* spp.** L.
Bg (1) великденче
De (–) Ehrenpreis
En (–) speedwell
Gr (2) βερόνικα
Ru (–) вероника
Sk (–) veronika

1791 ***Veronica acinifolia*** L.
Al (–) veronika me gjethe borziloku
De (–) Steinquendelblättriger Ehrenpreis
Hu (1) csomborlevelű veronika
Ru (0) вероника душевиколистная
Sk (–) veronika pamätníkolistá

1792 ***Veronica agrestis*** L.
Bg (1) посевно великденче
Cs (3) rozrazil polní
De (–) Acker-Ehrenpreis
En (–) green field-speedwell
Hu (1) fénytelen veronika
Pl (3) przetacznik rolny
Ru (1) вероника пашенная
Sh (1) njivska čestoslavica
Sk (–) veronika poľná

* ***Veronica anagallis*** auct. → *Veronica anagallis-aquatica*

1793 ***Veronica anagallis-aquatica*** L.
Al (–) veronika ujëse
Bg (1) изворно великденче
Cs (0) rozrazil drchničkovitý
De (–) Gauchheil-Ehrenpreis
En (–) blue water-speedwell
Hu (1) póléveronika
Pl (1) przetacznik bobownik
Ru (1) вероника ключевая
Sh (1) rečna čestoslavica; riječna čestoslavica
Sk (–) veronika drobničkovitá

1794 ***Veronica arvensis*** L.
Al (–) veronika e arave
Bg (1) полско великденче
Cs (3) rozrazil rolní
De (–) Feld-Ehrenpreis
En (–) wall speedwell
Hu (3) mezei veronika
Pl (3) przetacznik polny
Ru (2) вероника полевая
Sh (1) poljska čestoslavica
Sk (–) veronika roľná

1795 ***Veronica austriaca*** L.
Al (–) veronika e Austrisë
Bg (1) австрийско великденче
Cs (0) rozrazil ožankový
De (–) Österreichischer Ehrenpreis
En (–) large speedwell
Hu (0) fogaslevelű veronika
Pl (1) przetacznik pagórkowy
Ru (0) вероника австрийская
Sk (–) veronika rakúska

1796 ***Veronica beccabunga*** L.
Al (–) veronika e hutit
Bg (1) крайпоточно великденче
Cs (0) rozrazil potoční
De (–) Bach-Ehrenpreis
En (–) brooklime
Hu (–) deréceveronika
Ro (1) bobornic
Ru (0) вероника поточная
Sh (1) bubrežasta čestoslavica
Sk (–) veronika potočná

* ***Veronica buxbaumii*** Ten. → *Veronica persica*

* ***Veronica byzantina*** (Sibth. & Sm.) Degen → *Veronica persica*

1797 ***Veronica chamaedrys*** L.
Al (–) veronika dushk i vogël
Bg (1) храсталачно великденче
Cs (2) rozrazil rezekvítek
De (–) Gamander-Ehrenpreis
En (–) germander speedwell
Hu (0) ösztörűs veronika
Pl (2) przetacznik ożankowy
Ro (0) stejărel
Ru (1) вероника дубравная
Sh (1) dvostabljična čestoslavica; zmijska čestoslavica
Sk (–) veronika obyčajná

1798 ***Veronica cymbalaria*** Bodard
Bg (1) цимбаловидно великденче
De (–) Zymbelkraut-Ehrenpreis
En (–) pale speedwell
Ru (0) вероника цимбалярная
Sk (–) veronika cimbálok

* ***Veronica didyma*** auct. → *Veronica polita*

1799 ***Veronica dillenii*** Cran.
Cs (0) rozrazil Dillenův
De (–) Dillenius' Ehrenpreis
Hu (1) sovány veronika
Pl (2) przetacznik Dillena
Ru (0) вероника Диллена
Sk (–) veronika pekná

1800 ***Veronica hederifolia*** L.
Al (–) veronika gjetheurthi
Bg (2) бръшлянолистно великденче
Cs (3) rozrazil břečtanolistý
De (–) Efeublättriger Ehrenpreis
En (–) ivy-leaved speedwell
Hu (3) repkényveronika
Pl (3) przetacznik bluszczykowaty
Ro (3) doritoare
Ru (2) вероника плющелистная
Sh (1) bršljanolisna čestoslavica; bršljenasta čestoslavica
Sk (–) veronika brečtanolistá

1801 ***Veronica longifolia*** L.
Cs (–) rozrazil dlouholistý
De (–) Langblättriger Ehrenpreis
Ru (1) вероника длиннолистная
Sh (1) dugolisna čestoslavica
Sk (–) veronika dlholistá

1802 ***Veronica officinalis*** L.
Al (–) veronika mjekësore
Bg (1) лечебно великденче
Cs (0) rozrazil lékařský
De (–) Wald-Ehrenpreis
En (–) heath speedwell
Hu (1) orvosi veronika
Pl (2) przetacznik leśny
Ro (0) doritoare
Ru (1) вероника лекарственная
Sh (1) lekovita čestoslavica; ljekovita čestoslavica
Sk (–) veronika lekárska

1803 ***Veronica opaca*** Fries
Cs (1) rozrazil matný
De (–) Glanzloser Ehrenpreis
En (–) dark speedwell
Hu (1) fénytelen veronika
Pl (2) przetacznik ćmy
Ro (1) şopîrliţă nelucitoare
Ru (1) вероника тусклая
Sk (–) veronika tmavá

1804 ***Veronica persica*** Poir.
Al (–) veronika pjeshkë
Bg (1) персийско великденче
Cs (3) rozrazil perský
De (–) Persischer Ehrenpreis
En (–) common field-speedwell
Hu (3) perzsa veronika
Pl (3) przetacznik perski
Ro (3) ventrilică
Ru (3) вероника персидская
Sh (2) perzijska čestoslavica
Sk (–) veronika perzská

1805 ***Veronica polita*** Fries
Bg (1) изящно великденче
Cs (1) rozrazil lesklý
De (–) Glänzender Ehrenpreis
En (–) grey field-speedwell
Hu (3) fényes veronika
Pl (2) przetacznik lśniący
Ro (3) şopîrliţă nelucitoare
Ru (1) вероника двойчатая
Sh (1) glatka čestoslavica; ulaštena čestoslavica
Sk (–) veronika podvojná

1806 ***Veronica praecox*** All.
Al (–) veronika e hershme
Bg (1) ранно великденче
Cs (2) rozrazil časný
De (–) Frühblühender Ehrenpreis
En (–) Breckland speedwell
Hu (1) korai veronika
Pl (0) przetacznik wczesny
Ru (1) вероника ранняя
Sh (1) jamičasta čestoslavica; rana čestoslavica
Sk (–) veronika včasná

1807 ***Veronica prostrata*** L.
Al (–) veronika e shtrirë
Bg (1) полегнало великденче
Cs (1) rozrazil rozprostřený
De (–) Niederliegender Ehrenpreis
Hu (0) lecsepült veronika
Ro (0) coada mielului
Ru (0) вероника простёртая
Sk (–) veronika rozložitá

1808 ***Veronica scutellata*** L.
Al (–) veronika e kënetës; veronika në trajtë mburoje
Bg (1) щитовидно великденче
Cs (0) rozrazil štítkovitý
De (–) Schild-Ehrenpreis
En (–) marsh speedwell
Hu (1) pajzsos veronika
Pl (1) przetacznik błotny
Ru (1) вероника щитковая
Sh (1) končasta čestoslavica; štitasta čestoslavica
Sk (–) veronika štítkovitá

1809 ***Veronica serpyllifolia*** L.
Al (–) veronika me gjethe lisëre
Cs (2) rozrazil douškolistý
De (–) Quendelblättriger Ehrenpreis
En (–) thyme-leaved speedwell
Hu (1) kakukkveronika
Pl (3) przetacznik macierzankowy
Ru (1) вероника тимьянолистная
Sh (1) jajastolisna čestoslavica; sitna čestoslavica
Sk (–) veronika tymiánolistá

1810 ***Veronica spicata*** L.
Bg (1) класовидно великденче
Cs (1) rozrazil klasnatý
De (–) Ähriger Ehrenpreis
En (–) spiked speedwell
Hu (0) macskafarkú veronika
Pl (1) przetacznik kłosowy
Ro (0) băieţei
Ru (0) вероника колосистая
Sk (–) veronika klasnatá

* ***Veronica teucrium*** L. → *Veronica austriaca*

* ***Veronica tournefortii*** Gmel. → *Veronica persica*

* ***Veronica triloba*** Opiz → *Veronica hederifolia*

1811 ***Veronica triphyllos*** L.
Bg (1) трипръстно великденче
Cs (2) rozrazil trojlistý
De (–) Finger-Ehrenpreis
En (–) fingered speedwell
Hu (2) ujjaslevelű veronika

Pl (3) przetacznik trójlistkowy
Ru (1) вероника трёхлистная
Sh (1) troprsta čestoslavica
Sk (–) veronika trojzárezová

1812 ***Veronica verna*** L.
Bg (1) пролетно великденче
Cs (1) rozrazil jarní
De (–) Frühlings-Ehrenpreis
En (–) spring speedwell
Hu (1) tavaszi veronika
Pl (3) przetacznik wiosenny
Ru (1) вероника весенняя
Sk (–) veronika jarná

1813 ***Vicia* spp.** L.
Bg (1) глушина
De (–) Wicke
Gr (2) αγριοβίκος
Ro (2) măzăriche
Ru (1) горошек

* ***Vicia angustifolia*** L. → *Vicia sativa* ssp. *nigra*

1814 ***Vicia articulata*** Hornem.
Al (–) thjerrëz amerikane
Bg (1) нежна глушина
Cs (–) vikev článkovaná
De (–) Linsen-Wicke
En (–) oneflower tare
Pl (0) wyka jednokwiatowa
Sk (–) ľadník článkovaný

* ***Vicia biebersteinii*** Bess. → *Vicia grandiflora*

1815 ***Vicia bithynica*** (L.) L.
Al (–) grashinë e Bitinesë; mollira
Bg (1) битинска глушина
De (–) Bithynische Wicke
En (–) Bithynian vetch
Ru (0) горошек вифинский
Sk (–) ľadník maloázijský

* ***Vicia brachytropis*** Kar. & Kir. → *Vicia tenuifolia*

1816 ***Vicia cassubica*** L.
Al (–) grashinë kasube
Bg (0) касубска глушина
Cs (0) vikev kašubská
De (–) Kassuben-Wicke
Hu (0) vitézbükköny
Pl (1) wyka kaszubska
Ru (0) горошек кашубский
Sk (–) ľadník kašubský

1817 ***Vicia cracca*** L.
Al (–) grashina kraka
Bg (1) птича глушина
Cs (3) vikev ptačí
De (–) Vogel-Wicke
En (–) tufted vetch
Hu (2) kaszanyűg bükköny
Pl (3) wyka ptasia
Ru (2) горошек мышиный
Sh (2) ptičja grahorica
Sk (–) ľadník vtáčí

* ***Vicia dasycarpa*** auct. → *Vicia villosa* ssp. *varia*

1818 ***Vicia dumetorum*** L.
Al (–) grashinë kaçubore
Bg (1) храсталачна глушина
Cs (0) vikev křovištní
De (–) Hecken-Wicke
Hu (0) cserebükköny
Pl (1) wyka zaroślowa
Ru (0) горошек зарослевый
Sk (–) ľadník krovinatý

1819 ***Vicia ervilia*** (L.) Willd.
Al (–) urovi
Bg (1) бурчак
Cs (0) vikev čočková
De (–) Linsen-Wicke
En (–) bitter vetch

Ru (1) горошек четкообразный
Sk (–) ľadník šošovicovitý

* **Vicia glabrescens** (Koch) Heim. → *Vicia villosa* ssp. *villosa*

1820 **Vicia grandiflora** Scop.
Al (–) grashinë lulemadhe
Bg (2) едроцветна глушина
Cs (0) vikev velkokvětá
De (–) Großblütige Wicke
En (–) large yellow-vetch
Hu (2) szennyes bükköny
Pl (2) wyka brudnożółta
Ru (1) горошек крупноцветковый
Sh (1) krupnocvetna grahorica; velecvjetna grahorica
Sk (–) ľadník veľkokvetý pravý

* **Vicia heterophylla** Presl → *Vicia sativa* ssp. *sativa*

1821 **Vicia hirsuta** (L.) Gray
Al (–) grashinë me qime
Bg (1) космата глушина
Cs (3) vikev chlupatá
De (–) Rauhhaarige Wicke
En (–) hairy tare
Hu (3) borzas bükköny
Pl (3) wyka drobnokwiatowa
Ro (2) cosiţă
Ru (2) горошек волосистый
Sh (1) hrapava grahorica
Sk (–) ľadnik chlpatý

1822 **Vicia hybrida** L.
Al (–) grashinë hibride
Bg (1) хибридна глушина
En (–) hairy yellow-vetch
Ru (1) горошек гибридный
Sk (–) ľadník zkrížený

1823 **Vicia lathyroides** L.
Al (–) grashinë si vingjër
Bg (1) секирчеподобна глушина
Cs (1) vikev hrachorovitá
De (–) Platterbsen-Wicke
En (–) spring vetch
Hu (1) pici bükköny
Pl (1) wyka lędźwianowata
Ru (1) горошек чиновидный
Sk (–) ľadník hrachorovitý

* **Vicia lerchenfeldiana** Schur → *Vicia lutea*

* **Vicia lineata** Bieb. → *Vicia pannonica*

1824 **Vicia lutea** L.
Al (–) grashinë e verdhë
Bg (1) жълта глушина
Cs (0) vikev žlutá
De (–) Gelbe Wicke
En (–) common yellow-vetch
Hu (1) sárga bükköny
Ru (1) горошек жёлтый
Sk (–) ľadník žltý

1825 **Vicia melanops** Sibth. & Sm.
Bg (1) трицветна глушина
Cs (0) vikev černavá
De (–) Grünblütige Wicke
En (–) black vetch
Sk (–) ľadník černastý

* **Vicia monanthos** (L.) Desf. → *Vicia articulata*

1826 **Vicia narbonensis** L.
Al (–) greshinë e Narbones; momëleci
Bg (1) нарбонска глушина
Cs (0) vikev narbonská
De (–) Maus-Wicke
En (–) French vetch
Hu (0) fogaslevelű bükköny
Ru (1) горошек нарбонский
Sk (–) ľadník narbonský

1827 *Vicia onobrychioides* L.
Al (–) grashinë si esparsetë
Bg (1) еспарзетоподобна глушина
De (–) Esparsetten-Klee
Sk (–) ľadník ľadníčekovitý

1828 *Vicia pannonica* Cran. **ssp. *pannonica***
Al (–) grashinë e Hungarisë; grashinë e Panonisë
Bg (1) панонска глушина
Cs (1) vikev panonská
De (–) Ungarische Wicke
En (–) Hungarian vetch
Hu (1) pannon bükköny
Pl (0) wyka pannońska
Ru (0) горошек венгерский
Sh (1) panonska grahorica
Sk (–) ľadník panónsky pravý

1829 *Vicia pannonica* Cran. **ssp. *striata*** (Bieb.) Nyman
Bg (2) ребреста глушина
Hu (2) csíkos bükköny
Ro (1) măzăriche roşie
Ru (1) горошек паннонский
Sh (1) prugasta grahorica
Sk (–) ľadník panónsky pruhovaný

* ***Vicia parviflora*** Loisel. → *Vicia hirsuta*

1830 *Vicia peregrina* L.
Bg (1) чуждоземна глушина
De (–) Landstreicher-Wicke
Hu (1) idegen bükköny
Ru (1) горошек иноземный
Sk (–) ľadník cudzí

* ***Vicia polyphylla*** Desf. → *Vicia villosa* ssp. *villosa*

* ***Vicia purpurascens*** DC. → *Vicia pannonica* ssp. *striata*

1831 *Vicia sativa* L. **ssp. *nigra*** (L.) Ehrh.
Al (–) buxhaku i zi
Bg (2) теснолистна глушина
De (–) Schmalblättrige Wicke
En (–) narrow-leaved vetch
Hu (2) fekete bükköny
Pl (3) wyka wąskolistna
Ro (3) măzăriche
Ru (1) горошек посевной
Sh (1) crna grahorica; uskolisna grahorica
Sk (–) ľadník siaty čierny

1832 *Vicia sativa* L. **ssp. *sativa***
Al (–) grashinë e kultivuar
Bg (1) пролетен фий, пролетен урок
Cs (2) vikev setá
De (–) Futter-Wicke
En (–) common vetch
Gr (2) βίκος
Hu (1) takarmánybükköny
Pl (2) wyka siewna
Ro (2) măzăriche de primăvară
Ru (1) вика посевная
Sh (1) njivska grahorica; obična grahorica
Sk (–) ľadník siaty obyčajný

1833 *Vicia sepium* L.
Al (–) grashinë e gjerdheve; grozhël
Bg (1) горска глушина
Cs (3) vikev plotní
De (–) Zaun-Wicke
En (–) bush vetch
Hu (0) gyepűbükköny
Pl (3) wyka płotowa
Ru (1) горошек заборный
Sk (–) ľadník plotný

* ***Vicia serratifolia*** Jacq. → *Vicia narbonensis*

* ***Vicia sordida*** Waldst. & Kit. → *Vicia grandiflora*

* **Vicia striata** Bieb. → *Vicia pannonica* ssp. *striata*

1834 Vicia tenuifolia Roth
Al (–) grashinë gjethehollë
Bg (1) тънколистна глушина
De (–) Feinblättrige Wicke
En (–) fine-leaved vetch
Hu (1) keskenylevelű bükköny
Ru (1) горошек тонколистный
Sh (1) tankolisna grahorica
Sk (–) ľadník tenkolistý

1835 Vicia tetrasperma (L.) Schreb.
Al (–) grashinë me katër rara
Bg (1) четирисеменна глушина
Cs (3) vikev čtyřsemenná
De (–) Viersamige Wicke
En (–) smooth tare
Hu (1) négymagvú bükköny
Pl (3) wyka czteronasienna
Ru (1) горошек четырёхсемянный
Sh (1) četverozrna grahorica
Sk (–) ľadník štvorsemenný

* **Vicia tricolor** Sebast. & Mauri → *Vicia melanops*

* **Vicia varia** Host → *Vicia villosa* ssp. *varia*

1836 Vicia villosa Roth **ssp. varia** (Host) Corb.
Al (–) grashinë me frute leshatake
Bg (1) пъстроцветна глушина
De (–) Kahle Wicke
Pl (2) wyka pstra
Ru (1) горошек мохнатый
Sk (–) ľadník huňatý drsnoplodý

1837 Vicia villosa Roth **ssp. villosa**
Al (–) grashinë leshtore
Bg (1) вълнеста глушина
Cs (2) vikev huňatá
De (–) Zottige Wicke
En (–) fodder vetch
Hu (3) szöszös bükköny
Pl (2) wyka kosmata
Ro (3) măzăriche păroasă
Ru (1) вика озимая
Sh (–) maljava grahorica
Sk (–) ľadník huňatý pravý

1838 Vinca major L.
Al (–) bed i madhi; menekshe gomari më e madhe
Bg (1) голям зимзелен
De (–) Großes Immergrün
En (–) greater periwinkle
Gr (1) μεγάλη βίνκα
Pl (1) barwinek pospolity
Ru (0) барвинок большой
Sk (–) zimozel väčší

1839 Vinca minor L.
Al (–) bed i vogël; menekshe gomari më e vogël
Bg (1) малък зимзелен
Cs (1) brčál menší
De (–) Kleines Immergrün
En (–) lesser periwinkle
Gr (1) μικρή βίνκα
Hu (0) kis télizöld
Ro (0) saschiu
Ru (0) барвинок малый
Sk (–) zimozel menší

1840 Vincetoxicum hirundinaria Med.
Al (–) qenmbytsja helmmundëse
Cs (–) tolita lékařská
De (–) Weiße Schwalbenwurz
En (–) common vincetoxicum
Ru (0) ластовень лекарственный
Sh (1) obična lastavina
Sk (–) luskáč lekársky

1841 Viola spp. L.
Bg (1) теменуга

De (–) Veilchen
En (–) violet
Gr (1) αγριοπανσές
Ru (2) фиалка
Sk (–) fialka

1842 ***Viola ambigua*** Waldst. & Kit.
Cs (0) violka obojetná
De (–) Pontisches Veilchen
Hu (0) csuklyás ibolya
Pl (1) fiołek zmienny
Sk (–) fialka obojaká

1843 ***Viola arvensis*** Murray
Al (–) manushaqe e fushave
Bg (2) полска теменуга
Cs (3) violka rolní
De (–) Acker-Stiefmütterchen
En (–) field pansy
Hu (3) mezei árvácska
Pl (3) fiołek polny
Ro (3) trei frați pătați
Ru (2) фиалка полевая
Sh (2) poljska ljubičica
Sk (–) sirôtka roľná

1844 ***Viola biflora*** L.
Al (–) manushaqe dylulëshe
Bg (1) двуцветна теменуга
Cs (–) violka dvoukvětá
De (–) Zweiblütiges Veilchen
En (–) yellow wood-violet
Hu (0) kétvirágú ibolya; sárga ibolya
Pl (1) fiołek dwukwiatowy
Ro (0) viorele galbene
Ru (0) фиалка двухцветная
Sk (–) fialka dvojkvetá

1845 ***Viola dacica*** Borb.
Al (–) vjollca e Dacisë
Bg (1) дакийска теменуга
Pl (0) fiołek dacki
Ro (0) unghia păsării
Ru (–) фиалка дакийская
Sk (–) fialka dačska

1846 ***Viola kitaibeliana*** Schultes
Al (–) vjollca e Kitajbelit
Bg (1) Китайбелова теменуга
De (–) Kleines Stiefmütterchen
En (–) dwarf pansy
Hu (1) törpe árvácska
Pl (1) fiołek Kitaibela
Ru (0) фиалка Китайбелева
Sk (–) sirôtka najmenšia

1847 ***Viola odorata*** L.
Al (–) manushaqja; vjollca
Bg (1) миризлива теменуга
Cs (2) violka vonná
De (–) März-Veilchen
En (–) sweet violet
Hu (0) illatos ibolya
Pl (1) fiołek wonny
Ro (0) viorea
Ru (0) фиалка душистая
Sk (–) fialka voňavá

1848 ***Viola palustris*** L.
Al (–) manushaqe kënetash
Cs (0) violka bahenní
De (–) Sumpf-Veilchen
En (–) marsh violet
Pl (1) fiołek błotny
Ru (0) фиалка болотная
Sk (–) fialka močiarna

1849 ***Viola reichenbachiana*** Jord.
De (–) Wald-Veilchen
En (–) pale wood-violet
Ru (0) фиалка лесная
Sh (1) šumska ljubičica
Sk (–) fialka lesná

1850 ***Viola rhodopeia*** Beck.
Bg (1) родопска теменуга

Ru (–) фиалка родопская
Sk (–) fialka balkánska

1851 ***Viola rupestris*** Schm.
Bg (0) скална теменуга
Cs (0) violka skalní
De (–) Sand-Veilchen
En (–) Teesdale violet
Hu (0) homoki ibolya
Pl (1) fiołek skalny
Ru (0) фиалка скальная
Sk (–) fialka skalná

* ***Viola saxatilis*** Schm. → *Viola tricolor* ssp. *tricolor*

* ***Viola tricolor*** L. **ssp.** ***arvensis*** (Murray) Gaudin → *Viola arvensis*

1852 ***Viola tricolor*** L. **ssp.** ***tricolor***
Al (–) manushaqe; vjollcë tringjyrëshe
Bg (2) трицветна теменуга
Cs (3) violka trojbarevná
De (–) Gewöhnliches Stiefmütterchen
En (–) wild pansy
Hu (1) háromszínű árvácska
Pl (2) fiołek trójbarwny
Ro (1) trei frați pătați
Ru (1) фиалка трёхцветная
Sh (1) dan-i-noć ljubičica; trobojna ljubica
Sk (–) sirôtka trojfarebná

1853 ***Viscum album*** L.
Al (–) veshtulli
Bg (0) бял имел
Cs (1) jmelí bílé
De (–) Mistel
En (–) mistletoe
Gr (1) ιξός
Hu (–) fagyöngy
Pl (1) jemioła pospolita
Ro (1) vîsc
Ru (1) омела белая
Sk (–) imelo biele

1854 ***Vitex agnus-castus*** L.
Al (–) konopicë; marena
Bg (1) аврамово дърво
De (–) Keuschbaum
En (–) chaste tree
Gr (1) λυγαριά
Hu (0) barátcserje
Ro (0) mielărea
Ru (0) витекс священный

* ***Vogelia apiculata*** (Fisch., Meyer & Avé.) Vierh. → *Neslia paniculata* ssp. *thracica*

* ***Vogelia paniculata*** (L.) Hornem. → *Neslia paniculata* ssp. *thracica*

1855 ***Vulpia*** **spp.** Gmel.
Bg (1) вулпия
De (–) Fuchsschwingel
Gr (1) βούλπια
Ru (–) вульпия
Sk (–) mrvka

1856 ***Vulpia bromoides*** (L.) Gray
Bg (1) дертонска вулпия
Cs (0) mrvka sveřepovitá
De (–) Trespen-Fuchsschwingel
En (–) squirrel-tail fescue
Hu (1) déli egércsenkesz
Pl (0) wulpia stokłosowata
Sk (–) mrvka stoklasovitá

1857 ***Vulpia ciliata*** Dumort.
Al (–) vulpia me qepalla
Bg (1) ресничеста вулпия
Cs (0) mrvka brvitá
De (–) Wimperiger Fuchsschwingel
Ru (0) вульпия реснитчатая
Sk (–) mrvka kolienkatá

* ***Vulpia dertonensis*** (All.) Gola → *Vulpia bromoides*

1858 ***Vulpia myuros*** (L.) Gmel.
Al (–) vulpia bishtamini
Bg (1) миша вулпия
Cs (0) mrvka myší ocásek
De (–) Mäuseschwanz-Fuchsschwingel
En (–) rat's-tail fescue
Hu (1) vékony egércsenkesz
Pl (0) wulpia mysi ogon
Ro (0) fulfucă
Ru (0) вульпия мышехвостная
Sh (1) obični brčak; repasti brčak
Sk (–) mrvka myší chvost

W

* ***Wiborgia parviflora*** Humb., Bonpl. & Kunth → *Galinsoga parviflora*

1859 ***Wolffia arrhiza*** (L.) Hork.
Bg (–) безкоренчеста волфия
Cs (0) drobnička bezkorenná
De (–) Zwerglinse
En (–) rootless duckweed
Hu (1) vízi dara
Pl (1) wolffia bezkorzeniowa
Ru (0) вольфия бескорневая
Sk (–) wolffia bezkoreňová

X

* ***Xanthium americanum*** Walt. →
Xanthium strumarium ssp. *italicum*

* ***Xanthium brasilicum*** Vell. →
Xanthium strumarium ssp. *italicum*

* ***Xanthium canadense*** Gray →
Xanthium strumarium ssp. *italicum*

* ***Xanthium echinatum*** Murray →
Xanthium strumarium ssp. *italicum*

* ***Xanthium italicum*** Moret. →
Xanthium strumarium ssp. *italicum*

* ***Xanthium macrocarpum*** DC. →
Xanthium strumarium ssp. *italicum*

* ***Xanthium saccharatum*** Wallr. →
Xanthium strumarium ssp. *italicum*

1860 ***Xanthium spinosum*** L.
Al (–) ksanta me gjemba; sybardhi
Bg (1) казашки бодил
Cs (1) řepeň trnitá
De (–) Dornige Spitzklette
En (–) spiny cocklebur
Gr (1) ασπράγκαθο
Hu (1) szúrós szerbtövis
Pl (0) rzepień kolczasty
Ro (2) holeră
Ru (2) дурнишник игольчатый
Sh (1) trnovita boca
Sk (–) voškovník tŕnistý

1861 ***Xanthium strumarium*** L. **ssp.** ***italicum*** (Moret.) Löve
Al (–) ksanta e Italisë; rodhja
Bg (2) италианска свиница
Cs (2) řepeň durkoman pobřežni
En (–) Italian cocklebur
Hu (1) olasz szerbtövis
Ro (3) cornaci
Ru (2) дурнишник итальянский
Sh (1) zelena boca
Sk (–) voškovník taliansky

1862 ***Xanthium strumarium*** L. **ssp.** ***strumarium***
Al (–) ksanta e lungave
Bg (2) свиница
Cs (2) řepeň durkoman
De (–) Gemeine Spitzklette
En (–) rough cocklebur
Gr (1) αγριομελιτζάνα
Hu (2) bojtorján szerbtövis
Pl (1) rzepień pospolity
Ro (2) cornut
Ru (2) дурнишник обыкновенный
Sh (1) obična boca
Sk (–) voškovník durkoman pravý

* ***Xanthoxalis corniculata*** (L.) Small →
Oxalis corniculata

* ***Xanthoxalis dillenii*** (Jacq.) Holub →
Oxalis stricta

1863 ***Xeranthemum annuum*** L.
Al (–) bar kashte; kserantema njëvjeçare
Bg (1) едногодишно безсмъртниче
Cs (–) suchokvět roční
De (–) Einjährige Spreublume
En (–) pink everlasting
Hu (1) ékes vasvirág
Ro (2) plevaiță
Ru (1) сухоцвет однолетний
Sh (1) jednogodišnja poljska metla; jednogodišnja slamnica
Sk (–) suchokvet ročný

1864 ***Xeranthemum cylindraceum*** Sibth. & Sm.

Bg (1) цилиндрично безсмъртниче
De (–) Spreublume
Hu (1) hengeres vasvirág
Ru (1) сухоцвет цилиндрический
Sh (1) smrdljiva slamnica; valjkasta nevenka
Sk (–) suchokvet valcovitý

Z

* ***Zacintha verrucosa*** Gaertn. → *Crepis zacintha*

* ***Zannichellia major*** (Hartm.) Boenn. → *Zannichellia palustris*

1865 ***Zannichellia palustris*** L.
Al (–) xanxikelia kënetash
Bg (1) блатна цанихелия
Cs (0) šejdračka bahenní
De (–) Teichfaden
En (–) horned pondweed
Hu (0) tófonal
Pl (0) zamętnica błotna
Ro (1) mătriţă
Ru (0) дзаникеллия болотная
Sh (1) barska zanihelija
Sk (–) zanichelka močiarna

1866 ***Ziziphora capitata*** L.
Al (–) thrumja
Bg (1) главеста зизифора
Ru (0) зизифора головчатая

1867 ***Zostera marina*** L.
Al (–) leshteriku i detit
Bg (1) голяма морска трева
De (–) Gemeines Seegras
En (–) eelgrass
Ro (0) iarbă de mare
Ru (0) взморник морской
Sk (–) pásomnica morská

1868 ***Zygophyllum fabago*** L.
Ru (1) парнолистник обыкновенный

ALBANIAN INDEX

БЪЛГАРСКИ УКАЗАТЕЛ

CZECH INDEX

GERMAN INDEX

ENGLISH INDEX

GREEK INDEX

HUNGARIAN INDEX

POLISH INDEX

ROMANIAN INDEX

RUSSIAN INDEX

SERBO-CROAT INDEX

SLOVAK INDEX

www.ingramcontent.com/pod-product-compliance
Ingram Content Group UK Ltd.
Pitfield, Milton Keynes, MK11 3LW, UK
UKHW020226290926
15470UKWH00001B/400

9 780444 989697